Advanced Principles and Techniques in Tissue Engineering

Advanced Principles and Techniques in Tissue Engineering

Editor: Heidi Lee

R CALLISTO REFERENCE

www.callistoreference.com

Callisto Reference,
118-35 Queens Blvd., Suite 400,
Forest Hills, NY 11375, USA

Visit us on the World Wide Web at:
www.callistoreference.com

ISBN: 978-1-64116-543-3 (Hardback)

Cataloging-in-Publication Data

Advanced principles and techniques in tissue engineering / edited by Heidi Lee.
    p. cm.
Includes bibliographical references and index.
ISBN 978-1-64116-543-3
1. Tissue engineering. 2. Tissue culture. 3. Biomedical engineering. 4. Regenerative medicine. I. Lee, Heidi.
R857.T55 A38 2022
610.28--dc23

Table of Contents

Preface VII

Chapter 1   **Integration of Rabbit Adipose Derived Mesenchymal Stem Cells to Hydroxyapatite Burr Hole Button Device for Bone Interface Regeneration**................................1
Viswanathan Gayathri, Varma Harikrishnan and Parayanthala Valappil Mohanan

Chapter 2   **Application of Synthetic Polymeric Scaffolds in Breast Cancer 3D Tissue Cultures and Animal Tumor Models**................................10
Girdhari Rijal, Chandra Bathula and Weimin Li

Chapter 3   **Biomechanical Performances of Networked Polyethylene Glycol Diacrylate: Effect of Photoinitiator Concentration, Temperature, and Incubation Time**................................19
Morshed Khandaker, Albert Orock, Stefano Tarantini, Jeremiah White and Ozlem Yasar

Chapter 4   **Sol-Gel Derived Mg-Based Ceramic Scaffolds Doped with Zinc or Copper Ions: Preliminary Results on their Synthesis, Characterization, and Biocompatibility**................................27
Georgios S. Theodorou, Eleana Kontonasaki, Anna Theocharidou, Athina Bakopoulou, Maria Bousnaki, Christina Hadjichristou, Eleni Papachristou, Lambrini Papadopoulou, Nikolaos A. Kantiranis, Konstantinos Chrissafis, Konstantinos M. Paraskevopoulos and Petros T. Koidis

Chapter 5   **Nano-TiO$_2$ Doped Chitosan Scaffold for the Bone Tissue Engineering Applications**................................38
Pawan Kumar

Chapter 6   **Novel Vanadium-Loaded Ordered Collagen Scaffold Promotes Osteochondral Differentiation of Bone Marrow Progenitor Cells**................................45
Ana M. Cortizo, Graciela Ruderman, Flavia N. Mazzini, M. Silvina Molinuevo and Ines G. Mogilner

Chapter 7   **Preparation Methods for Improving PEEK's Bioactivity for Orthopedic and Dental Application**................................56
Davood Almasi, Nida Iqbal, Maliheh Sadeghi, Izman Sudin, Mohammed Rafiq Abdul Kadir and Tunku Kamarul

Chapter 8   **Intrinsic Osteoinductivity of Porous Titanium Scaffold for Bone Tissue Engineering**................................68
Maryam Tamaddon, Sorousheh Samizadeh, Ling Wang, Gordon Blunn and Chaozong Liu

Chapter 9   **Strain and Vibration in Mesenchymal Stem Cells**................................79
Brooke McClarren and Ronke Olabisi

Chapter 10   **Injectable Hydrogel versus Plastically Compressed Collagen Scaffold for Central Nervous System Applications** 92
Magdalini Tsintou, Kyriakos Dalamagkas and Alexander Seifalian

Chapter 11   **Magnesium Oxide Nanoparticles Reinforced Electrospun Alginate-Based Nanofibrous Scaffolds with Improved Physical Properties** 102
R. T. De Silva, M. M. M. G. P. G. Mantilaka, K. L. Goh, S. P. Ratnayake,
G. A. J. Amaratunga and K. M. Nalin de Silva

Chapter 12   **Tissue-Engineered Vascular Graft of Small Diameter based on Electrospun Polylactide Microfibers** 111
P. V. Popryadukhin, G. I. Popov, G. Yu. Yukina, I. P. Dobrovolskaya,
E. M. Ivan'kova, V. N. Vavilov and V. E. Yudin

Chapter 13   **Silicone Substrate with Collagen and Carbon Nanotubes Exposed to Pulsed Current for MSC Osteodifferentiation** 121
Daniyal Jamal and Roche C. de Guzman

Chapter 14   **Graphene Family Nanomaterials: Properties and Potential Applications in Dentistry** 130
Ziyu Ge, Luming Yang, Fang Xiao, YaniWu, Tingting Yu, Jing Chen, Jiexin Lin
and Yanzhen Zhang

Chapter 15   **Scaffolds for Pelvic Floor Prolapse: Logical Pathways** 142
Julio Bissoli and Homero Bruschini

Chapter 16   **Recent Progress of Fabrication of Cell Scaffold by Electrospinning Technique for Articular Cartilage Tissue Engineering** 148
Yingge Zhou, Joanna Chyu and Mimi Zumwalt

Chapter 17   **Oxidative Nanopatterning of Titanium Surface Influences mRNA and MicroRNA Expression in Human Alveolar Bone Osteoblastic Cells** 158
Maidy Rehder Wimmers Ferreira, Roger Rodrigo Fernandes,
Amanda Freire Assis, Janaína A. Dernowsek, Geraldo A. Passos, Fabio Variola
and Karina Fittipaldi Bombonato-Prado

Chapter 18   **Preparation and Evaluation of Gelatin-Chitosan-Nanobioglass 3D Porous Scaffold for Bone Tissue Engineering** 173
Kanchan Maji, Sudip Dasgupta, Krishna Pramanik and Akalabya Bissoyi

Chapter 19   ***In Vitro* and *In Vivo* Characterization of N-Acetyl-L-Cysteine Loaded Beta-Tricalcium Phosphate Scaffolds** 187
Yong-Seok Jang, Phonelavanh Manivong, Yu-Kyoung Kim, Kyung-Seon Kim,
Sook-Jeong Lee, Tae-Sung Bae and Min-Ho Lee

Permissions

List of Contributors

Index

Preface

Tissue engineering is a field that uses the principles of engineering, material methods and suitable biochemical and physiochemical factors in combination of cells to improve or replace biological tissues. It is related to applications that repair or replace whole tissues or a portion of tissues such as bone, cartilage, blood vessels, bladder and skin. It involves the use of a tissue scaffold for a medical purpose which helps in the formation of new viable tissue. It also uses living cells as engineering materials, for example in skin repair or replacement, it utilizes living fibroblasts, cartilage repaired with living chondrocytes, etc. It also aims to perform some biochemical functions using cells within an artificially-created support system such as the artificial pancreas or bio artificial liver. This book contains some path-breaking studies in the field of tissue engineering. It strives to provide a fair idea about this discipline and to help develop a better understanding of the latest advances within this field. This book is a resource guide for experts as well as students.

After months of intensive research and writing, this book is the end result of all who devoted their time and efforts in the initiation and progress of this book. It will surely be a source of reference in enhancing the required knowledge of the new developments in the area. During the course of developing this book, certain measures such as accuracy, authenticity and research focused analytical studies were given preference in order to produce a comprehensive book in the area of study.

This book would not have been possible without the efforts of the authors and the publisher. I extend my sincere thanks to them. Secondly, I express my gratitude to my family and well-wishers. And most importantly, I thank my students for constantly expressing their willingness and curiosity in enhancing their knowledge in the field, which encourages me to take up further research projects for the advancement of the area.

Editor

Integration of Rabbit Adipose Derived Mesenchymal Stem Cells to Hydroxyapatite Burr Hole Button Device for Bone Interface Regeneration

Viswanathan Gayathri, Varma Harikrishnan, and Parayanthala Valappil Mohanan

Biomedical Technology Wing, Sree Chitra Tirunal Institute for Medical Sciences and Technology, Poojappura, Thiruvananthapuram, Kerala 695 012, India

Correspondence should be addressed to Parayanthala Valappil Mohanan; mohanpvl0@gmail.com

Academic Editor: Sean Peel

Adipose Derived Mesenchymal Stem Cells, multipotent stem cells isolated from adipose tissue, present close resemblance to the natural *in vivo* milieu and microenvironment of bone tissue and hence widely used for in bone tissue engineering applications. The present study evaluates the compatibility of tissue engineered hydroxyapatite burr hole button device (HAP-BHB) seeded with Rabbit Adipose Derived Mesenchymal Stem Cells (ADMSCs). Cytotoxicity, oxidative stress response, apoptotic behavior, attachment, and adherence of adipose MSC seeded on the device were evaluated by scanning electron and confocal microscopy. The results of the MTT (3-(4,5-dimethylthiazol)-2,5-diphenyl tetrazolium bromide) assay indicated that powdered device material was noncytotoxic up to 0.5 g/mL on cultured cells. It was also observed that oxidative stress related reactive oxygen species production and apoptosis on cell seeded device were similar to those of control (cells alone) except in 3-day period which showed increased reactive oxygen species generation. Further scanning electron and confocal microscopy indicated a uniform attachment of cells and viability up to 200 μm deep inside the device, respectively. Based on the results, it can be concluded that the in-house developed HAP-BHB device seeded with ADMSCs is nontoxic/safe compatible device for biomedical application and an attractive tissue engineered device for calvarial defect regeneration.

1. Introduction

Burr hole cranial neurosurgical procedures often retain cosmetically abhorrent puckered scars on the scalp over burr hole sites [1]. Reconstructive surgeries using burr hole button fabricated from biomimetic bone substitutes are currently being used to renovate the residual depression on the scalp over the burr hole [2]. Even though the conventional method seems to be promising, strategies on improvising the local integration and healing process are always demanding [3].

The recent advancement in tissue engineering and regeneration technologies has paved the way for a conceptual shift in the otherwise conventional practices in orthopedic and reconstructive surgery of using prosthesis and other implantable devices [4]. Use of biodegradable scaffolds integrated with biological cells, particularly stem cells, has shown

to promote the repair and/or regeneration of tissues plus excellent integration to the surrounding tissues. The stem cell based therapy aims at integrating endogenous mesenchymal stem/progenitor cells (MSCs) with appropriate three-dimensional scaffolds along with other biochemical cues to enhance the healing process particularly the local integration of the construct to the surrounding tissues [5].

Adipose Derived Mesenchymal Stem Cells (ADMSCs) have become a focus of research due to their multipotential properties, high responsiveness to distinct environmental cues, and ease of isolation [6]. Adipose tissue can be easily harvested from subcutaneous tissue through percutaneous or minima open aspiration techniques [7]. The use of ADMSCs to treat calvarial defects is well documented [8]. Implanted poly(lactic-coglycolic acid) scaffolds seeded with ADMSCs promoted complete bone bridging in 12 weeks in a rat model

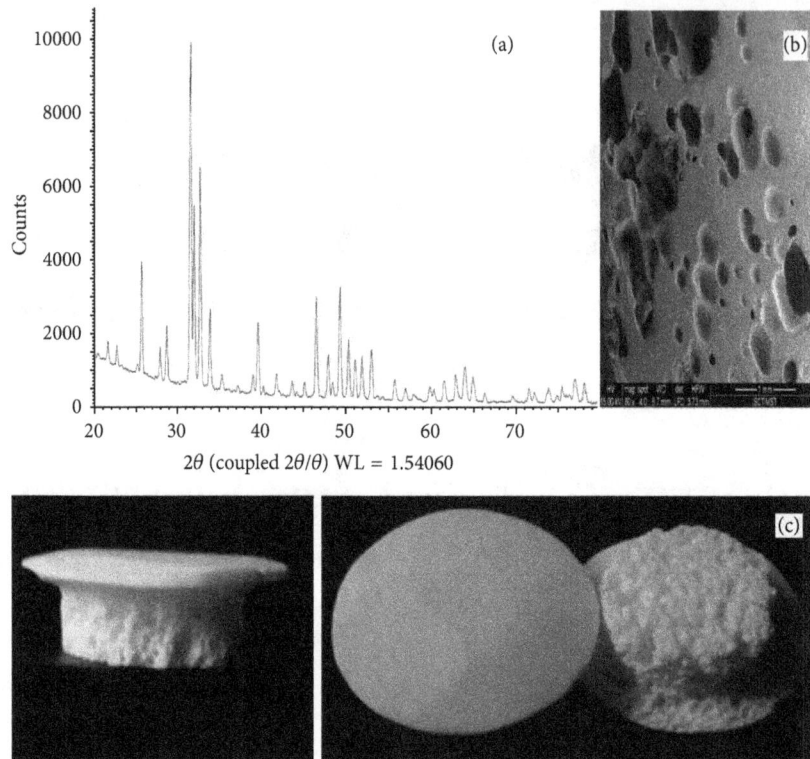

FIGURE 1: (a) Physicochemical characterization of HA burr hole buttons device using XRD spectrum of the powdered device material with the standard pattern of hydroxyapatite (PDF 9-432) superposed. (b) SEM micrograph of HA Burr Hole Buttons showing porosity distribution. (c) HA-Burr Hole buttons.

of calvarial defects [9]. The osteogenic potential of adipose stem cells on scaffolds has been examined in *in vitro* cell cultures and *in vivo* animal models, for successful healing and accelerated regeneration process in calvarial and other bone defects [10]. So, it is hypothesized that osteogenesis/ osteoinductive factors produced by the adipose stem cells along with a biomimetic bone substitute such as hydroxy apatite would aid in engineering an ideal bone substitute to heal burr hole or other calvarial defects.

As in any tissue engineering procedures, the possible clinical outcome depends on the cell survival in 3D scaffolds which is again defined by various factors such as adequate seeding density, cell attachment and proliferation, and diffusion of nutrients and oxygen supply. Also the immediate cell death after implantation due to hypoxia and other stress factors is generally thought to be the cause of failure of bone tissue engineering strategies. This could be taken by culturing the stem cells on the porous scaffold for a short period ensuring sufficient cell growth prior to implantation onto the defect site instead of the on the spot repair as seen in conventional treatment modalities. The size of the device will be manipulated depending on the size of the defect.

Hence, the objective of the present study is to evaluate the interaction and cytocompatibility of adipose mesenchymal derived stem cells (ADMSCs) and in-house developed hydroxyapatite burr hole device with potential implication on

treating burr holes and other calvarial defects using stem cell based tissue engineering strategies.

2. Materials and Methods

2.1. Preparation of Hydroxyapatite Burr Hole Button Device (HAP-BHB). The HAP-BHB device (Figure 1(c)) [will be referred to as "device" in the text] developed by the Biomedical Technology Wing of the Sree Chitra Tirunal Institute for Medical Sciences and Technology (Patent India 495/CHE/2006 dated 20/3/2006) consisted of two parts; a column made of porous hydroxyapatite (HA) which "dips" into the burr hole defect (pore size range: 100–300 mm), and a dome which "sits" on the surrounding bone made of dense HA (pore size less than 100 mm). It has a convex surface and could be contoured with mechanical drills. The thickness of the dome was 2 mm at the centre and 1 mm at the periphery. The column of the device was 3-4 mm long which was designed for easy insertion into the burr hole with the dome of the device being as superficial cover. The button could be implanted at any burr hole site irrespective of the location. Precipitation method was employed for the preparation of hydroxyapatite, consisting of calcium and phosphate salts, and the precipitate was washed thoroughly to remove ions adsorbed on the surface. The precipitate was later gel casted, dried slowly, and shaped according to requirements. Sintering was done at 1200°C to get the final ceramic. The Ca/P ratio was 1.67 mimicking mammalian bone. The implants were subjected to standard sterilization techniques [1].

2.2. Isolation, Culture, and Characterization of Rabbit Adipose Derived Mesenchymal Stem Cells (ADMSCs) Using Specific Surface Marker (CD+ 90).

Mesenchymal stem cells were isolated from rabbit's adipose tissue. The adipose tissues which were collected from the cadaver as per the approval from Institutional Animal Ethics Committee (CPCSEA, Committee for the Purpose of Control and Supervision on Experiments on Animals, India) and in accordance with the approved Institutional protocol. ADMSCs isolation was done as described elsewhere [11]. Briefly, 10 g of tissue was incubated with 30 mL of 1.5 mg/mL collagenase 1 (Invitrogen) at 37°C with continuous shaking for 30–45 min. Enzymatic dissociation of tissue was stopped by the addition of double the volume of serum-containing medium and the resultant suspension was passed through 180 μm nylon mesh (Millipore). Cells were washed by centrifugation. The cell pellet was resuspended in complete medium consisting of high glucose Dulbecco's Modified Eagle's Medium (DMEM; Gibco), 10% fetal bovine serum (FBS; Gibco), 100 units/mL penicillin, 100 μg/mL streptomycin, and 0.25 μg/mL Fungizone (Antibiotic-Antimycotic (100x) Gibco, USA) solution. The cells were seeded onto a 25-cm^2 tissue culture polystyrene dish (TCPS) and kept at 37°C under 5% CO_2. The medium was replenished at 72 h intervals [11]. The medium changes continued until the cells reach approximately 80% confluence. Thereafter, the cells were subjected to trypsinization, 0.25% trypsin-EDTA (Gibco, USA), as described elsewhere [12]. ADMSCs, in passage 2 or 3, were selected for all the in vitro experiments. In brief, the plastic-adherent cells after second passage were analyzed with a positive marker (CD+ 90). Finally, the cells were washed and imaged using fluorescent microscope (Leica DMIL, Germany).

2.3. Evaluation of Plastic Adherence of ADMSCs by Actin Staining.

ADMSCs were grown on TCPS and later stained for F-actin. After incubation for 24 h at 37°C, the cells were washed with PBS and fixed with freshly prepared 4% para formaldehyde for 20 min. Permeabilization of adhered cells was carried out by incubation with 0.1% Triton-X for 2 min at room temperature. Soon after, the cells were incubated in rhodamine phalloidin (Sigma, USA) diluted 1 : 100 in phosphate buffer saline (PBS) for 15 min, following which the cells were washed and treated with Hoechst (Sigma, USA) for staining of nuclei for 5 min. Finally, the cells were washed and imaged using fluorescent microscope (Leica DMIL, Germany).

2.4. Assessment of Material Mediated Cytotoxicity of ADMSCs by MTT (3-(4,5-Dimethylthiazol)-2,5-diphenyl tetrazolium bromide) Assay.

The cytotoxicity of suspensions of powdered material on ADMSCs cells was assessed by MTT (3-(4,5-dimethylthiazol)-2,5-diphenyl tetrazolium bromide) assay. Briefly, ADMSCs at a density of 20,000–40,000 cells/mL were seeded onto 96-well tissue culture plates and cultured in DMEM with 10% FBS. After 24 h of seeding, the medium was replaced by the medium containing device powdered material at different concentrations (0.0005 g/mL, 0.005 g/mL, 0.05 g/mL, and 0.5 g/mL, resp.) and cultured again for 24 h. Cells supplied with culture media alone served as negative control. After the incubation period, the metabolically active cells were quantified using MTT assay and compared with untreated control. Phenol (1%) served as positive control and ADMSCs alone as negative control. For MTT assay, 10 μL of MTT dye (5 mg/mL in PBS) was added to each well in dark and incubated for 3 h at 37°C. The optical density was measured spectrophotometrically following dissolution in dimethyl sulfoxide (DMSO) at 540 nm (Elx 808iu Ultra Microplate Reader, Bio-Tek Instruments, USA).

2.5. Evaluation of Cell Morphology of ADMSC Seeded on the Device (In Vitro) Using Scanning Electron Microscopy.

ADMSCs were seeded (1 × 10^6 cells/material) on all sides of preconditioned device and were placed in 35 mm^2 single well plate (Nunc, Germany) under static condition. The device was coated on all the sides by the ADMSCs in the medium. After being incubated at 37°C for 1 hour to allow cell attachment, DMEM with 10% FBS (Gibco, USA) was added for 24 h. The seeded device was maintained for 3, 7, and 14 days, independently [13]. The ADMSC seeded device was washed with PBS, fixed in 1% glutaraldehyde in Sorensen phosphate buffer for 24 hours and dehydrated in a graded ethanol series and the morphology of adhered and attached ADMSC on the device was visualized by Scanning Electron Microscopy (Hitachi 2500).

2.6. Measurement of Oxidative Stress Potential (Reactive Oxygen Species Level).

The generation of ROS was monitored by employing 2,7, Dichloro Dihydro Fluorescein Diacetate (DCF-DA) [Invitrogen, USA]. DCF-DA fluoresces when oxidized by the intracellular ROS generation. Briefly, ADMSCs (20,000–40,000 cells) were preincubated with DCF-DA at a concentration of 100 μM. After washing with serum-free medium, the cells were treated with varying concentrations of powdered device material (0.0005, 0.005, 0.05, and 0.5 g/mL, resp.) for 2 h, at 37°C. Hydrogen peroxide (0.09% H_2O_2) treated cells were used as positive control for DCF-DA analysis. Cells were then washed twice in serum-free medium and the resulting fluorescence was quantified using a fluorescence microplate reader with an excitation wavelength of 488 nm and emission wavelength of 53 nm. The values were normalized to the negative control (ADMSCs alone) and respective fold change was calculated.

2.7. Evaluation of Live/Dead ADMSC Seeded on the Device Using FDA and PI Staining by Confocal Microscopy.

Fluorescence-based live-dead assays were used to evaluate the viability of cells. Simultaneous use of two fluorescent dyes allows a two-color discrimination of the population of live cells from the dead-cell population. Seeded ADMSCs on the device were stained with fluorescein diacetate (FDA) and propidium iodide (PI), which stain viable cells and dead cells, respectively. Briefly, the staining solution (culture medium without fetal calf serum (FCS) 5 mL + FDA (5 mg/mL) 8 μL +

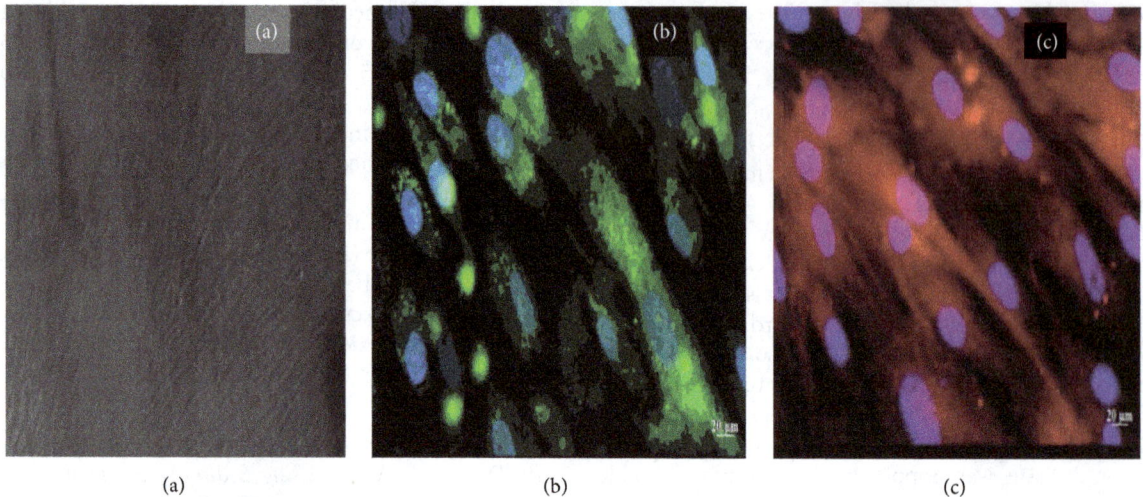

(a)   (b)   (c)

FIGURE 2: (a) Spindle-shaped morphology of rabbit adipose MSCs. (b) Staining of CD+ 90, specific cell surface marker. (c) Actin staining showing cytoskeletal arrangement.

PI (2 mg/mL) 50 μL) was added upon the device seeded with ADMSCs and was incubated at room temperature for 4 to 5 minutes in the dark. Later, staining solution was removed and subsequently PBS was added to wash the device seeded with ADMSCs. Samples were analyzed with confocal laser microscopy [14]. Future studies will concentrate to measure the bone regenerating ability of the HAP-BHB and to demonstrate the efficacy of the device.

2.8. Flow Cytometric Evaluation of Apoptosis by Annexin V/Propidium Iodide (PI) Staining. Briefly, ADMSCs were seeded on the device for different time periods (3, 7, and 14 days) independently. Later, the device seeded with the ADM-SCs were washed with PBS and trypsinized. The cell suspension was then double stained with Annexin V-Alexa Fluor and propidium iodide as per the manufacturer's protocol (Vybrant Apoptosis Assay Kit, Molecular Probes, Invitrogen, USA). The stained cells were then analyzed for apoptosis using BD FACSAria III flow cytometer.

2.9. Statistical Analysis. For all experiments, three separate devices were used (n = 3). All the measurements were done in triplicate in order to confirm the repeatability. Each parameter was expressed as mean of all values ± standard deviations. One-way analysis of variance (ANOVA) was employed to assess the statistical significance of results. p values less than 0.05 were considered significant.

3. Results

3.1. Physicochemical Characterization of HAP-Burr Hole Buttons. The phase purity of hydroxyapatite is ascertained by XRD compared with powder diffraction data (JCPDS 9-432) standard. The calcium to phosphorus ratio was estimated by EDTA titration and spectrophotometer method and observed as 1.66. The flexural strength is estimated in the three-point bending method and is found to be 139 MPa with a standard

deviation 25 (Figure 1(a)). Microstructure study has been carried out on a Hitachi scanning electron microscope. The figure is a SEM micrograph of the interface showing porosity as well as the high sintered characteristics (Figure 1(b)).

3.2. Isolation, Culture, and Expansion of ADMSCs and Characterization by CD+ 90 Marker. The ADMSCs were isolated from a heterogeneous population by virtue of their plastic adherence property and were successfully propagated in primary culture and in serial culture on TCPS. They attached in colonies and spread out from individual colonies. The ADMSCs exhibited their characteristic fusiform, spindle-shaped morphology from day 2 (Figure 2(a)). Medium given was DMEM high glucose supplemented with 10% fetal bovine serum. The culture attained 80% confluence by 7 days. Cells of passages 2-3 were used for all the studies. The ADMSCs were characterized by the specific surface marker CD+ 90 staining and viewed under fluorescent microscope (Figure 2(b)).

3.3. Evaluation of Cytoskeletal Arrangement and Plastic Adherence Property of ADMSCs by Actin Staining. ADMSCs are polystyrene/plastic adherent cells and are shown to spread quickly upon adhesion with an increase of cytoskeleton-associated actin (F-actin). Thus, the staining of cytoplasmic actin filaments was performed to demonstrate the plastic adherence property of ADMSCs. Fluorescent images of actin staining are shown in (Figure 2(c)).

3.4. Determination of Cytotoxicity of ADMSCs by MTT Assay. Cell viability was quantitatively estimated employing colorimetric MTT assay that detects mitochondrial activity of the cells. It is based on the reduction of the yellow tetrazolium dye 3-(4,5-dimethylthiazol)-2,5-diphenyl tetrazolium bromide (MTT) to a purple water insoluble formazan in cells bearing intact mitochondria and hence reflects the state of cultured cells. Results depicted that percentage viability of ADMSCs when treated with powdered device material

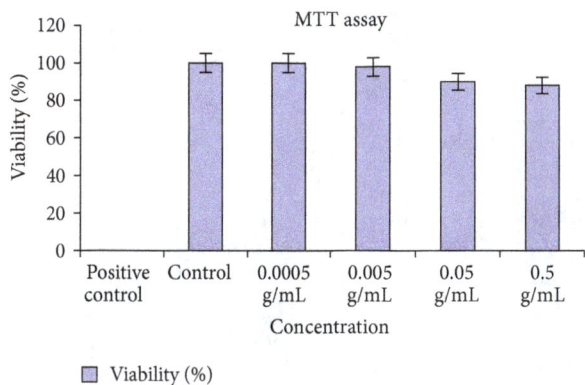

FIGURE 3: Cell viability by MTT assay. All values are expressed as mean ± Std. deviation ($n = 6$).

(Figure 3) was similar, when compared with negative control (cells alone). 1% phenol served as positive control. There was no significant difference in the values with the increase in concentration powdered device materials up to 0.5 g/mL that was found to be noncytotoxic when compared with control.

3.5. Evaluation of Cell Morphology and Attachment of ADMSC Seeded on the Device Using Scanning Electron Microscopy.
As evident from scanning electron micrographs, ADMSCs attached and adhered on the device from day 3 onwards. Three-day seeded ADMSCs on the device micrographs depicted typical ADMSCs morphology (Figures 4(a) and 4(b)). Eventually, ADMSCs were increased and distributed evenly over the device to give a cell sheet like appearance with filopodia (Figure 4(c)) on day 7. On day 14, ADMSCs formed sheets and layers of cells which were seen on both sides of the device (Figures 4(d) and 4(e)).

3.6. Measurement of ROS Generation of ADMSCs Seeded on the Device.
Figure 5 represents the intracellular ROS generation level of ADMSCs after exposure to powdered device material at different time periods (3, 7, and 14 days). The study showed a 10-percent increase in ROS generation on 3-day treated period when compared to control. Subsequently, this increase in ROS production on day 3 was neutralized at 7 and 14 days.

3.7. Evaluation of Live/Dead ADMSC Seeded on the Device Using FDA and PI Staining by Confocal Microscopy.
Confocal images showed that ADMSCs were significantly attached on porous surface of the device and exhibited viability (Figure 6(a)). Overlay images of porous surface is shown in Figure 6(b). Also, necrotic cells were absent on the porous surface. Similarly, dense surface of the device (Figure 6(c)) showed attachment of ADMSCs. Figure 6(d) shows overlay image of dense surface. Necrotic cells were very negligible on the dense surface also. Eventually, Figures 6(e) and 6(f) showed the depth code analysis of ADMSCs attachment and adherence on porous and dense surface of the device, respectively. It was noted that viable cells were attached up to 200 μm inside the devices.

3.8. Evaluation of Device Seeded with ADMSCs for Apoptosis Using Annexin V by Flow Cytometry.
Early and late apoptotic cells were detected using Annexin V FITC-propidium iodide kit by flow cytometry analysis (FACS BDAria). The data showed the presence of early apoptotic cells (16%) on the device seeded with ADMSCs on 3 days when compared to control (0.0%); the percentage of late apoptotic cells as well as necrotic cells was found to be negligible. This may be due to the initial stress experienced by the cell on acclimatization to the new substrate. However, following incubation, it was found that the apoptotic cells were not significant when compared with cells (ADMSCs) alone control. The data on day 7 and day 14 could not be correlated as events evaluated were not comparable with the 3rd day data. Figure 7 depicts graphical representation of percentage live/apoptotic/necrotic cells.

4. Discussion

Calvarial and craniotomy defect rebuilding is still a challenge to many tissue engineers. Bone has the extraordinary capacity to heal without scar formation, but this regenerative capacity is impaired in patients with large bone lesions and patients with impaired wound healing process. An ideal bone substitute is one which can repair or replace large bone defects especially calvarial segmented defects. Ideally bone substitute should show (a) biocompatibility, (b) osteoconduction, (c) osteointegration, (d) osteoinduction, (e) osteogenesis, and (f) vascularisation [15]. Available materials for the abovementioned applications display many of these properties, whereas properties like osteoinduction and vascularisation potential are exceptional in the available therapies. These processes are associated with the presence of growth factors or cells within the scaffold/material/device. Here lies the importance of tissue engineering, which utilizes a 3D porous biomaterial with cells and growth factors to provide structural support after neurosurgery procedures.

With hydroxy apatite being a natural constituent of bone and teeth, most of its applications are focused on the area of bone regeneration and dental restoration. Tissue engineered hydroxyapatites are particularly appealing because of the high biomimetic morphologies for applications such as orthopedic implant coating, bone substitute filler, and burr hole buttons during craniotomy [16].

Cytocompatible evaluation is the preliminary step in evaluating the toxicological behavior of any biomaterial. In this study, the results showed no evident abnormal morphological lesions in the cells (ADMSCs) exposed to device. Subsequently, the viability of ADMSCs obtained from MTT assay at all the exposed concentration of the powdered device material showed no cytotoxicity. Thus, mechanical damage to the cells was not observed.

The duration of culturing of cells on the materials *in vitro* prior to *in vivo* studies is a crucial feature that ensures the success of tissue engineered constructs. Some studies suggested cells to be cultured on the scaffolds for few hours, while, in other systems, the cells were cultured for one or more weeks before implantation [17]. Former approach gives an idea of cell attachment and proliferation time and the

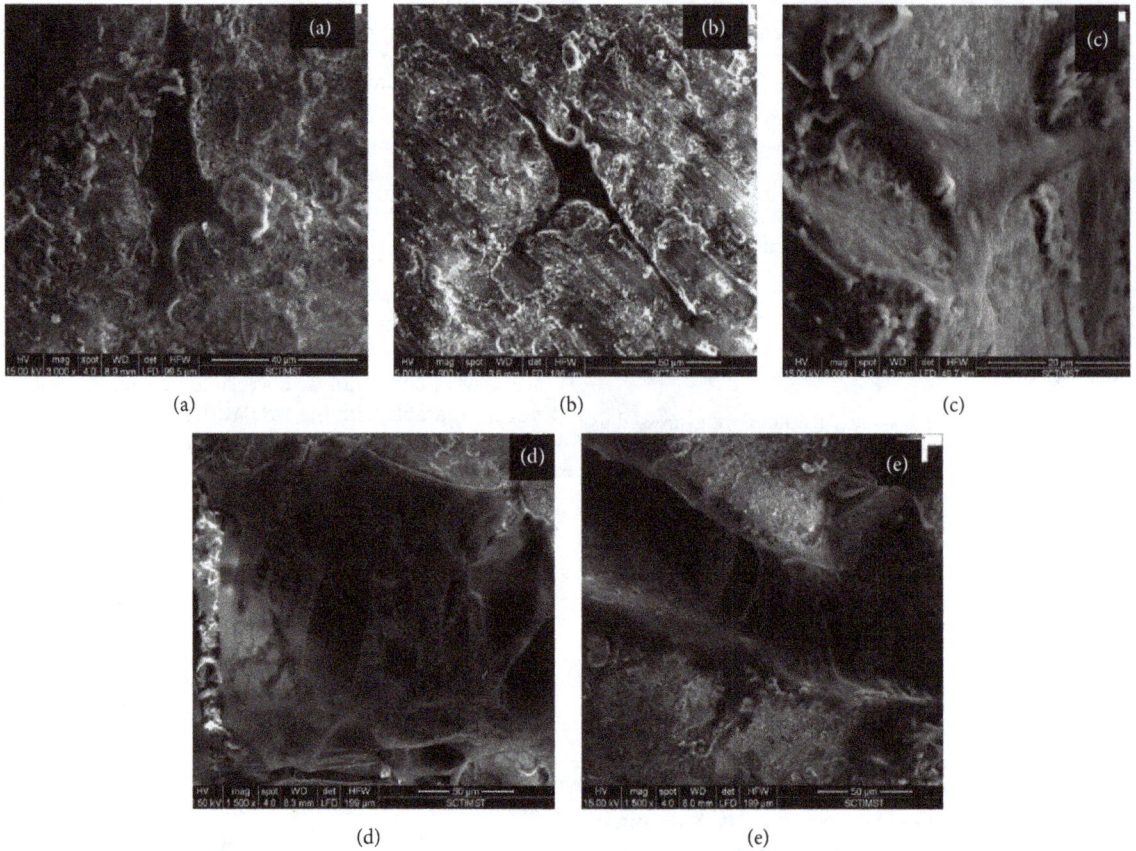

FIGURE 4: Scanning electron micrographs showing attachment and adhesion of ADMSCs on the device. (a and b) Day 3 attachment of ADMSCs on the device. (c) Day 7 attachment of ADMSCs on the devices showing filopodia formation. (d and e) Day 14 attachment of ADMSCs on the device showing layers of cells (ADMSCs) on the device.

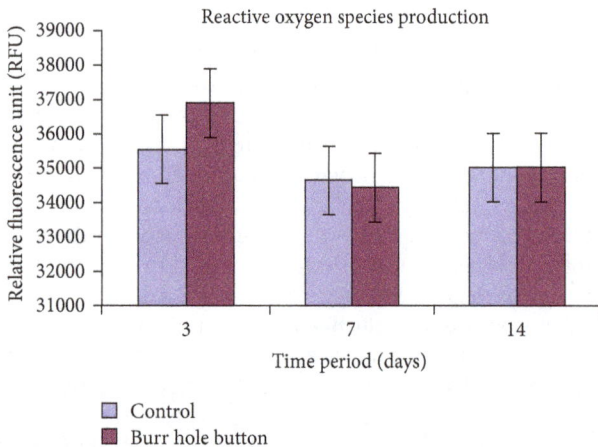

FIGURE 5: Measurement of the level of reactive oxygen species. All values are expressed as mean ± Std. deviation from three different replicates.

latter approach creates an extracellular matrix on the scaffold that provides signals towards the defect site. Several studies with rat bone marrow mesenchymal stem cells established that the bone formation was generated more effectively by culturing cells on the scaffold for a short *in vitro* culture period (days 1–8) [18]. Keeping the above facts in mind, the present study adopted both approaches to view cell attachment and adherence. Therefore, a period of 3, 7, and 14 days was selected to study the attachment and adherence of the cells (ADMSCs) on the device. The performance of cells (morphology, viability and attachment, and adherence) on the scaffolds was examined initially in an *in vitro* cell culture system to ensure that the cells can attach and adhere. Likewise, the SEM images and confocal microscopy depicted the morphology, attachment, proliferation, and viability of ADMSCs on the device, respectively.

The toxic effect of material at cellular level is evident by cellular death. Two distinct pathways adopted for cell death are apoptosis and necrosis. Acute cellular dysfunction in response to severe stress conditions leads to necrosis which is a relatively passive process coupled with rapid cellular depletion. The executor behind this fact is oxidative stress rising due to disturbance in the prooxidative, antioxidative balance resulting in cell damage [19]. Apoptosis is a form of programmed cell death that occurs during several pathological situations and is a complex process characterized by cell shrinkage, chromatin condensation, and formation of blebbing and apoptotic bodies [20]. However, 3-day result showed an increase in ROS level by the ADMSCs seeded

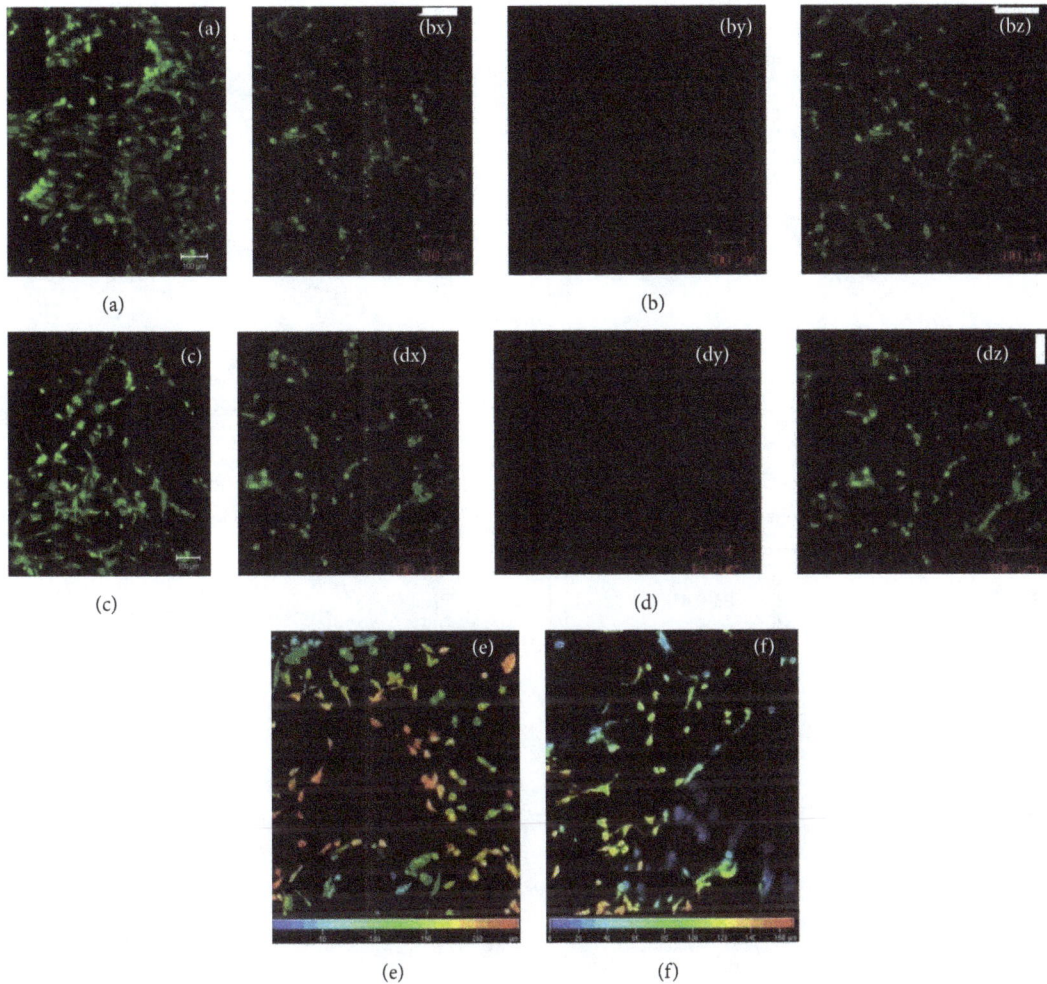

FIGURE 6: Confocal images showing ADMSCs attachment on device. (a) Porous surface of the material. (bx, by, and bz) Overlay image of live and dead cells on porous surface of the material (c) Dense surface of the material. (dx, dy, and dz) Overlay image of live and dead cells on dense surface of the material (e) Depth code analysis of cell viability on the porous surface. (f) Depth code analysis of cell viability on the dense surface.

on the device. But, this effect was seen neutralized in 7 and 14 days, respectively. Consequently, change in ROS level on the 3rd day could be the consequence of adapting to the new environment which subsequently subsided on further incubation.

Formation of new vasculature on the defect site is an important parameter for osteo induction and for osteogenesis to happen [21]. Various *in vivo* studies show the rapid vasculature with BMSc and ceramics [22]. Hartman et al. in 2004 [23] found that muscle recipient site could favour bone formation in a cell based bone graft substitute compared to subcutaneous recipient site due to high vascularity of the muscle tissue.

Tissue-specific differentiation of MSCs is dependent on the local microenvironment in which these cells reside and their interactions with the host tissue. Rat bone marrow mesenchymal cells covered on nanohydroxyapatite coated on silk fibroin Nano-HAp/silk. Fibroin sheets initiates adhesion and proliferation [24]. Similarly, the observations from

the current study suggest enhanced ADMSCs adherence, attachment, and viability on the device which is necessary for the bone regeneration and vasculature. Preliminary cytocompatibility data obtained from this study implies the use of device seeded with ADMSCs for bone tissue engineering applications. The efficacy of the device will be demonstrated on the rat model with a calvarial defect for 1-, 4-, and 12-week period.

5. Conclusions

In this study, the interaction and cytocompatibility of adipose mesenchymal derived stem cells (ADMSCs) and in-house developed device are evaluated as a preliminary step in projecting hydroxyapatite burr hole integrated with mesenchymal stem cells to improvise the healing process and local integration of burr hole button in neurosurgery patients. The device is found to be compatible with the stem cells

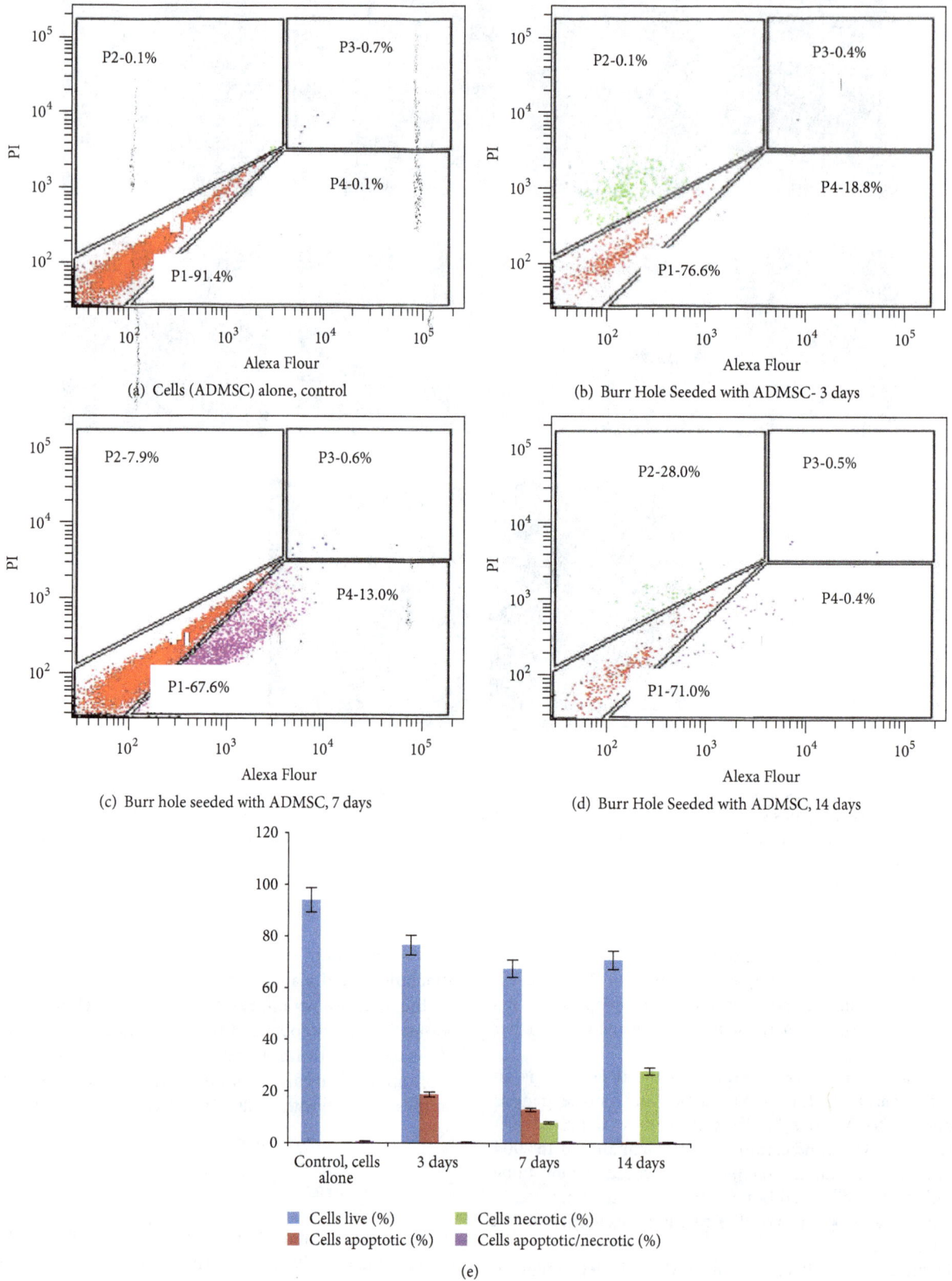

FIGURE 7: Flow cytometry evaluation for apoptosis of ADMSCs seeded onto the device. Annexin V assay. (a) Unstained ADMSCs control; (b) 3 days; (c) 7 days; (d) 14 days; (e) graphical representation of quantitative data ($n = 3$). All values are expressed as mean ± SD from three different replicates.

seeded and it can be concluded that the in-house developed HAP-BHB device seeded with ADMSCs are nontoxic/safe compatible device for biomedical application and may be an attractive tissue engineered device for calvarial defect regeneration in future.

Conflict of Interests

The authors declare that there is no conflict of interests.

Acknowledgments

The authors are thankful to the Director and Head, Biomedical Technology Wing, Sree Chitra Tirunal Institute for Medical Sciences and Technology, for providing the facilities to carry out the work. They also thank all the members who helped in this work.

References

[1] H. V. Easwer, A. Rajeev, H. K. Varma, S. Vijayan, and R. N. Bhattacharya, "Cosmetic and radiological outcome following the use of synthetic hydroxyapatite porous-dense bilayer burr-hole buttons," *Acta Neurochirurgica*, vol. 149, no. 5, pp. 481–485, 2007.

[2] H. Kashimura, K. Ogasawara, Y. Kubo, K. Yoshida, A. Sugawara, and A. Ogawa, "A newly designed hydroxyapatite ceramic burr-hole button," *Vascular Health and Risk Management*, vol. 6, no. 1, pp. 105–108, 2010.

[3] N. L. Fulmer, G. M. Bussard, T. J. Gampper, and R. F. Edlich, "Anorganic bovine bone and analogs of bone mineral as implants for craniofacial surgery: a literature review," *Journal of Long-Term Effects of Medical Implants*, vol. 8, no. 1, pp. 69–78, 1998.

[4] A. R. Amini, C. T. Laurencin, and S. P. Nukavarapu, "Bone tissue engineering: recent advances and challenges," *Critical Reviews in Biomedical Engineering*, vol. 40, no. 5, pp. 363–408, 2012.

[5] C. N. Salinas and K. S. Anseth, "Mesenchymal stem cells for craniofacial tissue regeneration: designing hydrogel delivery vehicles," *Journal of Dental Research*, vol. 88, no. 8, pp. 681–692, 2009.

[6] R. Hass, C. Kasper, S. Böhm, and R. Jacobs, "Different populations and sources of human mesenchymal stem cells (MSC): a comparison of adult and neonatal tissue-derived MSC," *Cell Communication and Signaling*, vol. 9, article 12, 2011.

[7] A. Park, M. V. Hogan, G. S. Kesturu, R. James, G. Balian, and A. B. Chhabra, "Adipose-derived mesenchymal stem cells treated with growth differentiation factor-5 express tendon-specific markers," *Tissue Engineering Part A*, vol. 16, no. 9, pp. 2941–2951, 2010.

[8] M. Barba, C. Cicione, C. Bernardini, F. Michetti, and W. Lattanzi, "Adipose-derived mesenchymal cells for bone regeneration: state of the art," *BioMed Research International*, vol. 2013, Article ID 416391, 11 pages, 2013.

[9] C. M. Cowan, Y.-Y. Shi, O. O. Aalami et al., "Adipose-derived adult stromal cells heal critical-size mouse calvarial defects," *Nature Biotechnology*, vol. 22, no. 5, pp. 560–567, 2004.

[10] C. Romagnoli and M. L. Brandi, "Adipose mesenchymal stem cells in the field of bone tissue engineering," *World Journal of Stem Cells*, vol. 6, no. 2, pp. 144–152, 2014.

[11] Y. Zhu, T. Liu, K. Song, X. Fan, X. Ma, and Z. Cui, "Adipose-derived stem cell: a better stem cell than BMSC," *Cell Biochemistry and Function*, vol. 26, no. 6, pp. 664–675, 2008.

[12] N. S. Remya, S. Syama, V. Gayathri, H. K. Varma, and P. V. Mohanan, "An in vitro study on the interaction of hydroxyapatite nanoparticles and bone marrow mesenchymal stem cells for assessing the toxicological behaviour," *Colloids and Surfaces B: Biointerfaces*, vol. 117, pp. 389–397, 2014.

[13] H.-T. Liao, J.-P. Chen, and M.-Y. Lee, "Bone tissue engineering with adipose-derived stem cells in bioactive composites of laser-sintered porous polycaprolactone scaffolds and platelet-rich plasma," *Materials*, vol. 6, no. 11, pp. 4911–4929, 2013.

[14] B. Gantenbein-Ritter, C. M. Sprecher, S. Chan, S. Illien-Jünger, and S. Grad, "Confocal imaging protocols for live/dead staining in three-dimensional carriers," *Methods in Molecular Biology*, vol. 740, pp. 127–140, 2011.

[15] V. Campana, G. Milano, E. Pagano et al., "Bone substitutes in orthopaedic surgery: from basic science to clinical practice," *Journal of Materials Science: Materials in Medicine*, vol. 25, no. 10, pp. 2445–2461, 2014.

[16] M. C. Kruyt, S. M. van Gaalen, F. C. Oner, A. J. Verbout, J. D. de Bruijn, and W. J. A. Dhert, "Bone tissue engineering and spinal fusion: the potential of hybrid constructs by combining osteoprogenitor cells and scaffolds," *Biomaterials*, vol. 25, no. 9, pp. 1463–1473, 2004.

[17] S. Seitz, K. Ern, G. Lamper et al., "Influence of in vitro cultivation on the integration of cell-matrix constructs after subcutaneous implantation," *Tissue Engineering*, vol. 13, no. 5, pp. 1059–1067, 2007.

[18] J. van den Dolder, J. W. M. Vehof, P. H. M. Spauwen, and J. A. Jansen, "Bone formation by rat bone marrow cells cultured on titanium fiber mesh: effect of in vitro culture time," *Journal of Biomedical Materials Research*, vol. 62, no. 3, pp. 350–358, 2002.

[19] J. Chandra, A. Samali, and S. Orrenius, "Triggering and modulation of apoptosis by oxidative stress," *Free Radical Biology & Medicine*, vol. 29, no. 3-4, pp. 323–333, 2000.

[20] S.-R. Yang, I. Rahman, J. E. Trosko, and K.-S. Kang, "Oxidative stress-induced biomarkers for stem cell-based chemical screening," *Preventive Medicine*, vol. 54, supplement, pp. S42–S49, 2012.

[21] M. Fröhlich, W. L. Grayson, L. Q. Wan, D. Marolt, M. Drobnic, and G. Vunjak-Novakovic, "Tissue engineered bone grafts: biological requirements, tissue culture and clinical relevance," *Current Stem Cell Research & Therapy*, vol. 3, no. 4, pp. 254–264, 2008.

[22] T. Tanaka, M. Hirose, N. Kotobuki, H. Ohgushi, T. Furuzono, and J. Sato, "Nano-scaled hydroxyapatite/silk fibroin sheets support osteogenic differentiation of rat bone marrow mesenchymal cells," *Materials Science and Engineering C*, vol. 27, no. 4, pp. 817–823, 2007.

[23] E. H. M. Hartman, J. W. M. Vehof, J. E. de Ruijter, P. H. M. Spauwen, and J. A. Jansen, "Ectopic bone formation in rats: the importance of vascularity of the acceptor site," *Biomaterials*, vol. 25, no. 27, pp. 5831–5837, 2004.

[24] M. Zandi, H. Mirzadeh, C. Mayer et al., "Biocompatibility evaluation of nano-rod hydroxyapatite/gelatin coated with nano-HAp as a novel scaffold using mesenchymal stem cells," *Journal of Biomedical Materials Research Part A*, vol. 92, no. 4, pp. 1244–1255, 2010.

2

Application of Synthetic Polymeric Scaffolds in Breast Cancer 3D Tissue Cultures and Animal Tumor Models

Girdhari Rijal, Chandra Bathula, and Weimin Li

Department of Biomedical Sciences, Elson S. Floyd College of Medicine, Washington State University, Spokane, WA 99210, USA

Correspondence should be addressed to Weimin Li; weimin.li@wsu.edu

Academic Editor: Jie Deng

Preparation of three-dimensional (3D) porous scaffolds from synthetic polymers is a challenge to most laboratories conducting biomedical research. Here, we present a handy and cost-effective method to fabricate polymeric hydrogel and porous scaffolds using poly(lactic-co-glycolic) acid (PLGA) or polycaprolactone (PCL). Breast cancer cells grown on 3D polymeric scaffolds exhibited distinct survival, morphology, and proliferation compared to those on 2D polymeric surfaces. Mammary epithelial cells cultured on PLGA- or PCL-coated slides expressed extracellular matrix (ECM) proteins and their receptors. Estrogen receptor- (ER-) positive T47D breast cancer cells are less sensitive to 4-hydroxytamoxifen (4-HT) treatment when cultured on the 3D porous scaffolds than in 2D cultures. Finally, cancer cell-laden polymeric scaffolds support consistent tumor formation in animals and biomarker expression as seen in human native tumors. Our data suggest that the porous synthetic polymer scaffolds satisfy the basic requirements for 3D tissue cultures both *in vitro* and *in vivo*. The scaffolding technology has appealing potentials to be applied in anticancer drug screening for a better control of the progression of human cancers.

1. Introduction

2D *in vitro* cell culture models have been instrumental in addressing various questions and providing invaluable knowledge in the field of cancer cell biology for decades. With the advancement of research technologies, some of the drawbacks of 2D cell culture models have been identified that include the lack of cell-ECM interactions and differences in cell morphology, proliferation rate, viability, polarity, motility, differentiation, and sensitivity to therapeutics compared to the characteristics of cells *in vivo* [1–6]. These limitations of 2D culture systems have become hindrance to the progress of our understanding of the mechanisms of cancer initiation and progression and of developing therapeutic approaches to treat human cancers, highlighting the needs for better culture platforms that are able to closely mimic tissue environments where native cancer cells live.

With the integration of the spatial concept, various 3D cell culture systems have been developed to overcome the limitations of 2D cultures. There is a remarkable increase in the use of 3D cultures over the past 10 years [7], resulting in many interesting findings that are distinct from the effects seen in

the traditional 2D cultures. For instance, cells grown in 3D cultures display changes in metabolic characteristics, such as increased glycolysis [8], in gene expression patterns, such as upregulation of VEGF and angiopoietin genes involved in angiogenesis [9–11], and in production of chemokines, such as interleukin-8 [12], compared to cells grown on 2D surfaces. It is noteworthy that genome wide gene expression analysis comparing gene expression patterns of U87 cells grown in 2D and 3D cultures with a cohort of 53 pediatric high grade gliomas revealed significant similarities between the 3D, but not the 2D, culture samples and the human brain tumors [13]. Moreover, several studies have shown increased chemoresistance of cancer cells grown in 3D systems compared to the cells in 2D cultures [14–16], recapitulating the responses of cancer cells to chemotherapeutics *in vivo*. Depending on scientific questions to be addressed and specific experimental design, 3D scaffolds applied in biomedical research are predominantly fabricated using either natural materials, such as native tissue proteins and algae, or synthetic polymers, such as PLGA, PCL, and poly(ethylene glycol) (PEG) [7, 17, 18]. The advantages of synthetic polymeric scaffolds are their abundant availability, low cost, suitability for large-scale

3D bioprinting and reconstruction of certain tissue structures, and flexibility to be tailored to meet specific physical requirements of different culture systems [19–24].

In this study, we focused on characterizing the efficacies of applying the synthetic polymer scaffolds fabricated using PLGA and PCL with a modified gas foaming approach for 3D tissue cultures and animal models in breast cancer research. The viability, morphology, proliferation, and receptor expression of breast cancer cells as well as their responses to anticancer drug and development into tumors in animals with the support of the 3D scaffolds were investigated.

2. Materials and Methods

2.1. Polymer Coating on Slides. Microscopic glass slides were cleaned with 70% ethanol, air-dried in a biological safety cabinet, coated with 2% of PCL (Sigma-Aldrich), PLGA (Sigma-Aldrich), or PCL and PLGA (1:1 ratio) dissolved in chloroform (Sigma-Aldrich) for 1 hour, air-dried in a biological safety cabinet, and rinsed twice with 1x PBS before cell seeding.

2.2. 3D Porous Scaffold Fabrication from Polymer. To fabricate porous scaffolds with similar pore sizes as decellularized mouse breast tissues ($\sim$100 μm) [16], 1.0 gram of PLGA or 0.5 gram of PCL was dissolved in 1 ml of chloroform followed by adding 1 gram of sodium bicarbonate ($NaHCO_3$, Sigma-Aldrich) into the solution. The solutions were slowly dispensed into the semispherical molds (4 mm in diameter) built in porcelain panels, which were kept in -80°C freezer for 1-2 hours and freeze-dried at -50°C for 48 hours as described previously [16]. The scaffolds were then washed in 0.1 N hydrochloric acid (HCl) solution for 6 hours (replacing the solution hourly) at room temperature to generate the pores after releasing CO_2 produced by the $NaHCO_3$ and HCl reactions from the scaffolds, followed by washing in distilled water for several times until the pH of the water became neutral. The scaffolds were soaked in 70% ethanol for 3–5 hours, washed three times in 1x PBS for 10 minutes, and kept in 1x PBS until use. The generation of PCL and PLGA combined scaffolds was achieved by mixing equal volume (1:1 ratio) of PCL and PLGA solutions and following the same procedures as described above.

2.3. In Vitro 2D and 3D Cultures. MCF10A cells (American Type Culture Collection, ATCC) were maintained in 1x DMEM/F12 50/50 (Mediatech) supplemented with 10 μg/ml insulin, 20 ng/ml EGF, 0.5 μg/ml hydrocortisone, 100 ng/ml cholera toxin, 5% horse serum, and 1% Penicillin-Streptomycin. MDA-MB-231 cells (ATCC) were maintained in 1x DMEM (Mediatech) supplemented with 10% FBS and 1% Penicillin-Streptomycin. The polymer-coated slides (circular, 12 mm in diameter; ThermoFisher Scientific) or the scaffolds were placed in 24-well or 96-well plates, washed several times with sterile 1x PBS, and preconditioned with culture medium. MCF10A or MDA-MB-231 cells suspended in the respective culture medium were seeded on the slides or scaffolds (1×10^5 per slide or scaffold) and allowed to attach to the matrices for 45 minutes. The cells were then cultured in the medium under

optimal conditions (37°C, 5% CO_2) and collected at indicated time points, analyzed, or used in downstream experiments. For longer period of culturing time, the culture medium was replenished every other day.

2.4. Live and Dead Cell Assay. The cell cultures on the polymer-coated slides or polymer scaffolds were briefly washed with 1x PBS (37°C) twice and incubated in the Live/Dead Cell Staining solution (2 μM of calcein-AM and 4 μM of EhtD-III in 1x PBS, PromoKine) at room temperature for 30 minutes. Images were captured using epifluorescence microscopy (Zeiss Axio Imager M2). Live cells take the calcein-AM stain and fluorescence green under EGFP filter, while dead cells take the EthD-III stain and fluorescence red under Texas Red filter.

2.5. Proliferation Assay. The proliferation of the cells grown on the coated slides and scaffolds was measured using the CCK-8 reagent (Sigma-Aldrich) at the time points indicated. Briefly, CCK-8 solution was added at a 1:10 dilution into the cultures and incubated (37°C, 5% CO_2) for 3 hours. The supernatants of the cultures were collected and the colorimetric reactions within the supernatants that reflect the proliferation status were measured using a Synergy 2 microplate reader (BioTek) for the absorbance at 490 nm. Error bars represent standard deviations (SD) of the means of three independent experiments.

2.6. Cell Surface Receptor Expression. The cells cultured on the polymer-coated slides at about 80% confluency were washed with cold 1x PBS twice and fixed in 4% cold paraformaldehyde. Immunofluorescence staining was performed as previously described [25] using primary antibodies against integrin-α2 (mouse, Santa Cruz Biotechnology, sc-74466) and collagen type 1 (rabbit, Abcam, ab34710) as set 1 as well as integrin α6 (rabbit, Invitrogen, 710201) and laminin-β3 (mouse, Santa Cruz Biotechnology, sc-33178) as set 2 along with Alexa Fluor® dye-conjugated anti-rabbit and anti-mouse (Thermo Fisher Scientific) secondary antibodies to detect the expression of the integrin receptors on the surface of the cells in response to the polymer matrices.

2.7. Response of Cells to Anticancer Drug. T47D breast cancer cells (ATCC, 1×10^5 cells per scaffold) were seeded on the 3D scaffolds and cultured in 96-well plates for 7 days. (Z)-4-Hydroxytamoxifen (4-HT, Abcam, ab1419430) at the final concentration of 1 μM was administered in alternate days from 7th to 14th day of culture. Cell survival experiment was performed using the Live/Dead assay as described above. Triplicate independent experiments were performed for statistical significance.

2.8. In Vivo Tumor Formation. MDA-MB-231 cells (1×10^5 cells/scaffold) were seeded on spherical porous PLGA scaffolds (4 mm-diameter) and cultured under optimal conditions (37°C, 5% CO_2) for 24 hours prior to implantation. The blank (without cells as negative controls) and cell-laden scaffolds were implanted into the right and the left 4th inguinal mammary fat pads, respectively, of 8-week-old

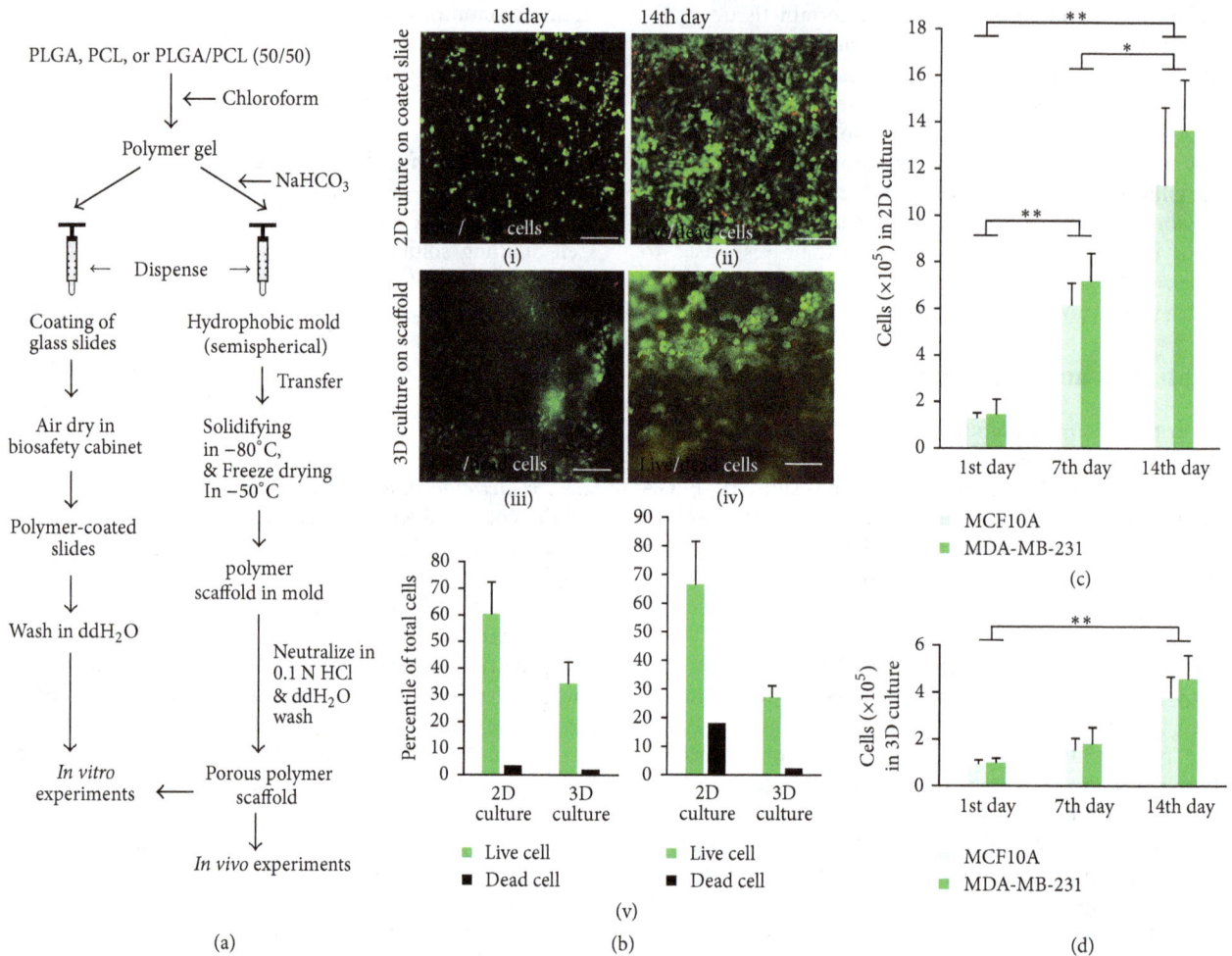

FIGURE 1: *Cell survival, morphology, and growth status on polymeric scaffolds.* (a) Main procedures of preparing PLGA-coated slides and 3D porous PLGA scaffolds. (b) Examination of MDA-MB-231 cell survival on PLGA-coated slides ((b)(i) and (b)(ii)) and porous PLGA scaffolds ((b)(iii) and (b)(iv)) using live and dead assays. Scale bars, 100 μm. The number of live and dead cells in the 2D and 3D cultures were quantified in ((b)(v)). (c) Proliferation rate of MCF10A and MDA-MB-231 cells on PLGA-coated slides. (d) Proliferation rate of MCF10A and MDA-MB-231 cells on porous PLGA scaffolds. $^{*}p < 0.05$, $^{**}p < 0.01$.

female NOD-SCID mice (Charles River Laboratories). Each implantation condition had six replicates. The growth of the tumors was monitored using spectrum computed tomography (CT) on an *in vivo* imaging system (IVIS, PerkinElmer). The tumors were collected into ice-cold 4% paraformaldehyde 4 weeks after implantation, paraffin embedded, cross-sectioned, antigen retrieved (1 mM EDTA solution, 10 mM Tris Base, and 0.05% Tween 20; pH 9.0), and stained with HER2 (rabbit, Cell Signaling Technology, 2165) and Ki-67 (mouse, Cell Signaling Technology, 9449) primary antibodies followed by Alexa fluorophore-conjugated secondary antibodies. Images were captured using fluorescence microscopy as described before [25].

2.9. *Statistical Analysis.* One-way ANOVA was performed using the StatPlus (Build 6.0.0/Core v5.9.92, AnalystSoft) software to analyze the statistical data. Error bars represent standard error of the mean (SEM) of three independent experiments unless otherwise indicated.

3. Results and Discussion

3.1. *Cell Survival, Morphology, and Proliferation on the Polymeric Scaffolds.* To examine the survival of cancer cells grown on the polymeric substrata, human triple (ER, PR, and HER2 receptor) negative breast cancer MDA-MB-231 cells were cultured on PLGA-coated microscopic glass slides (2D) and porous PLGA scaffolds (3D), respectively, as described in the methods and illustrated in Figure 1(a) for 14 days. The Day 1 and Day 14 culture samples were collected and stained with the Live/Dead Cell assay kit as described in the methods. This staining method labels live cells in green color and the dead cells in red color when observing the cells under fluorescence microscope. Our results showed that the number of dead cells detected on PLGA-coated glass slides (Figures 1(b)(i) and 1(b)(v)) or on PLGA 3D scaffolds (Figures 1(b)(iii) and 1(b)(v)) were negligible on Day 1. However, the number of dead cells detected on the PLGA-coated glass slides was markedly higher (Figures 1(b)(ii) and 1(b)(v)) than those on the 3D PLGA porous scaffolds (Figures 1(b)(iv) and 1(b)(v)) on Day 14. The reason for increased cell death in the 2D cultures could be due to the faster proliferation rate of MDA-MB-231 cells on flat surface compared to that of the cells on 3D scaffolds, consistent with the previous observations where other scaffolding materials were used in 3D cell cultures [16, 26]. Because of the nature of the staining and imaging method, some of the cells grown on 3D scaffolds appeared to be out of focus due to the growth of cells at different focal planes/depths of the 3D scaffolds (Figure 1(b)(iii) and 1(b)(iv)).

We next inspected the morphological differences between cancer cells grown on polymeric 2D surfaces and those on polymeric 3D scaffolds. The results showed that the MDA-MB-231 cells grown on the 2D PLGA surfaces were in spindle shapes and populated the surface areas in a more or less universal way (Figure 1(b)(ii)) while those cultured on the 3D PLGA scaffolds exhibited round shapes and expand as cell clusters (Figure 1(b)(iv)). These data are consistent with previous studies that showed distinct morphological features of cells grown on 3D scaffolds compared to those of cells gown on 2D cultures [27–29]. The morphological differences between the two types of cultures could be the results of two factors. One is the distinct physical features of the surfaces of the 2D flat and 3D porous scaffolds, with the former being smooth and even and the latter being rough and uneven because of the existing pores on the surfaces of the scaffolds. The other is the spatial interactions between the neighboring cells and between the cells and the substrata, with the 2D interactions being "bidirectional" at lateral and basal surfaces of the cells and the 3D interactions being "multidirectional" at most or all of the surfaces of the cells. The multidirectional interaction feature of the 3D cell cultures resembles the characteristics of the closed environment of native tissues, where the living cells attach to and interact with the surrounding matrices or/and other cells at all directions. Moreover, even though cancer cells grown in 2D culture can grow on top of each other when the cell population reaches confluency, they hardly form tumoroids as can be commonly achieved in 3D cultures. Therefore, the morphological properties of cancer cells within 3D microenvironments could be a fundamental factor contributing to cancer cell growth, motility, tumor development, and resistance to therapeutic drugs.

Cancer cell proliferation after adapting the living environment is essential for tumor growth. To assess the support of PLGA in cell proliferation both in 2D and in 3D cultures, MCF10A and MDA-MB-231 cells were grown on the PLGA-coated glass slides and the porous PLGA scaffolds for 14 days. Cell proliferation rates were measured using CCK-8 reagent on Day 1, Day 7, and Day 14 of culture. The results showed that MCF10A and MDA-MB-231 cells grown on the PLGA-coated glass slides proliferated rapidly during the culturing time (Figure 1(c)). Though a similar trend was observed in the 3D cultures, the proliferation rate of the cells was substantially lower than that of the 2D cultures (Figures 1(c) and 1(d)). In addition, the proliferation rate of MDA-MB-231 cells is slightly higher than that of MCF10A cells in both 2D and 3D cultures (Figures 1(c) and 1(d)). A similar discrepancy in cell proliferation between 2D and 3D cultures was observed in

our previous studies using native tissue ECM as scaffolding materials [16] and in studies using other materials [22, 30]. However, increased proliferation rate was observed in JIMT1 breast cancer cells grown on Matrigel compared to regular 2D cultures [31], suggesting a cell type- or/and culture system-related phenotype that should be taken into account for different experimental purposes. Overall, the proliferation rates of cell lines displayed in 3D culture models resemble those of tumor models *in vivo*.

3.2. Surface Receptor Expression of the Cells on Polymeric Scaffolds. Type I collagen is one of the major components of breast tissue ECM [16] and integrin $\alpha2\beta1$ is a primary receptor for type I collagen [32]. To investigate whether the synthetic polymers support the expression of ECM proteins and cell surface receptors, MCF-10A and MDA-MB-231 cells were cultured on the PLGA-coated slides for 24 hours, fixed with 4% paraformaldehyde, and stained with antibodies against type I collagen and integrin $\alpha2$ as described previously [25]. Immunofluorescence (IF) microscopy detected strong staining signals of type I collagen and integrin $\alpha2$ in both MCF10A and MDA-MB-231 cells (Figures 2(b), 2(c), 2(f), and 2(g)). Although the MDA-MB-231 cells appeared to be a bit smaller than the MCF10A cells on the PLGA-coated slides, the overall expression of type I collagen and integrin $\alpha2$ in the cancer cells is lower than in the normal MCF10A cells (Figures 2(b)–2(d) and 2(f)–2(h)). These data are consistent with a previous report of lower integrin $\alpha2$ (ITGA2) expression in primary breast cancers compared to normal breast tissues [33]. In addition, basal level expression of integrin $\alpha2\beta1$ may favor breast cancer cell migration and tumor growth [34, 35] since high levels of the integrin receptor inhibit cancer cell migration [36]. We observed a nice colocalization of integrin $\alpha2$ receptors with type I collagen in the cells especially around the edges of the cells, implicating local deposition of type I collagen on the slide surface and the binding of integrin $\alpha2$ receptors to the deposited collagen for the attachment and migration of the cells. Similarly, we examined the expression levels of type I collagen and integrin $\alpha2$ in MCF10A and MDA-MB-231 cells grown on PCL- or PLGA/PCL- (50/50) coated slides, and the results were coherent with those seen on the PLGA substratum (data not shown). Moreover, we did not notice significant differences in the morphology, viability, and cell proliferation of the cells grown on PLGA, PCL, or 3D PLGA/PCL (50/50) surfaces (data not shown). These data suggest that the synthetic polymeric surfaces or scaffolds could be used to study certain aspects of cancer biology. However, care needs to be taken into account in terms of the choices of different types of synthetic materials and fabrication methods to make 3D scaffolds for either biomedical or bioengineering applications owing to different advantages and limitations of the respective approaches compared to some biomaterial-based model systems [7].

3.3. Response of Cells Grown on the Polymeric Scaffold to Drugs. Traditionally, the efficacies of anticancer prodrugs were initially tested in 2D cell culture systems, and the promising drug candidates from these studies were further evaluated in animal experiments before entering clinical

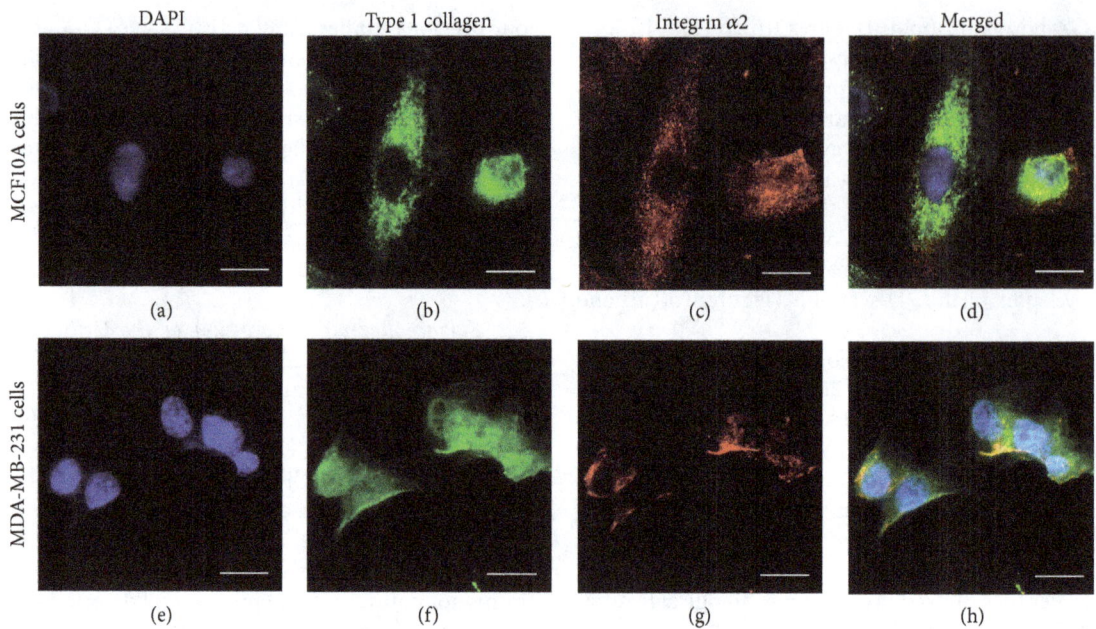

FIGURE 2: *Type I collagen and integrin 2 receptor expression in breast epithelial cells cultured on PLGA-coated surfaces.* The expression and deposition of type I collagen (green) as well as its cell surface receptor integrin $\alpha2$ (red) expression was inspected in MCF10A and MDA-MB-231 cells grown on the PLGA-coated glass slides using IF staining couple with confocal microscopy. The nuclei of the cells were stained with DAPI (Blue). Scale bars, 10 μm.

trials. However, the majority of drug candidates that were efficacious in 2D cultures failed in animal studies or clinical trials. One of the main reasons attributed to these drug testing failures is the inability of 2D culture systems to mimic the natural tissue microenvironment for cells living in them behave as they would be in native tissues. There is increasing evidence supporting 3D tissue cultures as better models to test efficacies of drug candidates [7].

In this study, we examined the effect of 4-hydroxyta-moxifen (4-HT), an active metabolite of the estrogen receptor (ER) antagonist tamoxifen, on the ER-positive luminal A type of breast cancer T47D cells grown on 3D PLGA scaffolds. The cells were treated with 1 μM 4-HT on alternate days starting on Day 7 through Day 13 after cell seeding on the scaffolds, and the images were taken on Day 8 and Day 14 time points. The viability of the cells was analyzed using live and dead cell assays as described before. In consistency with the drug testing results seen in biomaterial-based 3D tissue culture studies [16] and animal models [37–39], T47D cells grown on 3D PLGA scaffolds were less sensitive to 4-HT than those on PLGA-coated slides (data not shown). T47D cells treated with vehicle solvent only showed increased cell proliferation on Day 14 compared to Day 7 and did not show significant differences in cell viability on Day 7 or Day 14 in cells grown on the 3D scaffolds (Figures 3(a)–3(c) and 3(h)–3(j)). However, the proliferation of T47D cells on the scaffolds was markedly decreased after 4-HT treatment as exhibited in the Day 7 and Day 14 images (Figures 3(d)–3(g) and 3(k)–3(n)). A close to complete cell death was observed in 4-HT-treated samples on Day 14 (Figures 3(k)–3(n)). These data collectively support the notion that cancer cells cultured

in 3D environments develop chemoresistance as seen in native tumors and suggest that the polymeric scaffolds can be used as tissue-mimicry environments to study cancer cell responses to therapeutic drugs. The chemoresistance noticed in 3D cultures could be due to the heterogeneous populations of cancer cells and the complex physiochemical properties of the ECM environments deposited by the cells within the 3D structures that affect the permeability of the drugs and overexpression of multidrug resistance proteins [40–42].

3.4. The Polymeric Scaffold Support of Tumor Formation. To assess the capabilities of the polymeric porous scaffolds in supporting tumor formation in mice, porous PLGA scaffolds were coated with MDA-MB-231 cells, cultured for 24 hours, and implanted into mammary fat pads of NOD/SCID mice. Blank scaffolds without cells were used as negative control. Tumor growth was monitored by an *in vivo* imaging system (IVIS) spectrum CT, and tumor sizes were measured with caliper. The tumors were collected 4 weeks after implantation, paraffin embedded, and cross-sectioned for immunohisto-chemistry (IHC) staining and analyzed with IF microscopy. The animal whole body tomographic images taken at the experimental end point demonstrated that tumor lumps were developed from the cancer cell-laden scaffolds, but not the blank scaffold control groups, during the period of observation (Figures 4(a) and 4(f)). The IF images showed that, by end of week 4 after implantation, the scaffolds were occupied by cells as illustrated by DAPI staining of nuclei of the cells on the cross sections of the samples (Figures 4(b) and 4(g)). As expected, the proliferating cell nuclear antigen biomarker Ki-67 was nondetectable in the blank scaffold

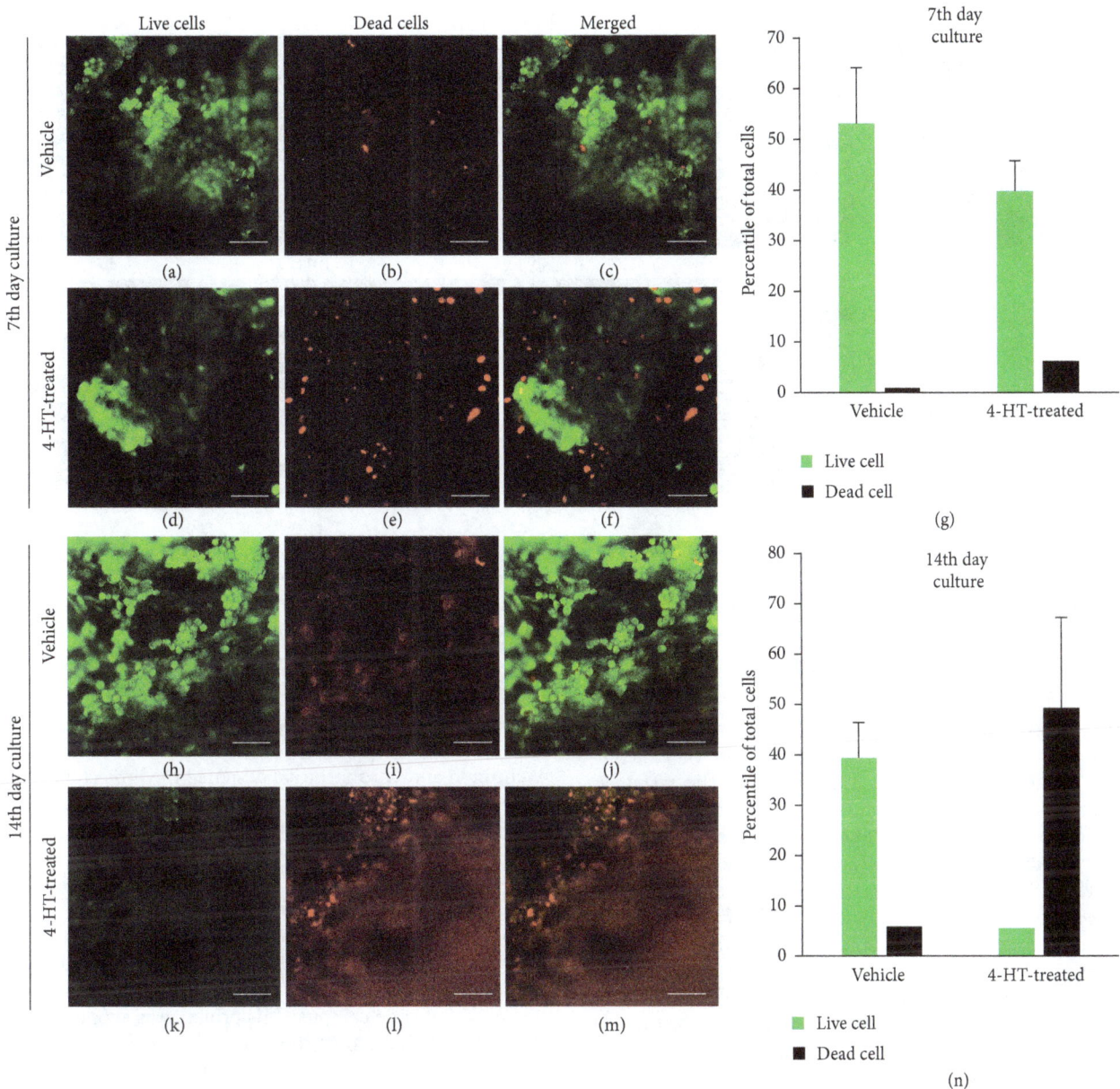

FIGURE 3: *Sensitivity of cancer cells grown on 3D PLGA scaffolds to anticancer drugs.* MDA-MB-231 cells cultured on the scaffolds for 7 days were treated with 4-HT every other day and examined for cell survival on Day 8 and Day 14, respectively. Live cells were indicated by green signals and dead cells in red signals. Scale bars, 100 μm.

implants and was detected at high levels within the tumors derived from the MDA-MB-231 cell-laden PLGA scaffolds (Figures 4(c) and 4(h)), indicating that fast proliferation of the cancer cell population was established in the scaffolds embedded within the native breast tissues. On the other hand, the HER2 receptors, which were negative in the MDA-MB-231 cells, were detected in some of the stromal cells within the normal and tumor tissues but not in the cancer cells (Figures 4(d) and 4(i)). We did not notice significant differences in tumor sizes when comparing the PLGA scaffold groups with the PCL or PLGA/PCL (50/50) scaffold groups in parallel animal experiments (data not shown). These data support

the feasibility of applying the porous polymeric scaffolds in animal tumor model generation with the advantage of consistent tumor formation within reasonable period of time.

4. Conclusions and Remarks

3D cell cultures have overcome many limitations of 2D culture models in cancer biology studies. Our data have added further insights into how the synthetic polymer scaffolds can be successfully used in 3D tissue cultures and in animal tumor models. The phenotypes of the cancer cells we observed in the 3D cultures with regard to survival,

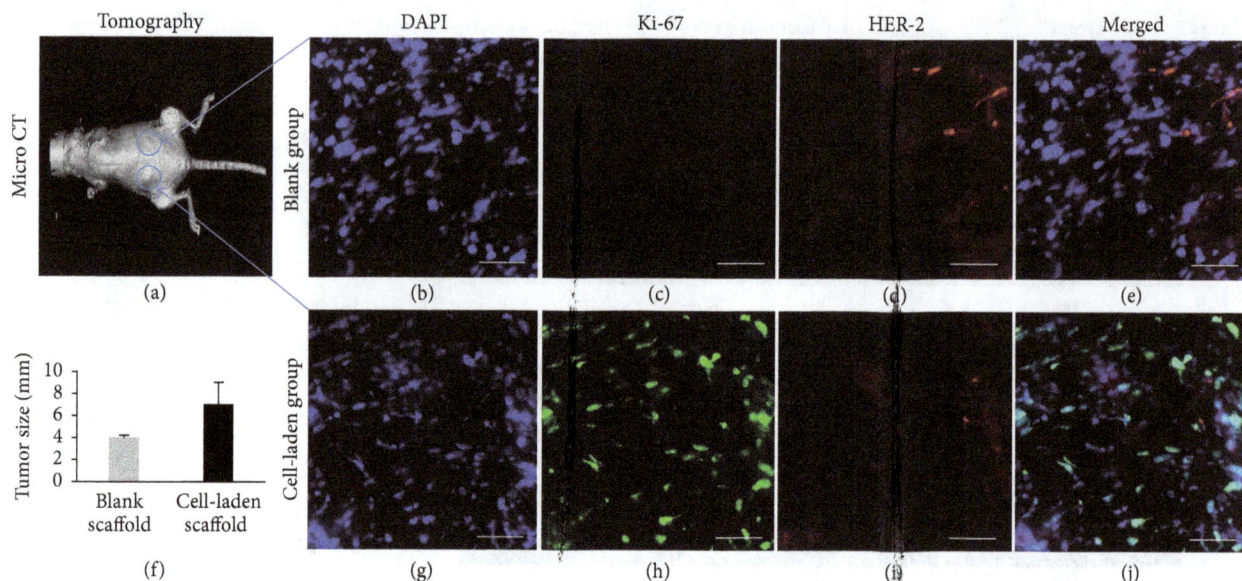

FIGURE 4: *Support of the polymeric scaffolds for tumor formation in mice.* Blank porous PLGA scaffolds (without MDA-MB-231 cells) and MDA-MB-231 cell-laden PLGA scaffolds were implanted into the mammary fat pads of the mice. Tumor growth were dynamically observed using IVIS during 4 weeks of the period ((a) and (f)). The cross sections of the tumors collected at the end point of the experiments were stained for DAPI (blue, (b) and (g)), Ki-67 (green, (c) and (h)), and HER2 (red, (d) and (i)). Scale bars, 100 μm.

morphology, proliferation, type I collagen and its receptor expression, and response to 4-HT treatment are very encouraging for additional research applications of the system in cancer research. For example, cancer cell migration and interaction with other types of cells within the 3D pores of the scaffolds can be studied. Because of the nonbiological features of the polymeric materials, nucleic acids and proteins can be extracted from the 3D cultures for further analysis without interference from biomolecules derived from native tissues.

Since the conventional tumor generation model, which injects cancer cells into the dorsal subcutaneous or mammary fat pads of animals, has big variations in tumor growth [43–45], our 3D porous scaffold-based animal tumor model can be very useful in consistently generating experimental tumors for both biomedical research and preclinical drug screening. Animal tumors produced using this scaffolding method can facilitate the observations of cancer biomarker expression, molecular regulation of cancer progression, and drug efficacies across tumors at similar sizes and developmental stages. Importantly, this easy and economically inexpensive scaffolding method could be adapted to bioengineering and other relevant fields. However, despite the rapid progress in the development of 3D culture models, there is not a one-for-all 3D system that could recapitulate all the features of native human tumors, and each model has its own advantages and disadvantages. Hence, it is important to select the 3D culture systems that best fit specific research purposes.

Conflicts of Interest

The authors declare that there are no conflicts of interest.

Authors' Contributions

Weimin Li and Girdhari Rijal designed the project. Girdhari Rijal performed the experiments. Girdhari Rijal, Chandra Bathula, and Weimin Li wrote and edited the manuscript.

Acknowledgments

The authors thank the colleagues in the Department of Biomedical Sciences for scientific discussion. They also greatly appreciate the consistent supports from the vivarium staff on the WSU Spokane campus. This project was supported by a WSU Startup Fund to Weimin Li.

References

[1] F. Pampaloni, E. G. Reynaud, and E. H. K. Stelzer, "The third dimension bridges the gap between cell culture and live tissue," *Nature Reviews Molecular Cell Biology*, vol. 8, no. 10, pp. 839–845, 2007.

[2] B. M. Baker and C. S. Chen, "Deconstructing the third dimension: how 3D culture microenvironments alter cellular cues," *Journal of Cell Science*, vol. 125, part 13, pp. 3015–3024, 2012.

[3] J. A. Hickman, R. Graeser, R. de Hoogt et al., "Three-dimensional models of cancer for pharmacology and cancer cell biology: capturing tumor complexity in vitro/ex vivo," *Biotechnology Journal*, vol. 9, no. 9, pp. 1115–1128, 2014.

[4] M. J. Bissell, A. Rizki, and I. S. Mian, "Tissue architecture: The ultimate regulator of breast epithelial function," *Current Opinion in Cell Biology*, vol. 15, no. 6, pp. 753–762, 2003.

[5] T. Mseka, J. R. Bamburg, and L. P. Cramer, "ADF/cofilin family proteins control formation of oriented actin-filament bundles in the cell body to trigger fibroblast polarization," *Journal of Cell Science*, vol. 120, part 24, pp. 4332–4344, 2007.

[6] G. D. Prestwich, "Evaluating drug efficacy and toxicology in three dimensions: using synthetic extracellular matrices in drug discovery," *Accounts of Chemical Research*, vol. 41, no. 1, pp. 139–148, 2008.

[7] G. Rijal and W. Li, "3D scaffolds in breast cancer research," *Biomaterials*, vol. 81, pp. 135–156, 2016.

[8] M. T. Santini, G. Rainaldi, R. Romano et al., "MG-63 human osteosarcoma cells grown in monolayer and as three-dimensional tumor spheroids present a different metabolic profile: a (1)H NMR study," *FEBS Letters*, vol. 557, no. 1–3, pp. 148–154, 2004.

[9] U. Cheema, R. A. Brown, B. Alp, and A. J. MacRobert, "Spatially defined oxygen gradients and vascular endothelial growth factor expression in an engineered 3D cell model," *Cellular and Molecular Life Sciences*, vol. 65, no. 1, pp. 177–186, 2008.

[10] M. Valcarcel, B. Arteta, A. Jaureguibeitia et al., "Three-dimensional growth as multicellular spheroid activates the proangiogenic phenotype of colorectal carcinoma cells via LFA-1-dependent VEGF: implications on hepatic micrometastasis," *Journal of Translational Medicine*, vol. 6, article 57, 2008.

[11] P. A. Kenny, G. Y. Lee, C. A. Myers et al., "The morphologies of breast cancer cell lines in three-dimensional assays correlate with their profiles of gene expression," *Molecular Oncology*, vol. 1, no. 1, pp. 84–96, 2007.

[12] C. Fischbach, J. K. Hyun, S. X. Hsiong, M. B. Evangelista, W. Yuen, and D. J. Mooney, "Cancer cell angiogenic capability is regulated by 3D culture and integrin engagement," *Proceedings of the National Acadamy of Sciences of the United States of America*, vol. 106, no. 2, pp. 399–404, 2009.

[13] S. J. Smith, M. Wilson, J. H. Ward et al., "Recapitulation of Tumor Heterogeneity and Molecular Signatures in a 3D Brain Cancer Model with Decreased Sensitivity to Histone Deacetylase Inhibition," *PLoS ONE*, vol. 7, no. 12, Article ID e52335, 2012.

[14] L. David, V. Dulong, D. Le Cerf, L. Cazin, M. Lamacz, and J.-P. Vannier, "Hyaluronan hydrogel: An appropriate three-dimensional model for evaluation of anticancer drug sensitivity," *Acta Biomaterialia*, vol. 4, no. 2, pp. 256–263, 2008.

[15] J. L. Horning, S. K. Sahoo, S. Vijayaraghavalu et al., "3-D tumor model for in vitro evaluation of anticancer drugs," *Molecular Pharmaceutics*, vol. 5, no. 5, pp. 849–862, 2008.

[16] G. Rijal and W. Li, "A versatile 3D tissue matrix scaffold system for tumor modeling and drug screening," *Science Advances*, vol. 3, no. 9, Article ID e1700764, 16 pages, 2017.

[17] Y. Y. Brahatheeswaran, T. Maekawa, and D. S. Kumar, "Polymeric scaffolds in tissue engineering application: a review," *International Journal of Polymer Science*, vol. 2011, Article ID 290602, pp. 1–19, 2011.

[18] M. S. Shoichet, "Polymer scaffolds for biomaterials applications," *Macromolecules*, vol. 43, no. 2, pp. 581–591, 2009.

[19] G. Rijal, B. S. Kim, F. Pati, D.-H. Ha, S. W. Kim, and D.-W. Cho, "Robust tissue growth and angiogenesis in large-sized scaffold by reducing H_2O_2-mediated oxidative stress," *Biofabrication*, vol. 9, no. 1, Article ID 015013, 2017.

[20] A. Naz, Y. Cui, C. J. Collins, D. H. Thompson, and J. Irudayaraj, "PLGA-PEG nano-delivery system for epigenetic therapy," *Biomedicine & Pharmacotherapy*, vol. 90, pp. 586–597, 2017.

[21] S. Taghavi, M. Ramezani, M. Alibolandi, K. Abnous, and S. M. Taghdisi, "Chitosan-modified PLGA nanoparticles tagged with 5TR1 aptamer for in vivo tumor-targeted drug delivery," *Cancer Letters*, vol. 400, pp. 1–8, 2017.

[22] Q. L. Loh and C. Choong, "Three-dimensional scaffolds for tissue engineering applications: role of porosity and pore size," *Tissue Engineering Part B: Reviews*, vol. 19, no. 6, pp. 485–503, 2013.

[23] P. X. Ma and J.-W. Choi, "Biodegradable polymer scaffolds with well-defined interconnected spherical pore network," *Tissue Engineering Part A*, vol. 7, no. 1, pp. 23–33, 2001.

[24] R. Izquierdo, N. Garcia-Giralt, M. T. Rodriguez et al., "Biodegradable PCL scaffolds with an interconnected spherical pore network for tissue engineering," *Journal of Biomedical Materials Research Part A*, vol. 85, no. 1, pp. 25–35, 2008.

[25] W. Li, M. Petrimpol, K. D. Molle, M. N. Hall, E. J. Battegay, and R. Humar, "Hypoxia-induced endothelial proliferation requires both mTORC1 and mTORC2," *Circulation Research*, vol. 100, no. 1, pp. 79–87, 2007.

[26] B. Weigelt, A. T. Lo, C. C. Park, J. W. Gray, and M. J. Bissell, "HER2 signaling pathway activation and response of breast cancer cells to HER2-targeting agents is dependent strongly on the 3D microenvironment," *Breast Cancer Research and Treatment*, vol. 122, no. 1, pp. 35–43, 2010.

[27] J. N. Li, P. Cao, X. N. Zhang, S. X. Zhang, and Y. H. He, "In vitro degradation and cell attachment of a PLGA coated biodegradable Mg-6Zn based alloy," *Journal of Materials Science*, vol. 45, no. 22, pp. 6038–6045, 2010.

[28] K. M. Hakkinen, J. S. Harunaga, A. D. Doyle, and K. M. Yamada, "Direct comparisons of the morphology, migration, cell adhesions, and actin cytoskeleton of fibroblasts in four different three-dimensional extracellular matrices," *Tissue Engineering Part A*, vol. 17, no. 5-6, pp. 713–724, 2011.

[29] E. Knight and S. Przyborski, "Advances in 3D cell culture technologies enabling tissue-like structures to be created in vitro," *Journal of Anatomy*, vol. 227, no. 6, pp. 746–756, 2015.

[30] C. M. Taylor, B. Blanchard, and D. T. Zava, "Estrogen receptor-mediated and cytotoxic effects of the antiestrogens tamoxifen and 4-hydroxytamoxifen," *Cancer Research*, vol. 44, no. 4, pp. 1409–1414, 1984.

[31] V. Hongisto, S. Jernström, V. Fey et al., "High-throughput 3D screening reveals differences in drug sensitivities between culture models of JIMT1 breast cancer cells," *PLoS ONE*, vol. 8, no. 10, Article ID e77232, 2013.

[32] C. G. Knight, L. F. Morton, D. J. Onley et al., "Identification in collagen type I of an integrin $\alpha2\beta1$-binding site containing an essential GER sequence," *The Journal of Biological Chemistry*, vol. 273, no. 50, pp. 33287–33294, 1998.

[33] W. Ding, X.-L. Fan, X. Xu et al., "Epigenetic silencing of ITGA2 by MiR-373 promotes cell migration in breast cancer," *PLoS ONE*, vol. 10, no. 8, Article ID e0135128, 2015.

[34] G. P. H. Gui, J. R. Puddefoot, G. P. Vinson, C. A. Wells, and R. Carpenter, "In vitro regulation of human breast cancer cell adhesion and invasion via integrin receptors to the extracellular matrix," *British Journal of Surgery*, vol. 82, no. 9, pp. 1192–1196, 1995.

[35] A. Lochter, M. Navre, Z. Werb, and M. J. Bissell, "$\alpha1$ and $\alpha2$ integrins mediate invasive activity of mouse mammary carcinoma cells through regulation of stromelysin-1 expression," *Molecular Biology of the Cell*, vol. 10, no. 2, pp. 271–282, 1999.

[36] M. M. Zutter, S. A. Santoro, W. D. Staatz, and Y. L. Tsung, "Reexpression of the $\alpha2\beta1$ integrin abrogates the malignant phenotype of breast carcinoma cells," *Proceedings of the National Acadamy of Sciences of the United States of America*, vol. 92, no. 16, pp. 7411–7415, 1995.

[37] M. Cekanova and K. Rathore, "Animal models and therapeutic molecular targets of cancer: utility and limitations," *Drug Design, Development and Therapy*, vol. 8, pp. 1911–1922, 2014.

[38] R. F. Song, X. J. Li, X. L. Cheng et al., "Paclitaxel-loaded trimethyl chitosan-based polymeric nanoparticle for the effective treatment of gastroenteric tumors," *Oncology Reports*, vol. 32, no. 4, pp. 1481–1488, 2014.

[39] H. Zhang, H. Hu, H. Zhang et al., "Effects of PEGylated paclitaxel nanocrystals on breast cancer and its lung metastasis," *Nanoscale*, vol. 7, no. 24, pp. 10790–10800, 2015.

[40] C. Sánchez, P. Mendoza, H. R. Contreras et al., "Expression of multidrug resistance proteins in prostate cancer is related with cell sensitivity to chemotherapeutic drugs," *The Prostate*, vol. 69, no. 13, pp. 1448–1459, 2009.

[41] X. Xu, M. C. Farach-Carson, and X. Jia, "Three-dimensional in vitro tumor models for cancer research and drug evaluation," *Biotechnology Advances*, vol. 32, no. 7, pp. 1256–1268, 2014.

[42] M. Håkanson, M. Textor, and M. Charnley, "Engineered 3D environments to elucidate the effect of environmental parameters on drug response in cancer," *Integrative Biology*, vol. 3, no. 1, pp. 31–38, 2011.

[43] C. Fischbach, R. Chen, T. Matsumoto et al., "Engineering tumors with 3D scaffolds," *Nature Methods*, vol. 4, no. 10, pp. 855–860, 2007.

[44] B. Kocatürk and H. H. Versteeg, "Orthotopic injection of breast cancer cells into the mammary fat pad of mice to study tumor growth," *Journal of Visualized Experiments*, no. 96, Article ID e51967, 2015.

[45] E. Iorns, K. Drews-Elger, T. M. Ward et al., "A new mouse model for the study of human breast cancer metastasis," *PLoS ONE*, vol. 7, no. 10, Article ID e47995, 2012.

Biomechanical Performances of Networked Polyethylene Glycol Diacrylate: Effect of Photoinitiator Concentration, Temperature, and Incubation Time

Morshed Khandaker,[1] **Albert Orock,**[1] **Stefano Tarantini,**[1]
Jeremiah White,[1] **and Ozlem Yasar**[2]

[1]*Department of Engineering and Physics, University of Central Oklahoma, Edmond, OK 73034, USA*
[2]*Department of Mechanical Engineering, New York City College of Technology, Brooklyn, NY 11201, USA*

Correspondence should be addressed to Morshed Khandaker; mkhandaker@uco.edu

Academic Editor: Feng-Huei Lin

Nutrient conduit networks can be introduced within the Polyethylene Glycol Diacrylate (PEGDA) tissue construct to enable cells to survive in the scaffold. Nutrient conduit networks can be created on PEGDA by macrochannel to nanochannel fabrication techniques. Such networks can influence the mechanical and cell activities of PEGDA scaffold. There is no study conducted to evaluate the effect of nutrient conduit networks on the maximum tensile stress and cell activities of the tissue scaffold. The study aimed to explore the influence of the network architecture on the maximum tensile stress of PEGDA scaffold and compared with the nonnetworked PEGDA scaffold. Our study found that there are 1.78 and 2.23 times decrease of maximum tensile stress due to the introduction of nutrient conduit networks to the PEGDA scaffold at 23°C and 37°C temperature conditions, respectively. This study also found statistically significant effect of network architecture, PI concentration, temperature, and wait time on the maximum failure stress of PEGDA samples (P value < 0.05). Cell viability results demonstrated that networked PEGDA hydrogels possessed increased viability compared to nonnetworked and decreased viability with increased photoinitiator concentrations. The results of this study can be used for the design of PEGDA scaffold with macrosize nutrient conduit network channels.

1. Introduction

Tissue engineering is a new field that allows the combination of engineering, biology, and material methods for developing new techniques with potential to create tissues and organs [1]. The ability of networked three-dimensional structure to elicit altered cell behaviors, including cell adhesion, has raised heightened interest in the scaffold materials for various biomedical applications, including orthopedic repair and regeneration [2]. Cells *in vitro* usually do not reproduce in a three-dimensional fashion unless being allowed to grow on scaffolding. The scaffolds should have appropriate characteristics such as pore size, shape, and mechanical properties to enable cells to grow in every dimension. The cells have to be able to attach, migrate, proliferate, and differentiate into various organs on the scaffold. Several engineered tissue grafts have been developed for the reconstruction of the injured hard and soft tissues [3]. Yasar et al. [4] used Lindenmayer systems, an elegant fractal-based language algorithm framework, in designing vasculature networks that could potentially be incorporated in hydrogel scaffolds like PEGDA. The reason for using PEGDA over other materials is that PEGDA is 3D networked structures that can be manufactured easily to allow for the cell growth at higher depth using photolithograph process. Photolithography is a process which is commonly used in microfabrication to produce the desired scaffolds with a high level of detail and precision. It has been found to be a valid method to manufacture multiple-layer scaffolds for allowing the constructions of channels within the scaffold to better distribute nutrients to the cells.

Yasar et al. [4] study also found that Polyethylene Glycol Diacrylate (PEGDA) tissue scaffolds having thickness higher than 1 mm were shown to have limited applications as a three-dimensional cell culture device due to the inability of cells to survive within the scaffolds. Without access to adequate nutrients, cells placed deep within the PEGDA tissue construct having thickness higher than 1 mm die out, leading to nonuniform tissue regeneration.

Photopolymerization system is usually comprised of three major parts: (1) a UV light source, (2) mold, and (3) a polymer solution. The role of the mold is to allow the PEGDA to polymerize in the desired shape. Tissue scaffolds, with nutrient conduit networks, need to be designed with intricate architecture, porosity, pore size and shape, and interconnectivity in order to provide the required structural strength, transport nutrients, and the microenvironment for cell and tissue ingrowth. By selecting the appropriate unit cell interior structures, structural properties such as the mechanical strength, ductility, and permeability and biological activities such as cell viability, degradation, and tissue generation of PEGDA structure can be controlled. The relationship between the interior nutrient conduit network structure and biomechanical properties (mechanical and biological properties) of PEGDA is not understood yet. Knowledge of the biomechanical properties of the networked PEGDA constructs with respect to the photoinitiator (PI) concentration, temperature, and incubation time is also necessary for adequate design and effective use of PEGDA for tissue engineering constructs.

To understand the effect of nutrient conduit networks on the PEGDA biomechanical performances, this study compared the failure stress of PEGDA flat dumbbell-shaped mold with nutrient conduit networks from the displacement controlled tension tests. Different photoinitiator concentration affects the materials properties due to the difference of crosslink density due to the difference of amount of free radicals created by the PI during the UV photopolymerization process. In addition, PEGDA hydrogels were susceptible to time and temperature dependent degradation, which in general negatively affect the mechanical strength. The failure assessments of the PEGDA molds were conducted in this study as a function of the PI concentration, temperature, and time. In addition, the cell viability assessments of with and without nutrient conduit networks cylindrical PEGDA molds were conducted as a function of the PI concentration. Flat dumbbell shaped PEGDA tension test samples with nutrient conduit networks at the gauge section were designed and fabricated by UV-photo polymerization process. PEGDA solution was added to different concentrations of PI (0.2% and 0.6%) solution to make the specimen. Tension tests were conducted on those samples at different temperatures (23°C and 37°C) and incubation time (0 and 7 days) after the fabrication of the specimen. Human DP147 mesenchymal fibroblasts cells were encapsulated within PEGDA hydrogel having nutrient conduit networks during the photopolymerization of the PEGDA. Cell viability experiments analyzed effects of conduit networks and photoinitiator concentration after 7 days of cell culture on the hydrogels.

2. Materials and Methods

2.1. Materials and Sample Types. Two solutions, PEGDA ($M_n = 700$; Sigma-Aldrich) with the Phosphate Buffer Solution (PBS) solvents, and the photoinitiator (PI), alpha-alpha-dimethoxy-alpha-phenylacetophenone ($M_w = 256.35$ g/mol; Sigma-Aldrich) with the 1-vinyl-2-pyrrolidone ($M_w = 111.14$ g/mol; Fluka) solvents, were used to fabricate the gel solutions. PBS was used instead of water in this study, since PBS is better biological solvent than water when preparing cell encapsulating PEGDA gel. Three samples were made for each experimental group of samples to evaluate the different concentrations of photoinitiator (0.2% and 0.6%) temperatures (23°C and 37°C) and incubation time (0 and 7 days) effect on the tension failure stresses of the samples. The reason for selecting 0.2% and 0.6% concentrations of photoinitiator for PEGDA solution was that in our earlier studies cells were successfully grown in PEGDA structures with significant cell viability differences with those concentrations of photoinitiator [5].

2.2. Mold Preparation to Prepare Mechanical Tests Samples. Figure 1 shows the steps used for the preparation of the mold, specimen, and mechanical tests. A silicon (Casting Craft Easymold Silicone Rubber, Environmental Technology Inc. Fields Landing, CA) mold (Figure 1-①-(a)) was fabricated to make ASTM E855-90 standard [6] flat dumbbell-shaped for mechanical experiments. Two additional ABS plastic pieces (Figure 1-①-(b and c)), fabricated using Dimension Elite 3D Printer (Stratasys, Inc.), were assembled with the silicon mold to fabricate nutrient conduit networked PEGDA specimen. Three pieces of mold were used in this study to generate same nutrient conduit networked PEGDA specimen for all group of test samples, while only silicon mold was used to prepare PEGDA samples without nutrient conduit network. Each plastic piece has an array of holes (diameter: ~2 mm and spacing: ~1 mm). The bottom piece has 14 (7 × 2) holes, whereas the side piece has 21 (7 × 3) holes. A total amount of 35 pins (0.8 mm diameter) were inserted through these holes in the gauge section of the flat dumbbell-shaped mold as shown in Figure 1-①. The purpose of the pins is to produce an array of nutrient conduit network channels at the gauge section of the flat dumbbell-shaped PEGDA sample after UV polarization of PEGDA solution in the mold (Figure 1-②).

2.3. Mold Preparation to Prepare Cell Viability Tests Samples. An open-ended sterile cylindrical borosilicate glass tube was used to prepare mold for cell viability tests samples. One side of the open-ended tube was cured on silicon rubber disc to prepare cylindrical shaped hydrogels on the tube. The glass tube functioned as hydrogel housing as well as providing a support for pins to create nutrient architecture channels on hydrogel inside the tube. In addition, the glass tube provided a novel way to acquire thin section of hydrogel samples for viability assays. To create networked channeled PEGDA hydrogels thin stainless steel pins were carefully placed into the open glass end and secured into the silicon disc (Figure 2).

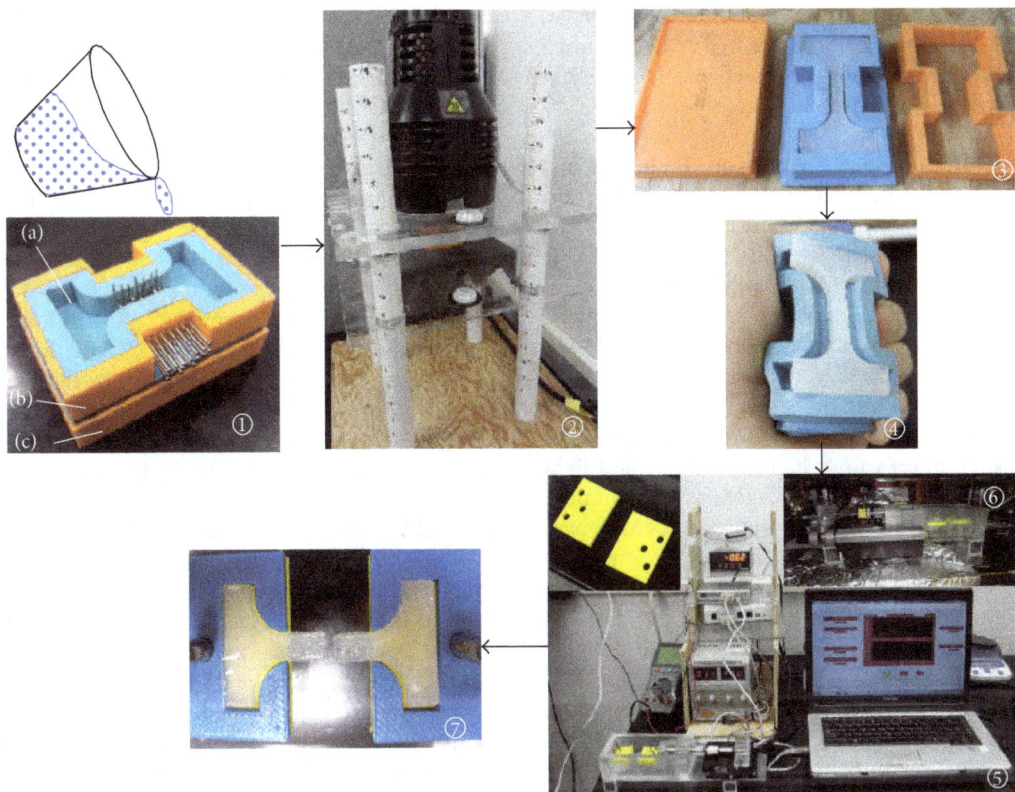

FIGURE 1: Steps that are performed for finding the failure stress of a networked PEGDA. Step 1: 20% PEGDA in PBS mixture was added to the desired concentration of photoinitiator mixture and poured in the custom-made mold to cure the mixture in flat dumbbell shape. Step 2: the solution was exposed to UV light for 3 min. Step 3: the mold was disassembled by the careful removal of pins. Step 4: the silicon mold was flexed to easily extract the specimen without damaging the PEGDA specimen. Step 5: tension test on PEGDA samples at room and incubator (body) conditions. Step 6: analysis of load and displacement data for the calculation of the failure stress of the specimen.

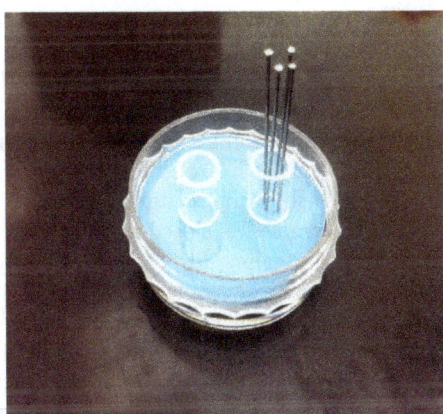

FIGURE 2: Networked and nonnetworked PEGDA hydrogels tissue culture dish with borosilicate glass hydrogel molds on silicon disc. PEGDA hydrogels with encapsulated cells were cured in the molds. Steel pins were inserted to create network conduit channels for networked PEGDA hydrogels (right), where nonnetworked PEGDA hydrogels curing was done without the presence of the pins (left).

2.4. Mechanical Tests Samples Preparation. The 20 wt% PEGDA solution was produced by mixing 2 mL of PEGDA with 8 mL of PBS. The PI solution was produced by mixing

0.3 g of PI with 1 mL of solvent in dark room to prevent premature crosslinking. The 0.2 wt% and 0.6 wt% PI gel solution was produced by mixing 4 μL and 12 μL of PI with 2 mL of PEGDA solution, respectively. The solution was poured in the custom-made mold to cure the mixture in a flat dumbbell-shaped gel. The solution was polymerized by exposure to 365 nm long wave UV (B-100SP Ultraviolet Lamp, UVP, LLC) light. In general, the biocompatibility of PEGDA hydrogel depends on complete polymerization of PEGDA solution while using minimal concentration of photoinitiator. The exposure to UV light causes photoinitiators to generate free radicals that initiate the polymerization to form the hydrogel. Since Mazzoccoli et al. [7] study found that 20 wt% and 40 wt% PEGDA having 0.6 wt% PI is biocompatible, therefore, this study used 0.2 wt% and 0.6 wt% PI gel solution to evaluate the PI concentration effect on failure stress of PEGDA samples. Due to the short-term UV exposure (3 to 5 minutes), photopolymerization is generally considered as a safe method to encapsulate cells [8]. Since encapsulation of cells is the main purpose of creating nutrient conduit networked in PEGDA gels, therefore, UV light was exposed for 3 min for all specimens to form the hydrogels and get the failure stresses of the samples. The pins were carefully removed from the solid plastic pieces after curing. The flexible silicon mold with hydrogel specimen was disassembled from the solid plastic pieces (Figure 1-③). The mold was bended to extract the PEGDA specimen (Figure 1-④). The specimen was stored in PBS solution before the mechanical tests. Eight groups of specimen were prepared.

2.5. Cell Viability Test Samples Preparation. DP147 dermal fibroblast cells, used in hydrogels, were acquired using standard techniques and protocols for culture and isolation. Cells used for culture were incubated at standard conditions, 37°C and 5% CO_2, in tissue culture dishes with Dulbecco's Modified Eagle's medium (DMEM) containing 10% fetal bovine serum (FBS) and 1% antibacterial/antimicrobial (ABAM). Cells for hydrogels were isolated by removing nutrient media, washing with DPBS, and adding trypsin to break up cell tissues and suspend in media for counting. Suspended cells were counted three times with a hemocytometer and light microscope and averaged. After counting the average number of cells per volume the current population density was calculated. Cell suspensions were centrifuged for 5 minutes until cell pellets formed, separating cells from the media. Liquid was suctioned from cell pellet in test tube followed by adding the two hydrogel solutions and mixing thoroughly directly before UV curing [9].

Cultured human DP147 fibroblasts, for hydrogel seeding, were trypsinized and counted to add to hydrogel solutions. The desired hydrogel mixtures were added to the cell pellet and vortexed to ensure thorough mixing. Cell infused PEGDA solution was photopolymerized by UV light. Under the aseptic conditions of a biological safety cabinet hood, custom-made molds, for networked and nonnetworked hydrogels, shown in Figure 1 were placed in covered 35 (mm) tissue dishes for sterile curing. Hydrogel cell solutions were pipetted into a mold and cured in layers under the lamp. Excess liquid was removed under hood after each layer cured, and the following layers were added and photopolymerized. For networked hydrogels, caution was taken during removal of secured pins, not to damage the delicate structures. Finished hydrogel samples were rinsed twice in DPBS to remove the noncured hydrogel liquid solution. Next cured hydrogels were directly placed in new tissue culture dishes containing nutrient media and incubated at standard conditions of 37 degrees Celsius and 5% CO_2. Nutrient media were removed and replenished every three days during incubation period.

2.6. Mechanical Tests. A custom-made tension test setup was designed and fabricated for determining the tensile failure stresses of the specimens at room (laboratory) and physiological (NuAire NU-4750 incubator) temperatures. The complete test setup for conducting mechanical tests of PEGDA samples at room temperature is shown in Figure 1-⑤. The specimens were placed in the holders in an unstressed state. Cover plates, same sizes as the holders, were placed above the specimen to restrict upward movement of the specimen. A precision actuator (Newport™ LTA-HL®) was used in the setup to extend the specimens at a rate of 0.01 mm/sec until failure of specimen. Force was measured using 1 lb load cell (Futek™ LRM200) consistently throughout extension. Load cell was calibrated before testing. The force and displacement data were recorded simultaneously by a user written LabVIEW program 10.0 (National Instruments) from the load cell and actuator, respectively. For conducting mechanical tests of PEGDA samples at physiological temperatures (37°C), an electric connection was developed using the utility side access port of the incubator (NuAire NU-4750) to operate the actuator and load cell inside incubator (Figure 1-⑥), while doing data acquisition outside the incubator. A sample after failure is shown in Figure 1-⑦. The stress-strain curves were developed for each of the samples. Stress was calculated by the magnitude of the force divided by the cross-sectional area (~5.5 mm × ~7.5 mm) at the center of the gage length. Gage length (~15.7 mm) for the specimen was determined as the distance between the holders at the initiation of positive load to the specimen. Stress was calculated by the magnitude of the displacement after the initiation of load force divided by the gage length.

2.7. Cell Viability Assay. Viability of cells infused in PEGDA hydrogels was assessed using the Invitrogen LIVE/DEAD Viability/Cytotoxicity Kit, for mammalian cells (Molecular Probes, Invitrogen), and the fluorescent microscopy techniques. Two probes, calcein AM, and ethidium homodimer-1 (EthD-1) were used in the assay. Invitrogen fluorescence microscopy protocol for the LIVE/DEAD assay was followed. Optical filters were selected for optimum observance of calcein and EthD-1. Two different bandpass filters were chosen for the individual probes resulting in two fluorescent images, red and green. A digital camera, attached to the UV microscope, and computer imaging software captured, saved, and merged the (10x) magnified images. The final third image, combined red and green, consisted of two merged saved images. Viability was calculated from the merged images by counting the number of live green cells and dead red cells. The equation for hydrogel cell viability ((number of live cells) ∗ 100%/(number of live and dead cells)) was applied to each merged image. Multiple assay samples were collected from each hydrogel during culture to show the percent change in cell viability over the hydrogel incubation period.

After the curing, thin sample sections of the incubated hydrogel were collected for hydrogel cell viability at 7 days. Figures 3(a) and 3(b) show the sectioning of cell imbedded hydrogel specimen samples, enclosed in glass tube housing for separated slices by scalpel, for analysis of hydrogel cell viability. Thin even sections ranging from 0.5 to 1.0 (mm) were sampled from the incubated hydrogels with difficulty. Acquiring desired sample sections with certain dimensions for the viability without damaging the hydrogel structure was a difficult task. This process was improved by designing a prototype device (Figure 3(a)) to hold the hydrogel in the glass tube securely. Once the glass was secured the built-in micrometer could be turned to move the hydrogel out of the glass housing in desired increments for samples. A scalpel was used to section thin disks from the hydrogel. After sectioning the sampled hydrogel was rinsed with DPBS and placed back in normal incubation conditions. Sample sections were rinsed twice with DPBS to remove media. Next, the viability assay LIVE/DEAD solution was pipetted onto sample sections, in 35 (mm) tissue culture dishes. Sampled sections were covered with aluminum foil and incubated

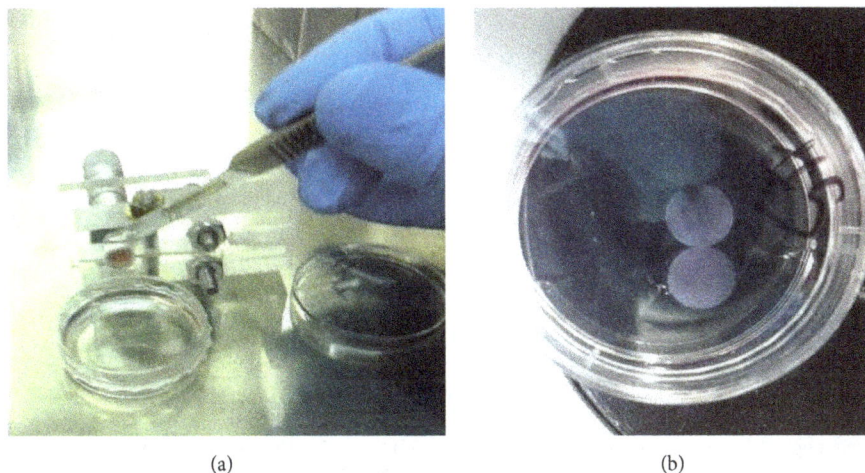

(a)   (b)

FIGURE 3: Sectioning of hydrogel specimen for cell viability experiments. (a) The cured cell infused hydrogel is placed in a custom-made holder equipped with micrometer. The micrometer allows for small increments of PEGDA hydrogel to be pushed outside of the tube to be sliced. (b) Thin hydrogel slices are obtained to be later analyzed under the microscope.

for 75 minutes. After the incubation period, sections were rinsed twice with DPBS, to remove excess stain, and placed on microscope slides that emerged in DPBS to prevent dehydration and remove unwanted liquids. Samples were assayed and viewed with two fluorescent microscope filters to produce an image for viability analysis.

2.8. Statistical Analysis. Statistical analyses were performed using Student's *t*-test for the different groups of specimen using Microsoft Excel 2000 statistical analysis toolkit. Datasets with a *P* value lower than 0.05 were considered significantly different.

3. Results and Discussion

Figure 4 compares the stress-strain curves between PEGDA hydrogels at variable PI concentrations (0.2% and 0.6%) and test temperatures (23°C and 37°C). The stress-strain response of all specimens is characterized as long elastic response, followed by a negligible inelastic region and then stable descending response. This result indicates that all PEGDA samples have brittle fracture behavior. This is reasonable since the conduit networks create three-dimensional voids on the samples.

Figure 5 and Table 1 showed a significant difference of maximum failure stress between the various PEGDA samples fabricated in this study due to photoinitiator (PI) concentration, incubation time, and temperature applied to the specimen during testing. This result clearly shows the higher PI concentration significantly increased the mechanical integrality of the PEGDA gel. This result can be explained with the fact that increasing PI concentration of the samples increased the crosslink density of the polymer matrix due to a larger number of reactive diacrylate groups, which, in turn, increased the failure stress of experimental samples. There is a thermodynamic relationship between the modulus (slope of stress and strain curve) and the crosslink density of

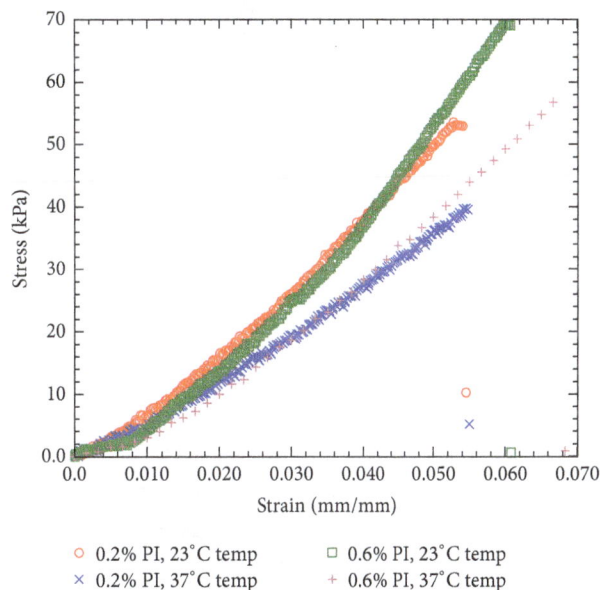

FIGURE 4: Typical stress versus strain diagram derived from the tension test on the flat dumbbell-shaped PEGDA sample having variable PI concentrations (0.2% and 0.6%) and test temperatures (23°C and 37°C). The tension tests were performed on these samples immediately after the preparation of the specimen at 0.01 mm/sec. strain rate.

a given polymer, where the modulus is directly proportional to changes in the crosslink density [10].

In each concentration, longer incubation (7 days versus 0 days) time and higher temperature (37°C versus 23°C) decreased the maximum failure stress of the PEGDA samples. This happens due to the fact that PEGDA undergoes small but significant degradation *in vitro* in PBS buffer solution as found by Xin et al. [11]. Such degradation lowers the mechanical integrity of PEGDA structures.

TABLE 1: Statistical parameters determined from the tensile tests of different kinds of PEGDA samples with nutrient conduit networks.

Test conditions	A	B	C	D	E	F	G	H
Failure stress								
Number of samples	4	6	3	6	6	3	6	4
Average	49.30	36.17	40.45	29.68	62.95	47.64	50.35	39.12
St. dev.	4.24	2.40	2.30	3.37	5.63	3.03	3.90	1.75
P values								
Temperature effect		0.007 (AB)		0.004 (CD)		0.003 (EF)		0.001 (GH)
Incubation time effect	0.031 (AC)		0.007 (BD)		0.003 (EG)		0.039 (FH)	
Photoinitiator concentration effect					0.005 (AE)	0.017 (BF)	0.006 (CG)	0.001 (DH)

PEGDA samples with nutrient conduit networks are represented by letters A to H, where samples A have photoinitiator concentration = 0.2%, test temperature = 23°C, and incubation time = 0 days, samples B have photoinitiator concentration = 0.2%, test temperature = 37°C, and incubation time = 0 days, samples C have photoinitiator concentration = 0.2%, test temperature = 23°C, and incubation time = 7 days, samples D have photoinitiator concentration = 0.2%, test temperature = 37°C, and incubation time = 7 days, samples E have photoinitiator concentration = 0.6%, test temperature = 23°C, and incubation time = 0 days, samples F have photoinitiator concentration = 0.6%, test temperature = 37°C, and incubation time = 0 days, samples G have photoinitiator concentration = 0.6%, test temperature = 23°C, and incubation time = 7 days, and samples H have photoinitiator concentration = 0.6%, test temperature = 37°C, and incubation time = 7 days.
P values from the t-tests of failure stresses of two groups of specimen are represented by ( ).

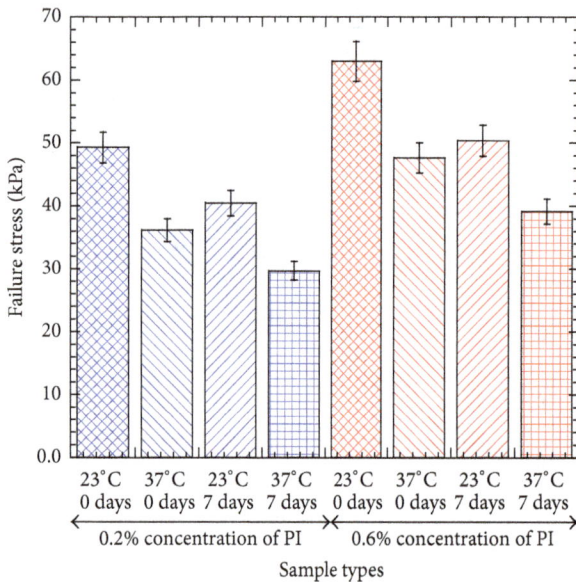

FIGURE 5: Tension test results of different PEGDA specimens showing the variation of the failure stress of the specimen due to photoinitiator (PI) concentration, incubation time, and temperature applied to the specimen during testing.

study, it was found that higher concentration of photoinitiator contributes to toughening the hydrogels, increasing their failure strength. PEGDA tissue constructs had the strongest maximum failure strength when they are at room temperature compared to body temperature. Also PEGDA hydrogel was found to lose strength when tested a week after production under incubation conductions. This result was expected since PEGDA is a biodegradable material. More network conduit channels for cells to be exposed to media nutrient flow should increase the cell viability. The degradation of strength over time could be a result of other factors. The samples may have been exposed to high heat causing damage in the curing process during the fabrication of the scaffold. A device for thin samples needs to be implemented to get more accurate results. Variations on curing times, adding collagen, and cell seeding encapsulation prior to curing could improve the strength in addition as suggested to increase the cell viability [14].

Cells were successfully grown in UV crosslinked PEGDA hydrogel structures with significant viability differences in networked architecture compared to nonnetworked architecture PEGDA samples for both PEGDA gels having 0.2% and 0.6 wt% PI as shown in Figure 6. Statistical significant differences of cell viability between 0.2% and 0.6% PI concentration PEGDA samples were found after 7 days of incubation times for both channel and w/o channel PEGDA samples (P values < 0.05).

Viability results were not as high as expected when compared to 80% fibroblast viability at 14 days observed by other research [1], although network conduit channels exposing cells to nutrient flow demonstrated increased viability. The difference of cell viability results is due to the fact that PEGDA molecular weight and/or UV light intensity in the curing process was different from the previous authors. Variations on curing times and UV intensity, lowering of PI concentration, adding nutrients to hydrogel solution, increasing the cell seeding density, and substituting higher molecular weights of PEGDA could improve the cell viability of PEGDA hydrogel.

This study suggested that failure stress of PEGDA samples was highly dependent on the amount of the PI concentration and the methods by which it was processed. The failure stress of the nutrient conduit networked PEGDA samples is significantly lower than the natural liver tissue tensile stress of 232 kPa [12] and breaking stress of 451 kPa [13]. Higher concentration of PI can be used to increase the failure stress of the PEGDA based tissue engineering scaffolds as liver implant materials. In our earlier studies, it was found that cell viability was not as high in PEGDA scaffolds [5]. The cell viability was found higher for a concentration of PI of 0.2% than 0.6% for without network conduit PEGDA structure, whereas, in this

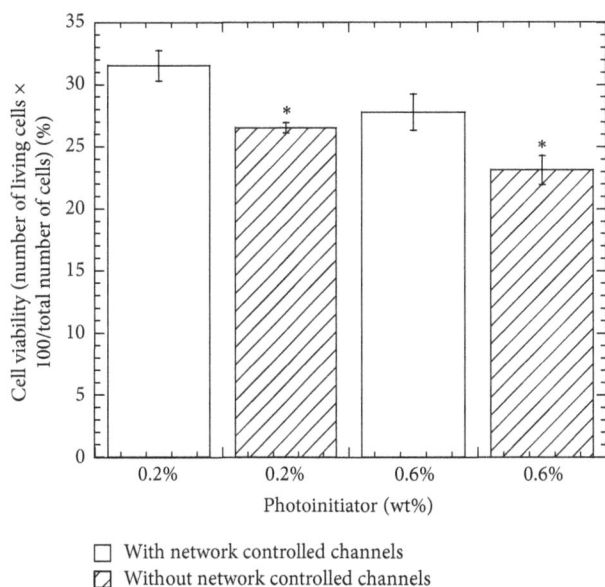

FIGURE 6: Cell viability test results of different PEGDA specimens showing the variation of the cell viability of the specimen due to the presence of network channel and photoinitiator (PI) concentrations of the specimen during testing.

This study is limited to determine the effect of photoinitiator concentration, temperature, and incubation time on the mechanical and cell viability properties of network conduit channel PEGDA. The chemical properties (e.g., degree of polymerization and polymer spectra), surface characteristics (e.g., scanning electron microscope), and physical properties (e.g., swelling ratio, density) are beyond the scope of this study. The authors used same test setup and condition to prepare the different test samples. Since the sample preparation for different groups of sample is identical, therefore, the study assumes the chemical, surface, and physical properties of different test samples groups are identical.

In the future, microsize network conduit channel will be created on PEGDA by adding layered degradable fiber mat inside the PEGDA gel. The effect of photoinitiator concentration, temperature, and incubation time on the mechanical and cell viability properties on such produced PEGDA samples will be determined and will be compared with the macrosize network conduit channeled PEGDA samples results.

4. Conclusion

Tensile failure assessment of nutrient conduit networked PEGDA was conducted as a function of incubation time, test temperature, and photoinitiator concentration. This study concludes that the maximum failure stress of PEGDA can be increased significantly by the degree of photocrosslinking concentration, although significant decrease of failure stress occurs within 7 days of incubation time and at 37°C incubation temperature. Cell assay results demonstrated networked PEGDA hydrogels possessed increased viability compared to nonnetworked and decreased viability with increased

photoinitiator concentrations. Further research using higher molecular weights of PEGDA, improved designs for networked molds, a device for attaining thin uniform assay samples, and the infusion of nutrients in hydrogels could increase the cell viability during incubation. Nutrient conduit networked PEGDA formed hydrogels can be tailored with adequate mechanical properties for various cell-based tissue engineering needs.

Conflict of Interests

The authors have no conflict of interests.

Acknowledgments

This publication was made possible by the funding provided by the Department of Engineering Physics of the University of Central Oklahoma and University of Central Oklahoma Office of Research and Grant. Special thanks go to Dr. Melville Vaughan, Professor in department of biology, for his assistance in cell culture.

References

[1] D. W. Hutmacher, "Scaffolds in tissue engineering bone and cartilage," *Biomaterials*, vol. 21, no. 24, pp. 2529–2543, 2000.

[2] K. Y. Lee and D. J. Mooney, "Hydrogels for tissue engineering," *Chemical Reviews*, vol. 101, no. 7, pp. 1869–1879, 2001.

[3] W. Sun, A. Darling, B. Starly, and J. Nam, "Computer-aided tissue engineering: overview, scope and challenges," *Biotechnology and Applied Biochemistry*, vol. 39, no. 1, pp. 29–47, 2004.

[4] O. Yasar, B. Starly, and S.-F. Lan, "A Lindenmayer systems based approach for the design of nutrient delivery networks in tissue constructs," *Journal of Biofabrication*, vol. 1, no. 4, Article ID 045004, 2009.

[5] O. Yasar, A. Orock, S. Tarantini, J. White, and M. Khandaker, "Mechanical characterization of polyethylene glycol diacrylate (PEGDA) for tissue engineering applications," in *Mechanics of Biological Systems and Materials, Volume 5*, B. C. Prorok, F. Barthelat, C. S. Korach et al., Eds., Conference Proceedings of the Society for Experimental Mechanics Series, pp. 189–195, Springer, New York, NY, USA, 2013.

[6] ASTM, *Annual Book of ASTM Standards: Section 3—Metals Test Methods and Analytical Procedures*, vol. 03.01, ASTM, 1994.

[7] J. P. Mazzoccoli, D. L. Feke, H. Baskaran, and P. N. Pintauro, "Mechanical and cell viability properties of crosslinked low- and high-molecular weight poly(ethylene glycol) diacrylate blends," *Journal of Biomedical Materials Research Part A*, vol. 93, no. 2, pp. 558–566, 2010.

[8] S. Varghese and J. H. Elisseeff, "Hydrogels for musculoskeletal tissue engineering," in *Polymers for Regenerative Medicine*, C. Werner, Ed., vol. 203 of *Advances in Polymer Science*, pp. 95–144, Springer, Berlin, Germany, 2006.

[9] M. B. Vaughan, R. D. Ramirez, S. A. Brown, J. C. Yang, W. E. Wright, and J. W. Shay, "A reproducible laser-wounded skin equivalent model to study the effects of aging in vitro," *Rejuvenation Research*, vol. 7, no. 2, pp. 99–110, 2004.

[10] L. H. Sperling, *Introduction to Physical Polymer Science*, John Wiley & Sons, New York, NY, USA, 2nd edition, 1992.

[11] A. X. Xin, C. Gaydos, and J. J. Mao, "In vitro degradation behavior of photopolymerized PEG hydrogels as tissue engineering

scaffold," in *Proceedings of the Annual International Conference of the IEEE Engineering in Medicine and Biology Society (EMBS '06)*, vol. 1, pp. 2091–2093, New York, NY, USA, September 2006.

[12] J. W. Melvin, R. L. Stalnaker, and V. L. Roberts, "Impact injury mechanisms in abdominal organs," SAE Transactions 730968, 1973.

[13] S. Seki and H. Iwamoto, "Disruptive forces for swine heart, liver, and spleen: their breaking stresses," *Journal of Trauma*, vol. 45, no. 6, pp. 1079–1083, 1998.

[14] C. A. Durst, M. P. Cuchiara, E. G. Mansfield, J. L. West, and K. J. Grande-Allen, "Flexural characterization of cell encapsulated PEGDA hydrogels with applications for tissue engineered heart valves," *Acta Biomaterialia*, vol. 7, no. 6, pp. 2467–2476, 2011.

Sol-Gel Derived Mg-Based Ceramic Scaffolds Doped with Zinc or Copper Ions: Preliminary Results on Their Synthesis, Characterization, and Biocompatibility

Georgios S. Theodorou,[1] Eleana Kontonasaki,[2] Anna Theocharidou,[2]
Athina Bakopoulou,[2] Maria Bousnaki,[2] Christina Hadjichristou,[2] Eleni Papachristou,[2]
Lambrini Papadopoulou,[3] Nikolaos A. Kantiranis,[3] Konstantinos Chrissafis,[1]
Konstantinos M. Paraskevopoulos,[1] and Petros T. Koidis[2]

[1]Department of Physics, Aristotle University of Thessaloniki, 54124 Thessaloniki, Greece
[2]Dentistry Department, Laboratory of Fixed Prosthesis and Implant Prosthodontics, Aristotle University of Thessaloniki,
  54124 Thessaloniki, Greece
[3]Department of Geology, Aristotle University of Thessaloniki, 54124 Thessaloniki, Greece

Correspondence should be addressed to Petros T. Koidis; pkoidis@dent.auth.gr

Academic Editor: Sang-Hoon Rhee

Glass-ceramic scaffolds containing Mg have shown recently the potential to enhance the proliferation, differentiation, and biomineralization of stem cells in vitro, property that makes them promising candidates for dental tissue regeneration. An additional property of a scaffold aimed at dental tissue regeneration is to protect the regeneration process against oral bacteria penetration. In this respect, novel bioactive scaffolds containing Mg^{2+} and Cu^{2+} or Zn^{2+}, ions known for their antimicrobial properties, were synthesized by the foam replica technique and tested regarding their bioactive response in SBF, mechanical properties, degradation, and porosity. Finally their ability to support the attachment and long-term proliferation of Dental Pulp Stem Cells (DPSCs) was also evaluated. The results showed that conversely to their bioactive response in SBF solution, Zn-doped scaffolds proved to respond adequately regarding their mechanical strength and to be efficient regarding their biological response, in comparison to Cu-doped scaffolds, which makes them promising candidates for targeted dental stem cell odontogenic differentiation and calcified dental tissue engineering.

1. Introduction

Research on "engineered tissues" is remarkably growing in recent years, as there is an increased demand of many clinical specialties for biomaterials able not only to substitute the lost or destroyed tissues but also to provide an environment that could induce its own regeneration. Extended research has started to emerge in the field of dental tissue regeneration based on the use of stem cells in combination with various scaffolds and relevant growth and differentiation factors which make the classical tissue engineering triad [1]. In literature, various types of biomaterial scaffolds have been developed as ECM analogs capable of supporting cell attachment, proliferation, and differentiation and, ultimately, forming new engineered tissues or organs [2, 3]. Although a wide range of biomaterials have been proposed for this purpose (ceramics, natural or synthetic polymers, etc.), to date, glass, glass-ceramic, and ceramic scaffolds present important advantages compared to polymeric scaffolds, such as porous structure and chemical texture that promotes mesenchymal cells differentiation and mineralization of the extracellular matrix, lack of toxic byproducts, and the formation of dentinal tubule-like structures [4–7]. Furthermore, ceramic scaffolds can be used as carriers of growth factors and angiogenetic agents, drugs, and cell differentiation products [8, 9]. Calcium-phosphate (β-TCP) or hydroxyapatite (HA) scaffolds are the most investigated compositions for dental tissues regeneration due to their chemical resemblance to the mineral component of natural dentin in mammals. However, other ceramic compositions may be more beneficial in triggering dental tissue formation, although research in this area is limited. Mg-doped phosphate glasses and ceramics have been shown to enhance the bioactivity of the scaffolds related to osteogenesis [10]. However, to date, the effect of Mg ions on dentin regeneration is largely unknown, although the use of Mg-containing ceramic scaffolds for dentin regeneration seems a reasonable concept due to the increased amount of Mg contained in dentin [11]. Based on the fact that magnesium plays a fundamental role in cellular processes [12, 13] and skeletal metabolism [14, 15], Mg-containing glass-ceramics with high porosity, suitable degradability, and bioactivity have been only recently proposed for dental tissue regeneration [16]. Huang et al. [17] compared akermanite ($Ca_2Mg(Si_2O_7)$) and β-tricalcium phosphate (β-TCP) in their ability to induce differentiation of human mesenchymal stem cells (MSCs), showing that the release of Si and Mg significantly facilitated stem cell proliferation and differentiation. Furthermore, Qu et al. [18] reported that the sustained release of Mg ions from magnesium-containing nanostructured hybrid scaffolds significantly enhanced the proliferation, differentiation, and biomineralization of human DPSCs in vitro.

New studies over the introduction of various metallic ions when synthesizing bioactive glasses report that when used in small amounts, they could be beneficiary [19], since by tailoring the composition and ionic dissolution process of bioactive glasses, the stimulation of specific cell behavior may be achieved. The addition of Cu ions has been proposed to show beneficial effects on angiogenesis [6] and to induce an increase in differentiation of MSCs [7] whereas the addition of Zn ion shows anti-inflammatory effects and stimulates bone formation in vitro by activating protein synthesis in osteoblasts [8, 9]. Recently biomaterials with antibacterial properties have been suggested in dental tissues engineering for the creation of a bacteria-free environment while healing and regenerating the defect area. This is particularly important for regenerating dental tissues which are prone to bacterial invasion from the oral cavity. The synthesis of three-dimensional porous scaffolds with interconnected porous structure, able to function as temporary 3D templates for cell attachment, proliferation, and differentiation when in contact under controlled environment and capable of releasing ions with antimicrobial or cariostatic properties [12], could constitute a suitable inductive carrier that could enhance dentin regeneration and induce the optimal formation of new dentin matrix. Although the last few years Mg, Zn, and Cu ions have shown promising results as additives or dopants to bioceramic scaffolds, the effect of their simultaneous presence in quaternary systems of SiO_2-CaO-MgO-CuO or SiO_2-CaO-MgO-ZnO has not been investigated, to the best of the authors' knowledge. Consequently, the aim of this work was to synthesize Mg-based glass-ceramic scaffolds with incorporated Zn/Cu ions and to investigate their physical, mechanical, and biological properties.

TABLE 1: Bioactive scaffold compositions in %wt.

	SiO_2	CaO	MgO	ZnO	CuO
ZnA2	60	30	7.5	2.5	—
CuA2	60	30	7.5	—	2.5

2. Materials and Methods

2.1. Scaffold Fabrication. Mg-based scaffolds of different composition as indicated in Table 1 were synthesized. Polyurethane (PU) foam was used as a sacrificial template in order to produce 3D porous scaffolds. The foam was cut into pieces of $10 \times 10 \times 5$ mm and used for the fabrication of the bioactive scaffolds through the foam replica technique as described by Chen et al. [20] while it was immersed in sol-gel. The sol-gel solution was prepared as described by Goudouri et al. [16]. Briefly, TEOS was added in the mixture of ultrapure H_2O and HNO_3 (2 N) and stirred—for approximately 30 min—until partial hydrolysis of TEOS occurred. Calcium nitrate tetrahydrate ($Ca(NO_3)_2 \cdot 4H_2O$), magnesium nitrate hexahydrate ($Mg(NO_3)_2 \cdot 6H_2O$), and zinc nitrate hexahydrate ($Zn(NO_3)_2 \cdot 6H_2O$) or cupric nitrate hemipentahydrate ($Cu(NO_3)_2 \cdot 2.5H_2O$) were added to the mixture allowing 50 min for the hydrolysis reaction to complete at 60°C. After the immersion of the foam in the sol-gel and mechanical stirring for 5 min, the samples (green bodies) were retrieved from the sol-gel and squeezed in order to remove the excess of sol from the pores and then left to dry out for at least 12 h. The thickness of the bioactive glass on the green bodies was adjusted by pouring droplets of sol-gel. The excess was removed after centrifuging the green bodies.

In order to understand the structural changes upon heating of the bioactive glasses, as well as their mass loss percentage, the Thermogravimetric (TG) and Differential Scanning Calorimetry (DSC) curves were received with heating rate of 10°C/min from room temperature to 1400°C, under nitrogen atmosphere. According to TG-DSC results, the synthesized bioactive scaffolds were sintered at 890°C (Zn-based) and 866°C (Cu-based) with 2 h annealing.

2.2. Characterization. Fourier Transform Infrared Spectroscopy (FTIR) and X-ray diffraction analysis (XRD) were used in order to examine thoroughly the scaffold's crystal structure. For FTIR measurements a Perkin-Elmer Spectrometer Spectrum 1000 in MIR region was used to determine the chemical composition of the fabricated scaffolds. Representative scaffolds from both groups were ground into powder and pellets with powder to KBr ratio of 1:100 were fabricated under pressure (7 tons). For the XRD analysis a Philips (PW1710) diffractometer with Ni-filtered CuKa wave radiation was used. The scaffolds were crushed and ground into powder for XRD analysis. Archimedes method was used for the determination of the scaffold porosity, p, as indicated by the equation

$$p = \frac{V_{\text{pores}}}{V_{\text{bulk}}}, \tag{1}$$

where V_{bulk} was calculated by the mass and dimensions of the scaffolds. The morphology and microstructure of the scaffolds was monitored by the use of scanning electron microscopy with associated energy dispersive spectroscopic analysis (SEM-EDS).

2.3. Compressive Strength Evaluation.

The mechanical properties of the synthesized scaffolds were tested by an Instron 3344 loading apparatus in compression at a crosshead speed of 0.5 mm/min. Ten prismatic samples (five from each group) with dimensions $8 \times 8 \times 4$ mm were tested. The compressive load was applied until 1 mm (12.5%) compressive strain was achieved in the 8 mm dimension (height). Further loading applied during pilot experiments resulted in off-axis loading and thus the received stress values were not considered as valid. The highest stress values included in the 1 mm stress-strain curve were recorded and mean values with standard deviations were determined.

2.4. In Vitro Degradation.

Degradation test was performed according to the ISO 10993-14: 2009 (extreme and simulation solution tests). More specifically, 12 scaffolds of each group were tested. Mass calculation was performed with an electronic balance (Kern ABS) with an accuracy of 0.0001 mg and the specific mass of each scaffold was recorded as the difference between the mass of the container with and without the scaffold. For the simulation solution test each container was filled with 100 mL of freshly prepared TRIS-HCL buffer, with pH 7.4 ± 0.1 at $37 \pm 1°C$, while for the extreme solution test each container was filled with 10 mL of the buffered citric acid solution, with pH 3.0 ± 0.2 at $37 \pm 1°C$. Then, all containers were placed in a controlled-temperature environment at $37 \pm 0.5°C$, for 120 h. The containers were agitated at 2 Hz with circular movement. After 120 h the containers with scaffolds were allowed to cool at room temperature. Remnants of scaffolds were removed under filtration. Reweighted filter paper was used for filtration. Remnants were rinsed and filtrated three times with small amounts of water grade 2. Then, filter paper and scaffold's remnants were dried in an oven overnight at $100 \pm 2°C$. Drying procedure was continued until mass changes less than 0,1% were recorded. The difference between the mass of the filter paper with and without the remnants was the actual mass of the nondegraded scaffold. Finally, the difference between the initial recorded mass of the scaffolds and the mass of the nondegraded scaffold remnants was recorded as the mass of the degraded scaffold. The % weight loss was determined with the following equation:

$$\text{Weight loss (\%)} = \left[\frac{(W_o - W_t)}{W_o} \right] \times 100. \tag{2}$$

2.5. Apatite Forming Ability in SBF.

The scaffolds were placed in sterilized reagent bottles and submerged in SBF solution with mass to solution ratio adjusted at 1.5 mg/mL [21]. Then they were placed in an incubator (Incucell 55) at $37°C$ under renewal conditions for various times after immersion (6 h, 24 h, and then after every 48 h) [22]. Finally the specimens were removed from the SBF at each time point (after 10 and 21 days of immersion), washed with distilled water, and dried at room temperature.

2.6. Evaluation of Cell Viability/Proliferation and Cell Attachment/Morphology of Dental Pulp Stem Cells (DPSCs) Seeded into the Biomimetic Scaffolds.

DPSC cultures were established from third molars of young healthy donors aged 16–18 years and extensively characterized for several stem cell markers, as previously published by our group [23]. The collection of the samples was performed according to the guidelines of the Institutional Review Board and the parents of all donors signed an informed consent form. For the establishment of cell cultures the enzymatic dissociation method was used [24]. Briefly, teeth were disinfected and cut around the cementum-enamel junction to expose the pulp chamber. The tissue was minced into small segments and digested in a solution of 3 mg/mL collagenase type I and 4 mg/mL dispase II (Invitrogen, Karlsruhe, Germany) for 1 h at $37°C$. Single cell suspensions were obtained by passing the cells through a 70 μm cell strainer (BD Biosciences, Heidelberg, Germany). Cells were expanded with α-MEM (Minimum Essential Media) culture medium (Invitrogen), supplemented with 15% FBS (EU-tested, Invitrogen), 100 mM L-ascorbic acid phosphate (Sigma-Aldrich, Taufkirchen, Germany), 100 units/mL penicillin, 100 mg/mL streptomycin, and 0.25 mg/mL Amphotericin B (all from Invitrogen) (Complete Culture Medium (CCM)), and incubated at $37°C$ in 5% CO_2. Cultured DPSCs in passage numbers from 3 to 6 were used for all experiments.

To analyze cell viability/proliferation the MTT assay was used. Scaffolds were first preimmersed into CCM for 30 min at $37°C$ and 5% CO_2 atmosphere in an incubator in order to adjust pH and create a more biomimetic microenvironment before cell seeding. Afterwards, the medium was removed and DPSCs were spotted at low volume (100 μL) into the scaffolds at 5×10^5 cells/scaffold in 48 well-plates and allowed to attach first for 45 min before being fully covered with 500 μL CCM. Cell viability/proliferation was evaluated after 1, 3, 7, and 14 days (d) by the MTT assay ($n = 4$). Medium change was performed every 3 days during the entire experimental period. At the end of each time point 50 μL of MTT (5 mg/mL in PBS) was added in each well and scaffolds/cell constructs were incubated for 3 h at $37°C$ and 5% CO_2. After this period, the medium containing the MTT solution was discarded, the scaffold/cell constructs were washed with PBS, and the insoluble formazan was dissolved with DMSO overnight at $37°C$. The absorbance was measured against blank (DMSO) at a wavelength of 545 nm and a reference filter of 630 nm by a microplate reader (Epock, Biotek, Biotek instruments, Inc, Vermont, USA). As controls, scaffolds (CuA2 and ZnA2) without cells were incubated under the same conditions and the optical density values were subtracted from values obtained by the corresponding scaffold/cell constructs. Finally, OD values were normalized to those of control DPSCs cultures beginning with the same cell number (5×10^5 cells/well) and the final results were expressed as % percentage of control.

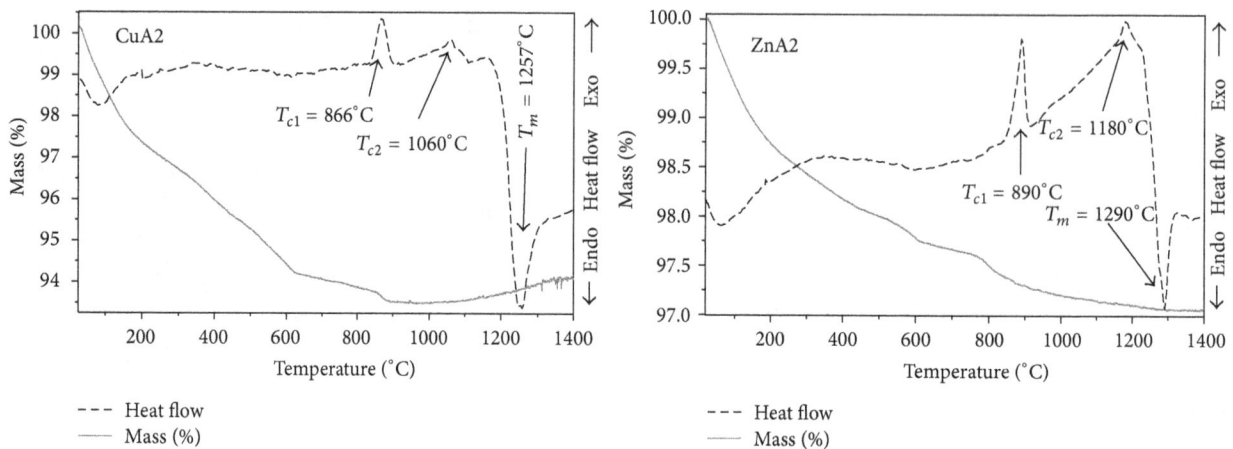

FIGURE 1: TG and DSC curves of CuA2 and ZnA2 glass powders.

In order to evaluate cell attachment and morphology of DPSCs seeded into the biomimetic scaffolds, samples of scaffold/cell constructs were processed for scanning electron microscopy (SEM). Cells were seeded into the biomimetic scaffolds, as described for the MTT assay. After 3, 7, and 14 d, the scaffold/cell constructs were washed twice with PBS and fixed with 3% glutaraldehyde (in 0.1 M sodium cacodylate, pH 7.4, containing 0.1 M sucrose). The specimens were subsequently dehydrated in a series of increasing concentrations of ethanol and hexamethyldisilazane. For SEM analysis they were carbon-coated and observed with a Jeol (Japan) electronic microscope at 20 kV.

3. Results and Discussion

3.1. TG – DSC Analysis. Thermal analysis of bioactive glasses can efficiently determine mass variations and thermal content changes as a function of temperature. Therefore, it is critically essential to determine these variations and changes for every material, which is going to receive heat treatment, in order to be able to predict its behavior at high temperatures. As already mentioned, ZnA2 scaffolds were sintered at 890°C and CuA2 scaffolds at 866°C, with 2 h annealing. Those temperatures were extracted from DSC curves (Figure 1, red lines) for each bioactive glass and represent exothermic peaks (T_c). The temperature at these peaks corresponds to the crystallization of the samples, while additional exothermic peaks are observed at higher temperatures (1060°C for CuA2 and 1180°C for ZnA2). Endothermic peaks assigned to the melting point (T_m) of each sample are observed at 1257°C for the CuA2 and 1290°C for the ZnA2 glasses. In this study, sintering temperatures were chosen at the first exothermic peak of each glass, in order to produce scaffolds with improved mechanical properties, as the crystallization of a glass provides a mechanically enhanced system without necessarily impairing the bioactive response of the glass-ceramic [25].

The TG curves (Figure 1, blue lines) of both samples indicate that the mass variations were insignificant, being under 8% for both of them. Mass loss for both bioactive

glasses takes place under 600°C and is caused because of the H_2O, CO, and CO_2 release from the samples, which were entrapped inside during the synthesis process.

3.2. Characterization of the Fabricated Scaffolds. The glass-ceramic scaffolds were successfully fabricated via the foam replica technique. Bioactive scaffolds in order to be applied for the development of calcified tissue should be able to favor cell penetration, vascularization, and nutrient and metabolic waste transportation [26, 27]. To achieve such a goal scaffolds should exhibit interconnected porous structure with pore sizes between 300 and 500 μm [18, 28]. The porous structure and morphology of the bioactive glass-ceramic scaffolds are shown in Figure 2.

SEM microphotographs revealed pore size of approximately 200–400 μm and interconnected pore structure. The ZnA2 and CuA2 scaffolds presented a mean porosity of 84% and 74%, respectively. A primary goal of dental tissue regeneration is the development of suitable scaffolding materials that could support dental stem cells attachment and proliferation. Scaffolds of similar porosity and interconnectivity as those of the scaffolds fabricated in this study have been shown to support the attachment and proliferation of human Dental Pulp Stem Cells [18, 28].

The FTIR spectra of the fabricated scaffolds are shown in Figure 3. FTIR spectra of both ZnA2 and CuA2 glass-ceramic scaffolds present the characteristic peaks of silicate glasses shown by a broad peak at 900–1200 cm^{-1} and the peak at ~ 470 cm^{-1} [29]. In addition, the spectra of the ZnA2 reveal the presence of a strong peak at 796 cm^{-1} indicating the existence of bridging oxygen, which are connected with the inability of a glass-ceramic material to exhibit bioactive behavior [30, 31]. This peak—though present—is not so intense in the spectra of the CuA2 scaffolds. For CuA2 scaffolds, the FTIR peaks at 646 cm^{-1}, 690 cm^{-1}, 902 cm^{-1}, and 946 cm^{-1} were attributed to wollastonite ($CaSiO_3$) [32]. XRD patterns (Figure 4) revealed that ZnA2 scaffolds consist mainly of an amorphous phase. On the other hand, XRD patterns of CuA2 glass-ceramic scaffolds indicate the existence of wollastonite (approximate percentage 40%wt), while 10%wt

FIGURE 2: Digital camera photographs (a, d), light microscope images of different magnifications (b, e), and SEM microphotographs (c, f) of the glass-ceramic scaffolds (a, b, c: CuA2, d, e, f: ZnA2).

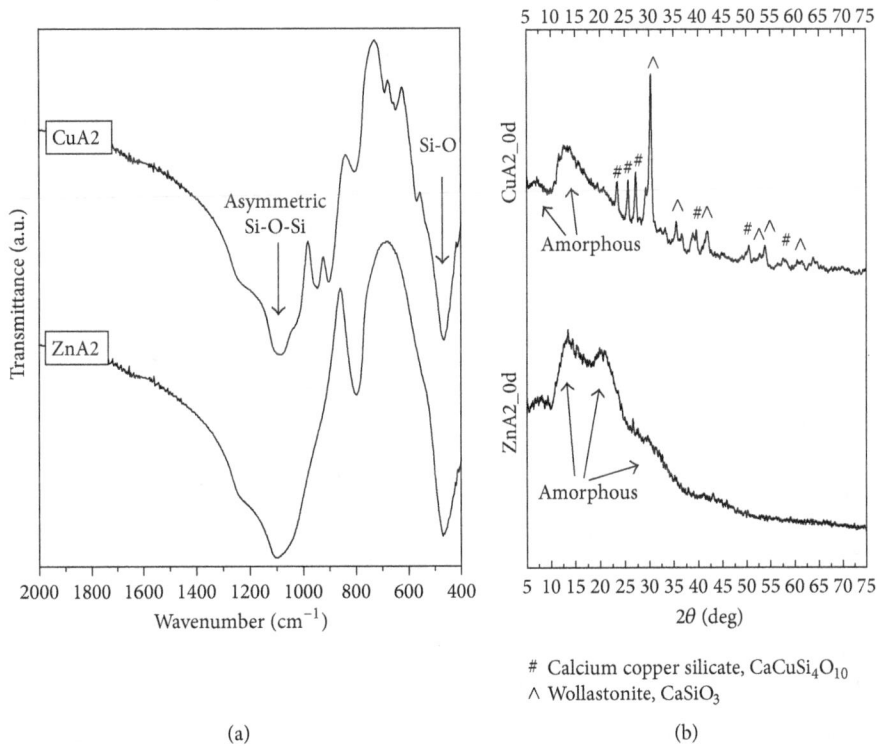

\# Calcium copper silicate, $CaCuSi_4O_{10}$
∧ Wollastonite, $CaSiO_3$

(a)　　　　　　　　(b)

FIGURE 3: (a) FTIR spectra and (b) XRD patterns of ZnA2 and CuA2 glass-ceramic scaffolds.

of calcium copper silicate ($CaCuSi_4O_{10}$) was also detected. These findings confirmed the FTIR results.

3.3. Compressive Strength.

The mechanical strength of both ZnA2 and CuA2 scaffolds under uniaxial compression stress was proven, as expected, rather low but in the range of values attained for ceramic scaffolds noncoated with gelatin or other polymeric materials [33, 34]. More specific, ZnA2 glass-ceramic scaffolds presented a mean compressive strength

at 0.10 (±0.06) MPa and CuA2 glass-ceramic scaffolds a mean compressive strength of 0.02 (±0.007) MPa. As it is shown from the typical stress-strain curves presented in Figure 4, a continuous section with peaks after isolating linear elastic regions and valleys, corresponding to the brittle crushing of the struts, was the dominant mode of fracture, as has been observed for brittle ceramic porous scaffolds [34, 35]. Further improvements of the mechanical properties of the scaffolds are necessary for the maintenance of their

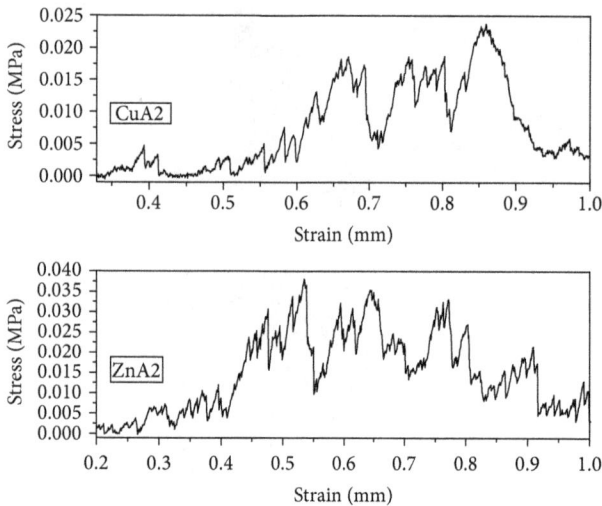

FIGURE 4: Indicative stress (MPa)-strain (mm) curves for each group of glass-ceramic scaffolds. The compressive load was applied until 1 mm (12.5%) compressive strain was achieved along the 8 mm dimension (height).

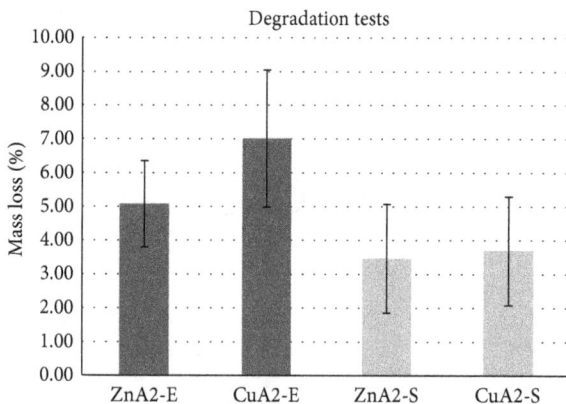

FIGURE 5: Results of degradation tests. Zn-S: ZnA2 conventional test, Cu-S: CuA2 conventional test, Zn-E: ZnA2 extreme test, and Cu-E: CuA2 extreme test.

structural integrity so as to allow time for the new calcified tissue to grow. The capability of improving the mechanical properties of ceramic scaffolds has been demonstrated in several composite polymer-ceramic formulations [36–39]. It is highly possible that coating these scaffolds with gelatin or alginate hydrogel could significantly improve their mechanical behavior and this is a subject of future research.

3.4. In Vitro Degradation. Results of degradation tests are presented in Figure 5. A mean degradation rate of 3.5% (ZnA2)-3.7% (CuA2) was recorded for the simulation test in Tris Buffer solution after 120 h immersion, while extreme test in citric acid solution revealed slightly higher degradation rate for the same time period (5% for ZnA2, 7% for CuA2). Both tests resulted in low solubility values. As it was expected the recorded degradation values were higher for the extreme test in comparison to the simulation test

for both ceramic scaffolds (ZnA2 and CuA2). Moreover, ceramic scaffolds of ZnA2 presented lower degradation values in extreme degradation test, in comparison with CuA2, while the two ceramic scaffolds presented almost equal degradation values in simulation degradation test. The higher degradation rate of the CuA2 compared to the ZnA2 scaffolds can be explained by the presence of wollastonite in the CuA2 scaffolds, as it has been found that the increase of wollastonite percentage rapidly increases the mass loss of composite poly(3-hydroxybutyrate-co-3-hydroxyvalerate) (PHBV)/wollastonite scaffolds [40]. This increased weight loss may be attributed to the dissolution of wollastonite when immersed in aqueous solution, the release of alkaline ions, and the subsequent destruction of the three-dimensional structure of the scaffolds. The release of alkaline ions of bioactive glasses is one of the basic mechanisms of apatite formation, as it leads to the formation of a high surface area of hydrated silica and finally to the crystallization of apatite through precipitation of P and Ca from the surrounding environment [41]. The increased degradation rate may be the reason for the bioactive behavior of CuA2 despite its lower porosity, while the lower degradation of ZnA2 that can be assigned to the presence of nonbridging oxygen as found with FTIR explains its inability for in vitro apatite formation.

Degradation of scaffolds is necessary during calcified tissue formation, as scaffolds are networks that assist initial cell attachment and proliferation but have to degrade simultaneously with the new tissue formation. Similar degradation rate values with those of the present study have been recorded for bioceramic scaffolds in literature [42, 43], although usually degradation rate of bioceramics is evaluated by measuring mass loss after immersion in solutions like SBF [44] or PBS [40] due to their resemblance with physiological body fluids. In this study degradation of ceramic scaffolds was evaluated according to ISO 10993-14: 2009 which is more appropriate for testing materials in contact with fluids of different pH. The simulation test is a mild, common test used to evaluate the degradation rate of most ceramic materials under physiological pH and temperature similar to physiological body fluids, while the extreme test is related to the more aggressive environment to which a material can be exposed to in the oral cavity due to low pH. Remarkable variation in degradation values was recorded among the scaffolds of both groups. Slight differences in scaffold's porosity or structure through the fabrication process could explain such variation [42, 45]. Greater degradation values were reported in literature, only for significantly longer immersion period (14–28 days) [43, 46]. Lower degradation values were recorded after 3 days of immersion in Tris Buffer for wollastonite/tricalcium phosphate macroporous scaffolds, with significantly lower porosity (50%) [45]. The degradation rate of Ca-P bioceramics is influenced by several parameters such as the sintering process, microstructure, crystallinity, and porosity [47]. The porosity plays a dominant role in the degradation of bioceramics as it is related to high specific surface area [48]. However the results of this study indicate that other mechanisms rather than porosity may be more crucial in determining the degradation of the scaffolds, such as composition and crystalline structure.

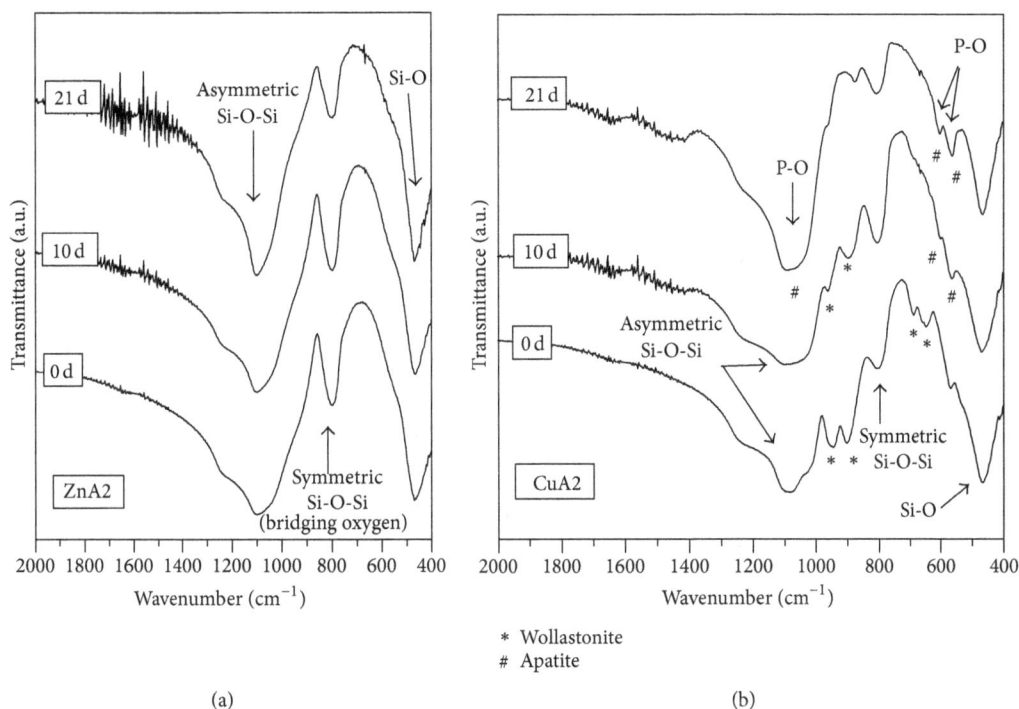

(a)   (b)

* Wollastonite
# Apatite

FIGURE 6: FTIR spectra of (a) ZnA2 and (b) CuA2 glass-ceramic scaffolds before and after immersion in SBF solution for 10 and 21 days.

3.5. Bioactivity Evaluation. FTIR spectra of ZnA2 glass-ceramic scaffolds could not reveal any differentiation in their chemical composition even after 21 days of immersion in SBF solution, as shown in Figure 6(a). This result may be attributed to the presence of bridging oxygen as shown by the strong peak at 796 cm^{-1}, as already mentioned.

On the contrary, CuA2 bioactive scaffolds presented bioactive behavior according to FTIR spectra (Figure 6(b)). More specific, FTIR spectra, after 10 days of immersion, revealed the formation of a weak double peak at 587 cm^{-1} and 603 cm^{-1}, which is attributed to the vibration of the P-O bond of the phosphate group. The high amount of wollastonite crystallized on the initial material could explain the delayed formation of apatite on the surface of CuA2 glass-ceramic scaffolds. This double peak is known to be associated with apatite formation. After 21 days of soaking in SBF solution, a stronger double peak at 587 cm^{-1} and 603 cm^{-1} was formed. Additionally, at the same immersion time, the broad peak at 900–1200 cm^{-1} shifted towards ~1100 cm^{-1} and became less wide. Therefore, after 21 days of soaking in SBF there is a strong indication of the formation of apatite on the surface of the glass-ceramic scaffolds of the CuA2 group.

These findings are in accordance with XRD patterns (Figure 7) for both groups of glass-ceramic scaffolds. Namely, ZnA2 samples did not show any compositional differentiation after 10 days of immersion in SBF solution, whereas CuA2 patterns revealed a peak corresponding to apatite after 10 days of soaking.

The effect of zinc incorporation on the structure of various bioactive glasses has resulted in different results concerning bioactivity depending on the microstructure and physicochemical properties of Zn-doped glasses. Although the acellular formation of calcium phosphate layer on the surface of bioactive silicate glasses doped with Zn have been shown to occur after soaking in biological fluids [49, 50], other studies have shown that Zn content reduces the overall leaching activity of the glass inhibiting the formation of the HCA layer on its surface [51]. Haimi et al. [52] reported a delayed formation of HCA which was related to the slower degradation profile of the Zn-doped bioactive glasses, in accordance with the results of this study. On the other hand, Cu^{2+}-doped 45S5 BG scaffolds exhibit high apatite forming ability, as proven by the rapid formation of a carbonated HA layer on their surface (3 days in SBF) [53]. Hoppe et al. [53] reported that Cu^{2+} addition (up to 2.5wt% CuO) had no effect on the reactivity of the undoped BG, as measured through immersion in SBF. Goudouri et al. [16] fabricated sol-gel Mg-based scaffolds with the foam replica technique and reported apatite formation on scaffolds sintered at 1350°C after 9 days in SBF. As the authors in the current study used the same starting glass formulation, the incorporation of copper resulted in a slight delay of apatite formation, taking into consideration the lower crystallization temperature of the scaffolds.

3.6. Evaluation of Cell Viability/Proliferation and Attachment/Morphology of DPSCs on the Biomimetic Scaffolds. DPSCs were able to attach and proliferate in both biomimetic scaffolds (Figures 8 and 9). However, both MTT assay and SEM analysis revealed a much better biological behavior of ZnA2 compared to the CuA2 scaffolds. ZnA2 supported a statistically significant higher viability of DPSCs compared

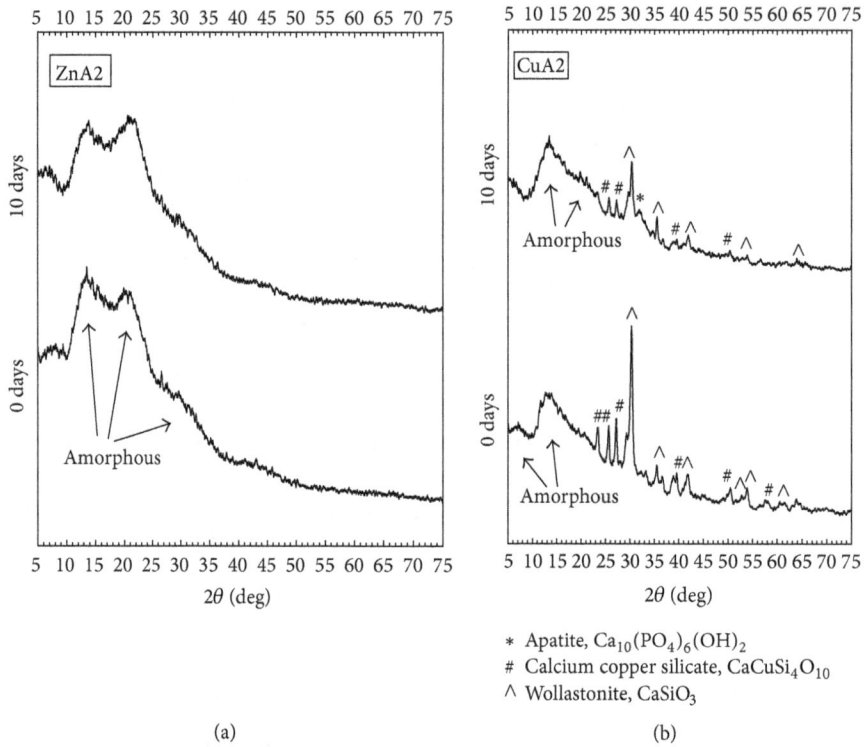

* Apatite, $Ca_{10}(PO_4)_6(OH)_2$
# Calcium copper silicate, $CaCuSi_4O_{10}$
∧ Wollastonite, $CaSiO_3$

(a)   (b)

FIGURE 7: XRD spectra of (a) ZnA2 and (b) CuA2 glass-ceramic scaffolds before and after immersion in SBF solution for 10 days.

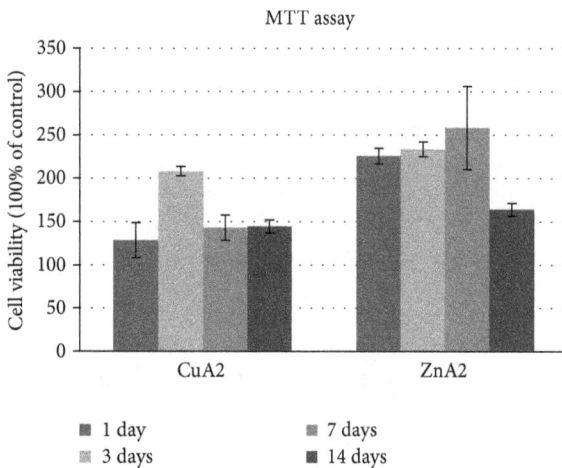

FIGURE 8: Evaluation of cell viability/proliferation of DPSCs seeded into the CuA2 and ZnA2 bioceramic scaffolds for 1, 3, 7, and 14 days (MTT assay).

to the CuA2 scaffolds at all time points tested ($p <$ 0.01). ZnA2 scaffolds showed an increase of cell viability/proliferation up to day 7 and decrease afterwards, which can be explained by the initiation of differentiation of DPSCs inside the biomimetic microenvironment, which is usually accompanied by a cease in proliferation, as already shown in preliminary experiments with real-time PCR analysis and western blotting (data under preparation). On the contrary

CuA2 scaffolds showed much lower OD values compared to the ZnA2 scaffolds, with viability/proliferation increasing until day 3 and significantly decreasing afterwards. Whether this inferior biological behavior of CuA2 scaffolds is due to a very high release of cytotoxic concentrations of Cu or any other elements needs further investigation.

The results of the MTT assay were also in accordance with the results obtained by the SEM analysis. Cells grown inside the ZnA2 scaffolds were more densely seeded, with an atractoid, spindle-shaped morphology, indicative of proper attachment and high viability of cells within the scaffold. Cells grown inside the CuA2 scaffolds, on the other hand, were fewer and with a rather rounded morphology, indicative of poor attachment and potential cytotoxicity. Zn-doped sol-gel derived glasses based on 58S have shown higher cellular viabilities than similar Cu-doped glasses, in a recent study by Bejarano et al. [54], although both were cytotoxic compared to undoped control 58S. The enhanced cell behavior recorded in the present study is probably attributed to a more stabilized Zn-derived glass structure that restricted mass glass dissolution and ion release that could exert cytotoxic behavior. Preliminary, unpublished data of the authors suggest that ZnA2 scaffolds combined with DPSCs and growth/morphogenetic factors such as Dentin Matrix Protein, DMP-1, and Bone Morphogenetic Protein, BMP-2, promote odontogenic differentiation and dentin-like tissue formation. These data need further investigation regarding the underlying molecular mechanisms of this biological response.

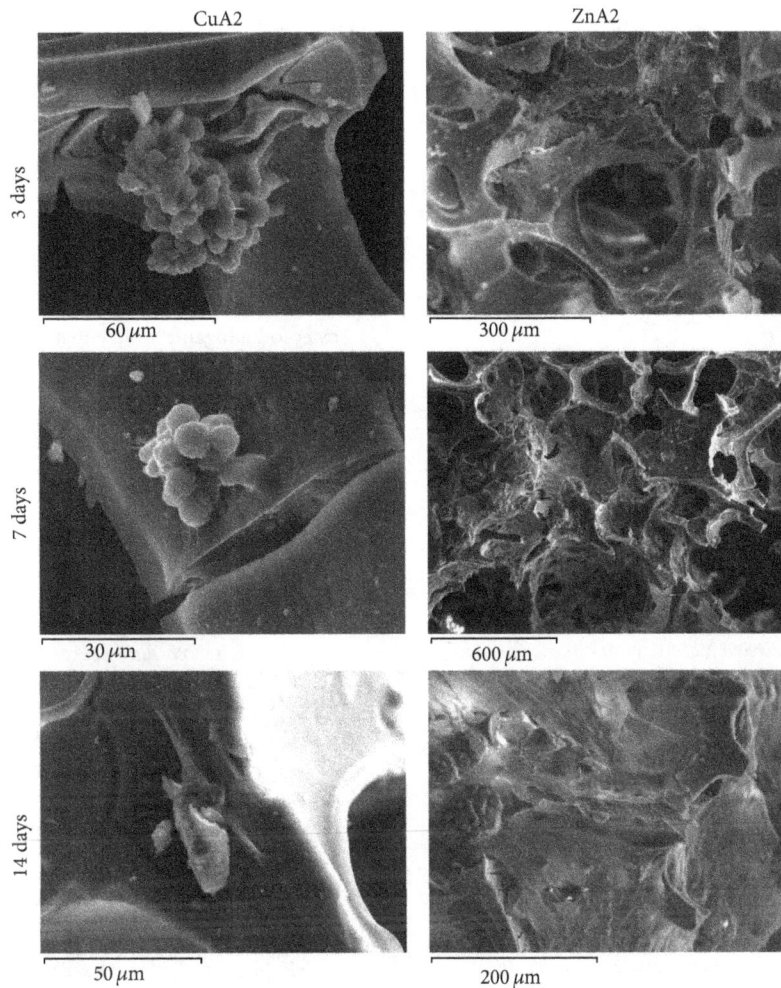

FIGURE 9: SEM microphotographs of CuA2 and ZnA2 scaffolds seeded with DPSCs after 3, 7, and 14 days.

4. Conclusions

Bioactive ceramic scaffolds, with adequate porosity, over 74%, and pore interconnectivity were produced by the foam replica technique. Cu-doped Mg-based scaffolds revealed apatite forming ability after 10 days immersion in SBF, while Zn-doped Mg-based scaffolds failed to develop apatite formation even after 21 days in SBF. Differences in structure are responsible for the different degradation profile, mechanical behavior, and bioactivity of the synthesized scaffolds. Despite failure to develop apatite ZnA2 scaffolds were proved very efficient to provide controlled degradation rate and a biomimetic environment for the long-term attachment and growth of DPSCs (up to 14 days), which makes them very promising for further research on their potential to induce odontogenic differentiation of DPSCs and calcified dental tissue production for targeted dentin regeneration.

Conflict of Interests

The authors declare that there is no conflict of interests regarding the publication of this paper.

Acknowledgments

This study was conducted under the action Excellence II (Project: 5105) and funded by the European Union (EU) and National Resources.

References

[1] F. Garcia-Godoy and P. E. Murray, "Status and potential commercial impact of stem cell-based treatments on dental and craniofacial regeneration," *Stem Cells and Development*, vol. 15, no. 6, pp. 881–887, 2006.

[2] L. Zhang, Y. Morsi, Y. Wang, Y. Li, and S. Ramakrishna, "Review scaffold design and stem cells for tooth regeneration," *Japanese Dental Science Review*, vol. 49, no. 1, pp. 14–26, 2013.

[3] R. Langer and D. A. Tirrell, "Designing materials for biology and medicine," *Nature*, vol. 428, no. 6982, pp. 487–492, 2004.

[4] A. El-Ghannam, P. Ducheyne, and I. M. Shapiro, "Formation of surface reaction products on bioactive glass and their effects on the expression of the osteoblastic phenotype and the deposition of mineralized extracellular matrix," *Biomaterials*, vol. 18, no. 4, pp. 295–303, 1997.

[5] M. Zayzafoon, "Calcium/calmodulin signaling controls osteoblast growth and differentiation," *Journal of Cellular Biochemistry*, vol. 97, no. 1, pp. 56–70, 2006.

[6] F. Mastrangelo, E. Nargi, L. Carone et al., "Tridimensional response of human dental follicular stem cells onto a synthetic hydroxyapatite scaffold," *Journal of Health Science*, vol. 54, no. 2, pp. 154–161, 2008.

[7] Y. Ando, M. J. Honda, H. Ohshima et al., "The induction of dentin bridge-like structures by constructs of subcultured dental pulp-derived cells and porous HA/TCP in porcine teeth," *Nagoya Journal of Medical Science*, vol. 71, no. 1-2, pp. 51–62, 2009.

[8] W. J. E. M. Habraken, J. G. C. Wolke, and J. A. Jansen, "Ceramic composites as matrices and scaffolds for drug delivery in tissue engineering," *Advanced Drug Delivery Reviews*, vol. 59, no. 4-5, pp. 234–248, 2007.

[9] T. Garg, O. Singh, S. Arora, and R. S. R. Murthy, "Scaffold: a novel carrier for cell and drug delivery," *Critical Reviews in Therapeutic Drug Carrier Systems*, vol. 29, no. 1, pp. 1–63, 2012.

[10] A. Hussain, K. Bessho, K. Takahashi, and Y. Tabata, "Magnesium calcium phosphate as a novel component enhances mechanical/physical properties of gelatin scaffold and osteogenic differentiation of bone marrow mesenchymal stem cells," *Tissue Engineering Part A*, vol. 18, no. 7-8, pp. 768–774, 2012.

[11] W. D. Armstrong and P. J. Brekhus, "Chemical constitution of enamel and dentine. I. Principal components," *The Journal of Biological Chemistry*, vol. 120, pp. 677–687, 1937.

[12] H. Oudadesse, S. Martin, A. C. Derrien, A. Lucas-Girot, G. Cathelineau, and G. Blondiaux, "Determination of Ca, P, Sr and Mg in the synthetic biomaterial aragonite by NAA," *Journal of Radioanalytical and Nuclear Chemistry*, vol. 262, no. 2, pp. 479–483, 2004.

[13] A. Hartwig, "Role of magnesium in genomic stability," *Mutation Research—Fundamental and Molecular Mechanisms of Mutagenesis*, vol. 475, no. 1-2, pp. 113–121, 2001.

[14] S. R. Kim, J. H. Lee, Y. T. Kim et al., "Synthesis of Si, Mg substituted hydroxyapatites and their sintering behaviors," *Biomaterials*, vol. 24, no. 8, pp. 1389–1398, 2003.

[15] R. K. Rude and M. Olerich, "Magnesium deficiency: possible role in osteoporosis associated with gluten-sensitive enteropathy," *Osteoporosis International*, vol. 6, no. 6, pp. 453–461, 1996.

[16] O. M. Goudouri, E. Theodosoglou, E. Kontonasaki et al., "Development of highly porous scaffolds based on bioactive silicates for dental tissue engineering," *Materials Research Bulletin*, vol. 49, no. 1, pp. 399–404, 2014.

[17] Y. Huang, X. Jin, X. Zhang et al., "In vitro and in vivo evaluation of akermanite bioceramics for bone regeneration," *Biomaterials*, vol. 30, no. 28, pp. 5041–5048, 2009.

[18] T. Qu, J. Jing, Y. Jiang et al., "Magnesium-containing nanostructured hybrid scaffolds for enhanced dentin regeneration," *Tissue Engineering Part A*, vol. 20, no. 17-18, pp. 2422–2433, 2014.

[19] A. Hoppe, N. S. Güldal, and A. R. Boccaccini, "A review of the biological response to ionic dissolution products from bioactive glasses and glass-ceramics," *Biomaterials*, vol. 32, no. 11, pp. 2757–2774, 2011.

[20] Q. Z. Chen, I. D. Thompson, and A. R. Boccaccini, "45S5 Bioglass®-derived glass-ceramic scaffolds for bone tissue engineering," *Biomaterials*, vol. 27, no. 11, pp. 2414–2425, 2006.

[21] T. Kokubo, H. Kushitani, S. Sakka, T. Kitsugi, and T. Yamamuro, "Solutions able to reproduce in vivo surface-structure changes in bioactive glass-ceramic A-W^3," *Journal of Biomedical Materials Research*, vol. 24, no. 6, pp. 721–734, 1990.

[22] Y. Zhang, M. Mizuno, M. Yanagisawa, and H. Takadama, "Bioactive behaviors of porous apatite- and wollastonite-containing glass-ceramic in two kinds of simulated body fluid," *Journal of Materials Research*, vol. 18, no. 2, pp. 433–441, 2003.

[23] T. Paschalidis, A. Bakopoulou, P. Papa, G. Leyhausen, W. Geurtsen, and P. Koidis, "Dental pulp stem cells' secretome enhances pulp repair processes and compensates TEGDMA-induced cytotoxicity," *Dental Materials*, vol. 30, no. 12, pp. e405–e418, 2014.

[24] A. Bakopoulou, G. Leyhausen, J. Volk et al., "Assessment of the impact of two different isolation methods on the osteo/odontogenic differentiation potential of human dental stem cells derived from deciduous teeth," *Calcified Tissue International*, vol. 88, no. 2, pp. 130–141, 2011.

[25] D. Bellucci, V. Cannillo, and A. Sola, "An overview of the effects of thermal processing on bioactive glasses," *Science of Sintering*, vol. 42, no. 3, pp. 307–320, 2010.

[26] D. W. Hutmacher, "Scaffolds in tissue engineering bone and cartilage," *Biomaterials*, vol. 21, no. 24, pp. 2529–2543, 2000.

[27] C. Wu, J. Chang, W. Zhai, S. Ni, and J. Wang, "Porous akermanite scaffolds for bone tissue engineering: preparation, characterization, and in vitro studies," *Journal of Biomedical Materials Research Part B: Applied Biomaterials*, vol. 78, no. 1, pp. 47–55, 2006.

[28] K.-T. Lim, H.-W. Choung, A.-L. Im et al., "Novel composite scaffolds for tooth regeneration using human dental pulp stem cells," *Tissue Engineering and Regenerative Medicine*, vol. 7, no. 5, pp. 473–480, 2010.

[29] E. Kontonasaki, T. Zorba, L. Papadopoulou et al., "Hydroxy carbonate apatite formation on particulate bioglass in vitro as a function of time," *Crystal Research and Technology*, vol. 37, no. 11, pp. 1165–1171, 2002.

[30] F. Gervais, A. Blin, D. Massiot, J. P. Coutures, M. H. Chopinet, and F. Naudin, "Infrared reflectivity spectroscopy of silicate glasses," *Journal of Non-Crystalline Solids*, vol. 89, no. 3, pp. 384–401, 1987.

[31] N. Koga, Z. Strnad, J. Šesták, and J. Strnad, "Thermodynamics of non-bridging oxygen in silica bio-compatible glass-ceramics," *Journal of Thermal Analysis and Calorimetry*, vol. 71, no. 3, pp. 927–938, 2003.

[32] V. C. Farmer, *The Infrared Spectra of Minerals*, The Mineralogical Society, London, UK, 1974.

[33] D. Desimone, W. Li, J. A. Roether et al., "Biosilicate®-gelatine bone scaffolds by the foam replica technique: development and characterization," *Science and Technology of Advanced Materials*, vol. 14, no. 4, Article ID 045008, 2013.

[34] M. Kim, I.-H. Park, Y.-H. Kim, H.-Y. Song, Y.-K. Min, and B.-T. Lee, "Fabrication and characterization of strengthened BCP scaffold through infiltration of PCL in the frame," *Bioceramics Development and Applications*, vol. 1, Article ID D110118, 4 pages, 2011.

[35] T. M. O'Shea and X. Miao, "Preparation and characterisation of PLGA-coated porous bioactive glass-ceramic scaffolds for subchondral bone tissue engineering," in *Proceedings of the 9th International Symposium on Ceramic Materials and Components for Energy and Environmental Applications*, Shanghai Institute of Ceramics, Chinese Academy of Sciences, Shanghai, China, November 2008.

[36] C. Wu, Y. Ramaswamy, P. Boughton, and H. Zreiqat, "Improvement of mechanical and biological properties of porous CaSiO$_3$ scaffolds by poly(d,l-lactic acid) modification," *Acta Biomaterialia*, vol. 4, no. 2, pp. 343–353, 2008.

[37] X. Miao, L.-P. Tan, L.-S. Tan, and X. Huang, "Porous calcium phosphate ceramics modified with PLGA–bioactive glass," *Materials Science and Engineering C*, vol. 27, no. 2, pp. 274–279, 2007.

[38] O. Bretcanu, S. Misra, I. Roy et al., "In vitro biocompatibility of 45S5 Bioglass®-derived glass-ceramic scaffolds coated with poly(3-hydroxybutyrate)," *Journal of Tissue Engineering and Regenerative Medicine*, vol. 3, no. 2, pp. 139–148, 2009.

[39] M. M. Erol, V. Mouriňo, P. Newby et al., "Copper-releasing, boron-containing bioactive glass-based scaffolds coated with alginate for bone tissue engineering," *Acta Biomaterialia*, vol. 8, no. 2, pp. 792–801, 2012.

[40] H. Li and J. Chang, "In vitro degradation of porous degradable and bioactive PHBV/wollastonite composite scaffolds," *Polymer Degradation and Stability*, vol. 87, no. 2, pp. 301–307, 2005.

[41] L. L. Hench, "Bioceramics: from concept to clinic," *Journal of the American Ceramic Society*, vol. 74, pp. 1487–1510, 1991.

[42] W. Qianbin, W. Qiguang, and W. Changxiu, "The Effect of porosity on the structure and properties of calcium polyphoshate bioceramics," *Ceramics-Silikáty*, vol. 55, pp. 43–48, 2011.

[43] C. Wu and J. Chang, "Degradation, bioactivity, and cytocompatibility of diopside, akermanite, and bredigite ceramics," *Journal of Biomedical Materials Research Part B: Applied Biomaterials*, vol. 83, no. 1, pp. 153–160, 2007.

[44] P. Feng, P. Wei, C. Shuai, and S. Peng, "Characterization of mechanical and biological properties of 3-D scaffolds reinforced with zinc oxide for bone tissue engineering," *PLoS ONE*, vol. 9, no. 1, Article ID e87755, 2014.

[45] F. Zhang, J. Chang, K. Lin, and J. Lu, "Preparation, mechanical properties and in vitro degradability of wollastonite/tricalcium phosphate macroporous scaffolds from nanocomposite powders," *Journal of Materials Science: Materials in Medicine*, vol. 19, no. 1, pp. 167–173, 2008.

[46] Q. Wang, Q. Wang, and C. Wan, "Preparation and evaluation of a biomimetic scaffold with porosity gradients in vitro," *Anais da Academia Brasileira de Ciencias*, vol. 84, no. 1, pp. 9–16, 2012.

[47] J. X. Lu, M. Descamps, J. Dejou et al., "The biodegradation mechanism of calcium phosphate biomaterials in bone," *Journal of Biomedical Materials Research*, vol. 63, no. 4, pp. 408–412, 2002.

[48] C. P. A. T. Klein, A. A. Driessen, K. De Groot, and A. Van den Hooff, "Biodegradation behavior of various calcium phosphate materials in bone tissue," *Journal of Biomedical Materials Research*, vol. 17, no. 5, pp. 769–784, 1983.

[49] M. Bini, S. Grandi, S. Capsoni, P. Mustarelli, E. Saino, and L. Visai, "SiO_2–P_2O_5–CaO glasses and glass-ceramics with and without ZnO: relationships among composition, microstructure, and bioactivity," *The Journal of Physical Chemistry C*, vol. 113, pp. 8821–8828, 2009.

[50] L. Courthéoux, J. Lao, J.-M. Nedelec, and E. Jallot, "Controlled bioactivity in zinc-doped sol–gel-derived binary bioactive glasses," *Journal of Physical Chemistry C*, vol. 112, no. 35, pp. 13663–13667, 2008.

[51] V. Aina, G. Malavasi, A. Fiorio Pla, L. Munaron, and C. Morterra, "Zinc-containing bioactive glasses: surface reactivity and behaviour towards endothelial cells," *Acta Biomaterialia*, vol. 5, no. 4, pp. 1211–1222, 2009.

[52] S. Haimi, G. Gorianc, L. Moimas et al., "Characterization of zinc-releasing three-dimensional bioactive glass scaffolds and their effect on human adipose stem cell proliferation and osteogenic differentiation," *Acta Biomaterialia*, vol. 5, no. 8, pp. 3122–3131, 2009.

[53] A. Hoppe, R. Meszaros, C. Stähli et al., "In vitro reactivity of Cu doped 45S5 Bioglass® derived scaffolds for bone tissue engineering," *Journal of Materials Chemistry B*, vol. 1, no. 41, pp. 5659–5674, 2013.

[54] J. Bejarano, P. Caviedes, and H. Palza, "Sol-gel synthesis and *in vitro* bioactivity of copper and zinc-doped silicate bioactive glasses and glass-ceramics," *Biomedical Materials*, vol. 10, no. 2, Article ID 025001, 2015.

5

Nano-TiO$_2$ Doped Chitosan Scaffold for the Bone Tissue Engineering Applications

Pawan Kumar ⓘD

Department of Materials Science and Nanotechnology, Deenbandhu Chhotu Ram University of Science and Technology, Murthal-131039, India

Correspondence should be addressed to Pawan Kumar; pawankamiya@yahoo.in

Academic Editor: Carlo Galli

The present focus is on the synthesis of highly effective, porous, biocompatible, and inert scaffold by using ceramic nanoparticles and natural polymer for the application in tissue engineering. Freeze-drying method was used to fabricate nano-TiO$_2$ doped chitosan sample scaffold. Nano-TiO$_2$/chitosan scaffold can considered as an effective solution for damaged tissue regeneration. The interaction between chitosan (polysaccharide) and nano-TiO$_2$ makes it highly porous and brittle that could be an effective substitute for bone tissue engineering. The TiO$_2$ nanoparticles have a great surface area and inert properties while chitosan is highly biocompatible and antibacterial. The physiochemical properties of TiO$_2$ nanoparticles and scaffold are evaluated by XRD and FTIR. The nanoparticles doped scaffold has given improved density (1.2870g/cm^3) that is comparatively relevant to the dry bone (0.8 - 1.2 gm/cm^3). The open and closed porosity of sample scaffold were measured by using Brunauer–Emmett–Teller analyzer (BET) and scanning electron microscopy (SEM). The mechanical properties are examined by stable microsystem (Texture Analyzer). The in vitro degradation of scaffold is calculated in PBS containing lysozyme at pH 7.4. Electron and fluorescence microscopy are used to study morphological characteristics of the scaffolds and TiO$_2$ nanoparticles. The growth factor and drug-loaded composites can improve osteogenesis and vascularization.

1. Introduction

In the human body, bone is an extremely dynamic and diverse tissue (structurally and functionally). The human skeletal system is a collection of long to short, flat, and irregular bones. The average pore sizes of bone are 100-300 μm. The active diffusion of nutrients occurs within 150-200 μm pore sizes from blood supply. The nonmineralized (type-1 collagen) and mineralized phases (plate like apatite) are main composition of bone extracellular matrix [1–3]. Trauma, injury, infections, and bone extracellular matrix (ECM) loss are among the most human health threatening problems. Bone tissue engineering easily eliminates the issues of surgical treatments including donor site morbidity, inadequate availability, immune response of the body, and infections. The different substitutes including hydroxyapatite, different composition of bioactive glass, a number of synthetic and natural polymers, and their composites were fabricated for the bone tissue engineering. Still, none of the above materials could meet all the characteristics of bone graft substitutes

[4, 5]. The three major approaches of tissue engineering include isolated bone cells (osteogenic) transplantation or injection to the desired injured site, application of isolated tissue inducing molecules, or growth factors to the defected site and designing of 3D scaffolds [6]. Chitosan (natural polysaccharide) is the derivative of chitin (most abundant in nature after cellulose), used as graft material in tissue restoration because of its high biocompatibility and rapid degradation without toxicity. The well-designed 3D scaffold having the capability to mimic the extracellular matrix of bone promotes cell adhesion and cell proliferation without producing toxicity and helps in the promotion of new tissue [7, 8]. The progress of any scaffold depends on its porosity and biomechanical and physical properties that allows fast vascularization without producing cytotoxicity. The chitosan scaffold exhibits cationic nature, which is liable for exchange of negatively, charged molecules like proteoglycans, glycosaminoglycan (GAG), and other nutrients [9]. GAG is the chief constituent of ECMs that allow cell accumulation or adhesion and proliferation on the surface of scaffold.

FIGURE 1: Lyophilized sample scaffold of chitosan/TiO$_2$.

Many studies stated that the porous structure [10] gels [11], thin films [12], membranes [13], and fibres [14] favour more bone cell growth with chitosan. However, chitosan has numerous shortcomings such as lack of mechanical strength, fast degradation, and missing of cell signaling molecules that are most important for growth of damaged tissue [15, 16]. Titanium oxide (TiO$_2$) is available in the form of nanocrystals having a high surface area. Some researcher [17] observed the antiseptic and antibacterial activity of titanium oxide. Some researcher also proposed that TiO$_2$ can be used as good filler materials in biodegradable polymers because the presence of titanium oxide enhances cell attachment and cell proliferation [18]. Due to its inert property, it has medical and health applications. Both chitosan and TiO$_2$ nanoparticles are biocompatible, inert, and chemically stable so that they can be useful for hard tissue engineering also. They support cell adhesion and proliferation without producing toxicity. A lightweight and cost-effective TiO$_2$/chitosan scaffold can be fabricated by freeze-drying method that fulfills many biomechanical requirements of a hard tissue graft. The present work includes TiO$_2$ nanoparticles doped chitosan for scaffold synthesis.

2. Materials and Methods

2.1. Materials.
Analytical grade low molecular weight chitosan (75-85% deacetylated), titanium tetra isopropoxide (TTIP, 97% assay), sodium borohydride, glutaraldehyde (25% in H$_2$O), acetic acid, and NaOH were purchased Sigma-Aldrich. Lysozyme was procured from Thermo Fisher Scientific.

2.2. Synthesis of TiO$_2$ Nanoparticles.
The hydrothermal method is used for the TiO$_2$ nanoparticles synthesis. Firstly, 1M titanium tetra isopropoxide (TTIP) is used to make colloidal solution by hydrolysis, mixed with acetic acid (4M), and allowed for one hour of stirring. After an aging period of 24 hours, 25 ml of this solution transferred to Teflon lined tube of stainless steel autoclave and placed in oven at 180°C for 12 hours. After that, precipitates were washed three times with distilled water. The mixture solution was filtered and placed in oven for drying at 100°C to get TiO$_2$ crystal [19].

2.3. Scaffold Synthesis.
For the synthesis of chitosan/TiO$_2$ sample scaffold, firstly make 2% (w/v) solution of chitosan by the addition of 1% (w/v) acetic acid solution with 6 h stirring. The calculated amount (1g) of titanium oxide nanoparticles is mixed with 100 ml distilled water to make slurry. Add this slurry in viscus solution of chitosan with continuous stirring for 12 h at room temperature. 1 M solution of NaOH is used dropwise to adjust pH 10 of the mixture. 0.5 ml solution of glutaraldehyde (0.25% v/v) is used to make cross-linking between chitosan and TiO$_2$. After 15 hours of stirring, degas the suspension centrifugally and dispense it into Petri plates. These Petri plates are placed at - 80°C in an ultralow temperature freezer (LFZ-86L series, LABFREEZ) for 72 hours. After 72 hours, these frozen samples were transferred to the chamber of lyophilizer (ALLID FROST, Macflow Engg. Pvt. Ltd.), where the ice was removed by direct sublimation and the unfrozen water removed by desorption in a secondary drying process [20]. The freeze-dried sample neutralized with distilled water to remove acetate residue. The free glutaraldehyde (uncross-linked) was removed by using 5% sodium borohydride (reducing agent) solution and then washed with distilled water. After that, lyophilize these samples again for 48 h and collect pale yellow color scaffolds as shown in Figure 1.

2.4. XRD Analysis.
The graphical X-ray powder diffraction patterns of the TiO$_2$ and chitosan/TiO$_2$ samples were noted down by Rigaku Ultima IV X-Ray diffractometer, using Cu-Ka (1.5406 Å) radiation at room temperature in the range of 10° to 60° at 2 theta degree scale.

2.5. FTIR Analysis.
The Fourier transform infrared (FTIR) spectra (from 4000 cm^{-1} to 400 cm^{-1}) of the hydrothermally synthesized TiO$_2$ nanoparticles and chitosan/TiO$_2$ scaffold were recorded in Perkin Elmer Spectrum RX1 spectrometer. FTIR gives the information related to the presence of different chemical or functional groups in the samples.

2.6. SEM Observation.
Scanning electron microscope (SEM, JEOL) has given the information about the surface and the fracture section of scaffold. For the SEM analysis, thin section of sample scaffold was used with gold prior coating.

2.7. TEM Observation. Transmission Electron Microscope produces high-resolution images by the transmission of high-energy electron through the specimen. Structure, composition, and size of nanoparticles were analyzed by TEM (Hitachi H 7500) results. These nanoparticles were used for doping with chitosan to make scaffold.

2.8. Bulk Structure Analysis. Fluorescence Microscope (Leica DM 250) studied the surface topography and distribution of TiO_2 nanoparticles in chitosan gel. The chitosan/TiO_2 gel was used to understand the connectivity and branched structure existing between chitosan and TiO_2.

2.9. Porosity Measurement. The pore volume, micropore radius, and pore specific surface area of the sample were examined by BET (Quantachrome® Nova Station). The small sized pieces of sample scaffold were loaded in the sample tube and set measurement conditions. These results are mostly concentration- and viscosity-dependent.

2.10. Mechanical Properties. The sample size of $2.5 \times 4 \times 4$ mm^3 was cut down for the fracture strength measurement. The values of results can be uttered as the mean ± standard error. The force versus time graph is used to explain the hardness of the chitosan/TiO_2 scaffold.

2.11. Density. Density measurement is important for physical property evaluation. Equation (1) was used to measure the density of scaffold that give the calculated ratio of the mass by volume of sample [21].

$$\rho = \frac{W}{\pi \times (D/2)^2 \times H} \tag{1}$$

where D is the diameter, ρ is density, H is the thickness, and W is the weight of the sample, respectively.

2.12. Swelling Behavior. PBS solution was used to check the swelling or water retention capability of the sample scaffold. The swelling capacity depends on the porosity of the sample and nature of materials. The swelling capacity of the sample was calculated by the following [21]:

$$Water\ Retention\ (\%) = \frac{W_w - W_d}{W_d} \times 100 \tag{2}$$

where W_d is initial weight and W_w is the weight of the sample after swelling.

2.13. Biodegradation. The measurement of *in vitro* weight degradation of the chitosan/TiO_2 scaffold was required to estimate the bioavailability of materials. The pieces of the sample were incubated in the PBS solution (pH 7.4) containing 1×10^4 U/ml of lysozyme at room temperature for 14 days. After the interval of 7 and 14 days, degradation of the sample was recorded by using the following [21]:

$$Weight\ degraded\ (\%) = \frac{W_f - W_0}{W_0} \times 100 \tag{3}$$

where W_0 is initial weight and W_f is the final weight after degradation sample.

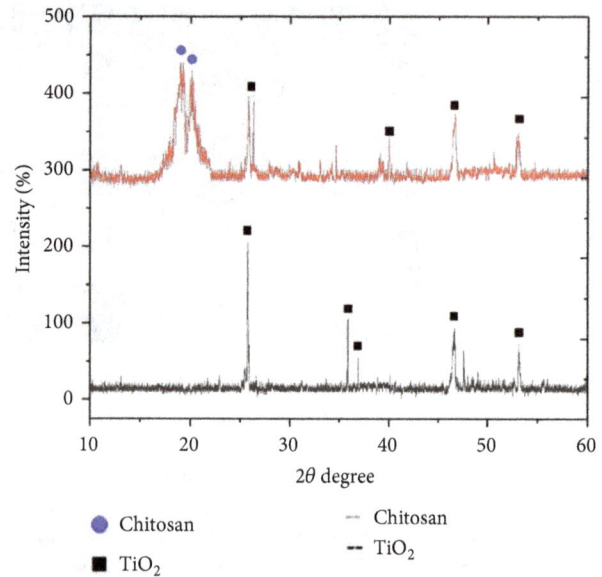

Chitosan · ·
TiO_2 ■

Chitosan —
TiO_2 --

FIGURE 2: XRD analysis of nano-TiO_2.

2.14. Cytotoxicity. To check the cytotoxicity and biocompatibility of chitosan/TiO_2 scaffolds, fibroblast cell lines were maintained in the cell culture facility in MEM with 10% FBS and 100 U/ml penicillin–streptomycin. Before cell seeding, all the sample scaffolds were sterilized and placed in an incubator with cell culture for 2 hours with 5% CO_2 and 85% humidity. The detached cells (1×10^5 cells/100 μl) were seeded dropwise on the surface of scaffolds for the investigation of cytocompatibility. The cell seeded scaffolds were placed in a humidified incubator at 37°C for 4 h for the cell attachment [22].

3. Results and Discussion

The formation of nano-TiO_2 was confirmed in XRD spectra (Figure 2). The particle size (4.48 nm) of TiO_2 was determined by using Scherrer equation. The chitosan/TiO_2 scaffold has shown a crystalline nature. The XRD of chitosan/TiO_2 scaffold and TiO_2 nanoparticles showed peaks at 25.7°, 35.8°, 36.9°, 40.2°, 46.6°, and 53.0° corresponding to TiO_2 (crystalline) and broad phase from 18 to 21° corresponding to chitosan (slightly amorphous). The FTIR spectrum has confirmed the presence of relevant functional groups in the sample (Figure 3). There is a band present from 2800 to 3400 cm^{-1} in the sample due to OH stretching vibrations. The band present from 600 to 711 cm^{-1} represents Ti–O–Ti stretching bonding [23]. The peak present in the sample near 1630 cm^{-1} shows Ti-OH bending vibrations of adsorbed H_2O molecules and peak at 1380 cm^{-1} indicating Ti-O [24]. The presence of C-H, C=O, and CH-OH groups was confirmed at 2924, 1656, and 1422 cm^{-1}, respectively [25].

The hydrothermally synthesized nano-TiO_2 is examined by TEM as shown in Figure 4. The TEM image of TiO_2 is shown around 2 nm sized particles with irregular shape. TiO_2 slurry does an interfacial interaction between chitosan and nanoparticles, which causes a nanoscale dispersion of TiO_2

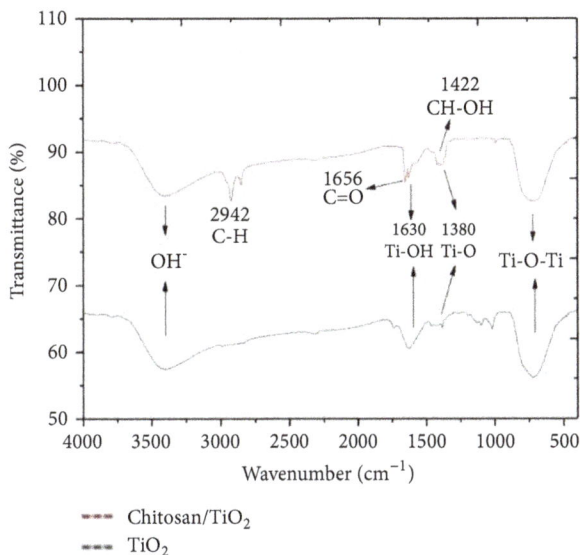

FIGURE 3: FTIR analysis of nano-TiO$_2$.

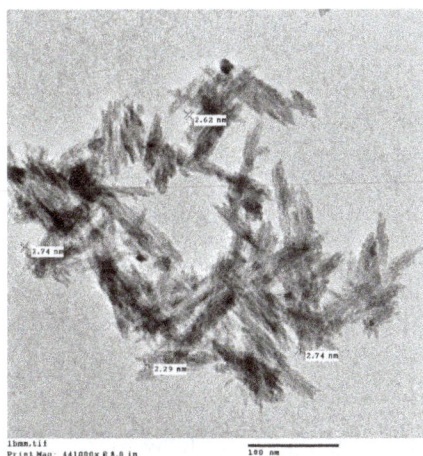

FIGURE 4: TEM image of TiO$_2$ nanoparticles.

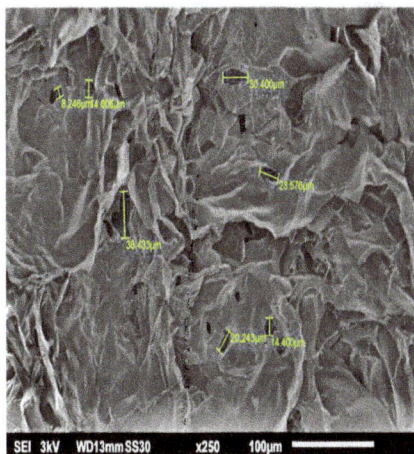

FIGURE 5: SEM images chitosan/TiO$_2$ sample.

in the matrix. The interaction between TiO$_2$ nanoparticles and chitosan depends on the charge state of the interface. Cationic chitosan is adsorbed on the surface of TiO$_2$, driven by electrostatic interactions and steric effects; adsorption is strictly pH-dependent and creation of stable TiO$_2$–chitosan composite [26]. Glass transition temperature (Tg) of chitosan (dry state) was found to be 118C confirmed by conventional DSC. The polymer-nanoparticles interactions play a key role in controlling the local dynamics of matrix and glass transition value of sample. The uniform dispersion of TiO$_2$ in chitosan matrix improved the glass transition temperature (Tg) chitosan/TiO$_2$ sample [27]. The well-interconnected, heterogeneous pore microstructures in sample scaffold are shown in Figure 5. Stretched pores were generated in the scaffold during lyophilization whose formation might be due to hydrogen bonding formation between polymer and nanoparticles and parallel ice crystal growth. SEM images of scaffold revealed mixed size of pores from 8.24 to 38.43μm were found in the chitosan/TiO$_2$ scaffold which are more

relevant for tissue engineering because the pore sizes of bone, muscle, and skin vary from 20 to 300 μm [20]. The pore size lower than 300μm helps to proliferate osteoblast cell easily through the scaffold [28]. Porosity of scaffold examined by BET revealed a specific surface area (2.7752 m^{2}g^{-1}), pore specific surface area (3.8751 m^{2}g^{-1}), pore volume (0.0030 cm^{3}g^{-1}), and pore diameter 2.86 nm. Previously examined chitosan and chitosan-gelatin scaffolds [20] show less porosity than chitosan/TiO$_2$ scaffold. The addition of TiO$_2$ nanoparticles increased porosity of scaffold. The addition of nano-f-CNT in chitosan increased porosity [29], while the addition of nanosilica reduces the porosity [30]. The sample of chitosan/TiO$_2$ as observed by fluorescence microscopy (Figure 6) revealed a low surface density of biomolecules and interconnected branched structure clearly shown in the image. The experimental density of chitosan/TiO$_2$ scaffold found 1.2870g/cm^3, which is more than chitosan and chitosan/gelatin scaffolds [20] and comparatively nearby the normal dry bone density (0.8 - 1.2 g/cm^3). The density of TiO$_2$ is 4.23 g/cm^3 which is less than stainless steel 316L (7.9g/cm^3), tantalum (16.6), and gold (19.3g/cm^3); however, it is better than fat (0.94g/cm^3), soft tissues (1.01-1.06g/cm^3), glass (2.4-2.8g/cm^3), bone (1-2 g /cm^3), and aluminum oxide (3.98g/cm^3). The density of chitosan is very low (0.15-0.30 g/cm^3) so TiO$_2$ is selected as a dopant to increase the density of scaffolds. The addition of nanobioglass and nanosilica enhanced the density of scaffolds [22, 30] but in this work we found that nano-TiO$_2$ reduces the density of scaffold. The change in density of a scaffold depends upon the concentration of chitosan in a composition. The force-time graph shows the fracture strength with respect to breaking time of sample as shown in Figure 7. The doping of titanium oxide makes effective improvements in the strength of scaffold without producing toxicity, proved by International Agency for Research on Cancer (IARC). The force bearing capacity of the sample was 1347.5 N/m^2 and breaking time was 0.58 sec that is better than chitosan scaffold, i.e., as previously described in [20], but less than CS/Alg, CS/Alg/nSiO$_2$, chitosan/CNT, chitosan-PPy-Alg, and chitosan-gelatin scaffolds

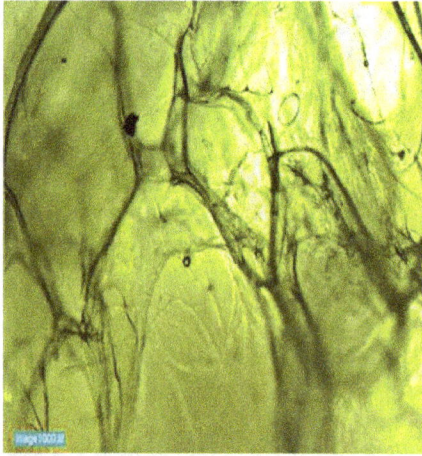

FIGURE 6: Microscopic image of chitosan/TiO$_2$ sample.

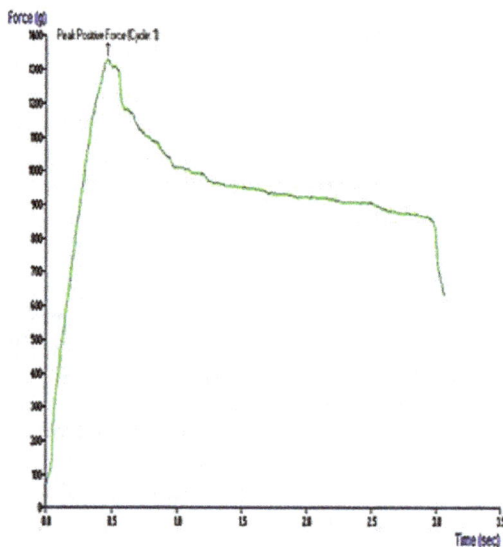

FIGURE 7: Force versus time graph for fracture strength.

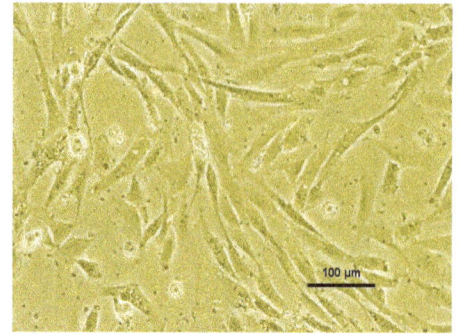

FIGURE 8: Fibroblast proliferation on chitosan/TiO$_2$ scaffold.

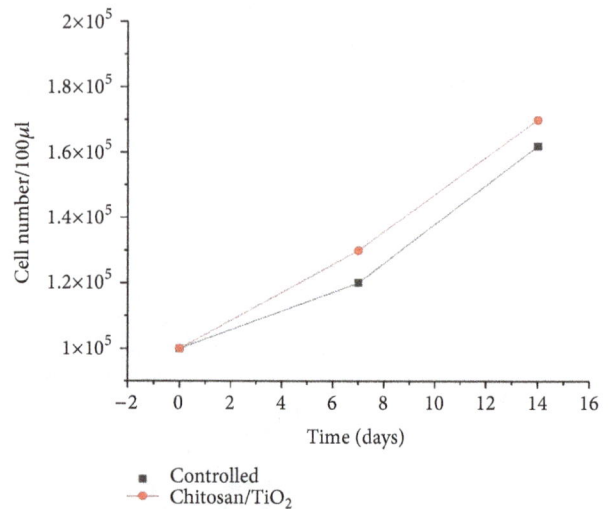

FIGURE 9: Cell proliferation on chitosan/TiO$_2$ scaffolds as a function of time.

[20, 29–32]. Hence, the scaffold can be used for preparing implants of bioinert and lightweight, which makes brittle composite with low density. The physical and mechanical characteristics of scaffold depend on the viscosity, ingredient concentration, pH of matrix, temperature, and lyophilization. At room temperature, the observed water retention capacity of chitosan-TiO$_2$ scaffold is found to be 24%, which is less than chitosan, chitosan-gelatin [20], chitosan-PPy-Alg [31], chitosan-alginate, chitosan-silica, [32], and chitosan-bioglass [22] scaffolds. The retention is also depending upon the porosity and nature of materials contained by the scaffolds. The percentage degradation of chitosan/TiO$_2$ scaffold was very slow compared to other chitosan based scaffolds [20, 29–32], 12.24% after 7 days and 14.80% after 14 days. The degree of crystallinity is the major factor, which controls the hydrolysis rate. However, TiO$_2$ supports apatite formation when it is encountered with Simulated Body Fluid [33]. TiO$_2$ nanoparticles have shown similarity with nanostructured nature of microenvironment. They have ability to pass the biological barriers, enter into the cell nuclei, and affect the cell functions such as proliferation and differentiation. From results, we found that the incorporation of TiO$_2$ in chitosan did not show any influence on fibroblast proliferation. Cytotoxicity and cell attachment studies showed the nanocomposites are nontoxic to an array of fibroblast cell line. The incorporation of TiO$_2$ into biodegradable scaffolds may enhance cell seeding and hence the subsequent tissue growth [34]. After preselected time intervals (7 and 14 days), the number of cells increased as shown in Figures 8 and 9. Fibroblast cells were successfully grown on the surface of chitosan/TiO$_2$ scaffolds.

4. Conclusion

The lyophilized sample of the scaffold was pale yellow in color, brittle, and inelastic. The scaffold fabricated using the freeze-drying technique exhibited high porosity and relatively low density with good mechanical properties. The remarkable adsorption or water retention capacity of TiO$_2$ doped chitosan scaffold was noticed during investigation. The addition of TiO$_2$ reduces the fast degradation of scaffold.

The nanoparticles restricted the formation of the strong bond during sample preparation. Hence, it can be concluded that nanoparticles doped scaffolds can be used for preparing a biocompatible implant with lighter weight. The natural origin and biocompatible nature of chitosan support bone cells attachment, proliferation, and mineralization. Titanium oxide is inert in nature, so the combination of chitosan/TiO$_2$ can be a good substitute for tissue engineering. Still a lot of research needs to be conducted for optimizing different parameters of scaffolds without compromising biodegradability and toxicity. Multidisciplinary approach for fabrication of new scaffolds with improved properties is highly desired.

Conflicts of Interest

The author declare that they have no conflicts of interest.

References

[1] F. Anagnostou, R. Bizios, and H. Petite, "Engineering bone: challenges and obstacles," *Journal of Cellular and Molecular Medicine*, vol. 9, no. 1, pp. 72–84, 2005.

[2] J. Zeltinger, J. K. Sherwood, D. A. Graham, R. Müeller, and L. G. Griffith, "Effect of pore size and void fraction on cellular adhesion, proliferation, and matrix deposition," *Tissue Engineering Part A*, vol. 7, no. 5, pp. 557–572, 2001.

[3] D. W. Hutmacher, "Scaffold design and fabrication technologies for engineering tissues—state of the art and future perspectives," *Journal of Biomaterials Science, Polymer Edition*, vol. 12, no. 1, pp. 107–124, 2001.

[4] R. J. O'Keefe and J. Mao, "Bone tissue engineering and regeneration: from discovery to the clinic-an overview," *Tissue Engineering—Part B: Reviews*, vol. 17, no. 6, pp. 389–392, 2011.

[5] C. Wang, H. Shen, Y. Tian et al., "Bioactive Nanoparticle–Gelatin Composite Scaffold with Mechanical Performance Comparable to Cancellous Bones," *ACS Applied Materials & Interfaces*, vol. 6, no. 15, pp. 13061–13068, 2014.

[6] H. Chen, R. Truckenmüller, C. Van Blitterswijk, and L. Moroni, "Fabrication of nanofibrous scaffolds for tissue engineering applications BT-Nanomaterials in Tissue Engineering," in *Woodhead Publishing Series in Biomaterials*, pp. 158–183, Woodhead Publishing, 2013.

[7] W. W. Thein-Han, J. Saikhun, C. Pholpramoo, R. D. K. Misra, and Y. Kitiyanant, "Chitosan-gelatin scaffolds for tissue engineering: physico-chemical properties and biological response of buffalo embryonic stem cells and transfectant of GFP-buffalo embryonic stem cells," *Acta Biomaterialia*, vol. 5, no. 9, pp. 3453–3466, 2009.

[8] A. M. Martins, C. M. Alves, F. Kurtis Kasper, A. G. Mikos, and R. L. Reis, "Responsive and in situ-forming chitosan scaffolds for bone tissue engineering applications: an overview of the last decade," *J. Mater. Chem.*, vol. 20, no. 9, pp. 1638–1645, 2010.

[9] A. di Martino, M. Sittinger, and M. V. Risbud, "Chitosan: a versatile biopolymer for orthopaedic tissue-engineering," *Biomaterials*, vol. 26, no. 30, pp. 5983–5990, 2005.

[10] T. Jiang, C. M. Pilane, and C. T. Laurencin, "Fabrication of Novel Porous Chitosan Matrices as Scaffolds for Bone Tissue Engineering," *Materials Research Society - Proceedings*, vol. 845, 2004.

[11] M. Yamamoto, Y. Takahashi, and Y. Tabata, "Controlled release by biodegradable hydrogels enhances the ectopic bone formation of bone morphogenetic protein," *Biomaterials*, vol. 24, no. 24, pp. 4375–4383, 2003.

[12] N. Sultana, "Study of in vitro degradation of biodegradable polymer based thin films and tissue engineering scaffolds," *African Journal of Biotechnology*, vol. 10, pp. 18709–18715, 2011.

[13] C. E. Orrego and J. S. Valencia, "Preparation and characterization of chitosan membranes by using a combined freeze gelation and mild crosslinking method," *Bioprocess and Biosystems Engineering*, vol. 32, no. 2, pp. 197–206, 2009.

[14] F. Pati, H. Kalita, B. Adhikari, and S. Dhara, "Osteoblastic cellular responses on ionically crosslinked chitosan-tripolyphosphate fibrous 3-D mesh scaffolds," *Journal of Biomedical Materials Research Part A*, vol. 101, no. 9, pp. 2526–2537, 2013.

[15] M. Okada and T. Furuzono, "Hydroxylapatite nanoparticles: fabrication methods and medical applications," *Science and Technology of Advanced Materials*, vol. 13, no. 6, p. 064103, 2012.

[16] J. R. Porter, T. T. Ruckh, and K. C. Popat, "Bone tissue engineering: a review in bone biomimetics and drug delivery strategies," *Biotechnology Progress*, vol. 25, no. 6, pp. 1539–1560, 2009.

[17] S. M. Gupta and M. Tripathi, "A review of TiO$_2$ nanoparticles," *Chinese Science Bulletin*, vol. 56, no. 16, pp. 1639–1657, 2011.

[18] J. K. Savaiano and T. J. Webster, "Altered responses of chondrocytes to nanophase PLGA/nanophase titania composites," *Biomaterials*, vol. 25, no. 7-8, pp. 1205–1213, 2004.

[19] R. Vijayalakshmi and V. Rajendran, "Synthesis and characterization of nano-TiO$_2$ via different methods," *Scholars Research Library*, vol. 4, pp. 1183–1190, 2012.

[20] P. Kumar, B. S. Dehiya, and A. Sindhu, "Comparative study of chitosan and chitosan–gelatin scaffold for tissue engineering," *International Nano Letters*, vol. 7, no. 4, pp. 285–290, 2017.

[21] J. S. Mao, L. G. Zhao, Y. J. Yin, and K. D. Yao, "Structure and properties of bilayer chitosan-gelatin scaffolds," *Biomaterials*, vol. 24, no. 6, pp. 1067–1074, 2003.

[22] M. Peter, N. S. Binulal, S. V. Nair, N. Selvamurugan, H. Tamura, and R. Jayakumar, "Novel biodegradable chitosan-gelatin/nano-bioactive glass ceramic composite scaffolds for alveolar bone tissue engineering," *Chemical Engineering Journal*, vol. 158, no. 2, pp. 353–361, 2010.

[23] H. Pan, X. Wang, S. Xiao, L. Yu, and Z. Zhang, "Preparation and characterization of TiO2 nanoparticles surface-modified by octadecyltrimethoxysilane," *Indian J. Eng. Mater. Sci*, vol. 20, pp. 561–567, 2013.

[24] A. León, P. Reuquen, C. Garín et al., "FTIR and Raman Characterization of TiO2 Nanoparticles Coated with Polyethylene Glycol as Carrier for 2-Methoxyestradiol," *Applied Sciences*, vol. 7, no. 1, p. 49, 2017.

[25] S. Yasmeen, M. Kabiraz, B. Saha, M. Qadir, M. Gafur, and S. Masum, "Chromium (VI) Ions Removal from Tannery Effluent using Chitosan-Microcrystalline Cellulose Composite as Adsorbent," *International Research Journal of Pure and Applied Chemistry*, vol. 10, no. 4, pp. 1–14, 2016.

[26] A. E. Wiącek, A. Gozdecka, and M. Jurak, " Physicochemical Characteristics of Chitosan–TiO ," *Industrial & Engineering Chemistry Research*, vol. 57, no. 6, pp. 1859–1870, 2018.

[27] F. A. Al-Sagheer and S. Merchant, "Visco-elastic properties of chitosan-titania nano-composites," *Carbohydrate Polymers*, vol. 85, no. 2, pp. 356–362, 2011.

[28] M. C. Nerantzaki, I. G. Koliakou, M. G. Kaloyianni et al., "New N-(2-carboxybenzyl)chitosan composite scaffolds containing nanoTiO2 or bioactive glass with enhanced cell proliferation for bone-tissue engineering applications," *International Journal of Polymeric Materials and Polymeric Biomaterials*, vol. 66, no. 2, pp. 71–81, 2017.

[29] J. Venkatesan, B. Ryu, P. N. Sudha, and S.-K. Kim, "Preparation and characterization of chitosan-carbon nanotube scaffolds for bone tissue engineering," *International Journal of Biological Macromolecules*, vol. 50, no. 2, pp. 393–402, 2012.

[30] K. C. Kavya, R. Jayakumar, S. Nair, and K. P. Chennazhi, "Fabrication and characterization of chitosan/gelatin/nSiO$_2$ composite scaffold for bone tissue engineering," *International Journal of Biological Macromolecules*, vol. 59, pp. 255–263, 2013.

[31] K. M. Sajesh, R. Jayakumar, S. V. Nair, and K. P. Chennazhi, "Biocompatible conducting chitosan/polypyrrole-alginate composite scaffold for bone tissue engineering," *International Journal of Biological Macromolecules*, vol. 62, pp. 465–471, 2013.

[32] J. A. Sowjanya, J. Singh, T. Mohita et al., "Biocomposite scaffolds containing chitosan/alginate/nano-silica for bone tissue engineering," *Colloids and Surfaces B: Biointerfaces*, vol. 109, pp. 294–300, 2013.

[33] T. Kasuga, H. Kondo, and M. Nogami, "Apatite formation on TiO2 in simulated body fluid," *Journal of Crystal Growth*, vol. 235, no. 1-4, pp. 235–240, 2002.

[34] R. Jayakumar, R. Ramachandran, V. V. Divyarani, K. P. Chennazhi, H. Tamura, and S. V. Nair, "Fabrication of chitin-chitosan/nano TiO2-composite scaffolds for tissue engineering applications," *International Journal of Biological Macromolecules*, vol. 48, no. 2, pp. 336–344, 2011.

6

Novel Vanadium-Loaded Ordered Collagen Scaffold Promotes Osteochondral Differentiation of Bone Marrow Progenitor Cells

Ana M. Cortizo,[1] Graciela Ruderman,[2] Flavia N. Mazzini,[1] M. Silvina Molinuevo,[1] and Ines G. Mogilner[2]

[1]LIOMM, Dto. Ciencias Biológicas, Facultad de Ciencias Exactas, Universidad Nacional de La Plata, 1900 La Plata, Argentina
[2]IFLYSIB, CONICET, Facultad de Ciencias Exactas, Universidad Nacional de La Plata, 1900 La Plata, Argentina

Correspondence should be addressed to Ana M. Cortizo; cortizo@biol.unlp.edu.ar

Academic Editor: Ravin Narain

Bone and cartilage regeneration can be improved by designing a functionalized biomaterial that includes bioactive drugs in a biocompatible and biodegradable scaffold. Based on our previous studies, we designed a vanadium-loaded collagen scaffold for osteochondral tissue engineering. Collagen-vanadium loaded scaffolds were characterized by SEM, FTIR, and permeability studies. Rat bone marrow progenitor cells were plated on collagen or vanadium-loaded membranes to evaluate differences in cell attachment, growth and osteogenic or chondrocytic differentiation. The potential cytotoxicity of the scaffolds was assessed by the MTT assay and by evaluation of morphological changes in cultured RAW 264.7 macrophages. Our results show that loading of VOAsc did not alter the grooved ordered structure of the collagen membrane although it increased membrane permeability, suggesting a more open structure. The VOAsc was released to the media, suggesting diffusion-controlled drug release. Vanadium-loaded membranes proved to be a better substratum than C0 for all evaluated aspects of BMPC biocompatibility (adhesion, growth, and osteoblastic and chondrocytic differentiation). In addition, there was no detectable effect of collagen or vanadium-loaded scaffolds on macrophage viability or cytotoxicity. Based on these findings, we have developed a new ordered collagen scaffold loaded with VOAsc that shows potential for osteochondral tissue engineering.

1. Introduction

Osteochondral damage is a frequent consequence of traumatic and degenerative alterations of joints and bones [1]. As articular cartilage is an avascular tissue in which differentiated cells are embedded in an organic matrix, it has a very limited potential for spontaneous healing [2, 3]. Different tissue engineering-based strategies have been used to improve the regenerative capacity of osteochondral lesions. In particular, a microfracture procedure combined with scaffold implantation with or without the addition of mesenchymal stem cells and/or growth factors has shown promising results [4–7]. Filling the osteochondral defect with an adequate matrix substitute (i.e., with functional characteristics similar to those of the original tissue) can decrease reparation time and thus patient morbidity [7–9]. The use of mesenchymal stem cells is of particular interest since in response to specific factors they present the ability to differentiate into osteoblasts or chondroblasts and since they can be obtained from the same patient [10]. Despite these encouraging aspects, an important number of patients submitted to these procedures show worsening of their lesions, underscoring the importance of perfecting the design of an adequate biomaterial.

The inclusion of different bioactive molecules for bone and/or cartilage regeneration has also been reported, leading to the design of scaffolds that can function as a controlled delivery system [11, 12]. This could improve the osteoinductive properties of the matrix and even reduce the cytotoxicity of a certain drug, thus reducing the time required for tissue repair. In addition, designing a functionalized or "intelligent" biomaterial could optimize the degradation rate and dosification of the bioactive molecule.

Although many materials have been developed for the repair of osteochondral defects, none has equalled the properties of collagen [13]. This natural polymer has been used extensively because of its inherent biocompatibility, biodegradability, osteoconductivity, and cost-effective availability [14]. Recently, we have prepared and characterized a microscale ordered collagen scaffold and shown its efficacy for bone tissue engineering [15]. Our results suggest that the orientation of collagen fibers positively regulates osteoblastic growth and development. When compared to a randomly oriented collagen scaffold, the ordered collagen matrix enhanced the in vitro osteogenic differentiation of MC3T3E1 preosteoblasts increasing markers of osteoblast differentiation and extracellular matrix mineralization [16].

In addition, we have previously reported the osteogenic properties of a vanadyl(IV) ascorbate (VOAsc) complex as well as its possible mechanisms of action, on two osteoblastic cell lines in culture [17]. VOAsc significantly stimulated osteoblastic proliferation and type I collagen production and increased the formation of mineralization nodules. We demonstrated that this complex inhibited several phosphatases, while simultaneously activating the ERK pathway and regulating intracellular calcium levels and the PI3-kinase pathway. Altogether our observations suggest that the vanadium(IV) ascorbate complex could be a useful pharmacological tool for bone tissue regeneration. On the other hand, we have evaluated a delivery system for vanadium using poly(β-propiolactone) films [18]. In this system, vanadium was liberated at a controlled rate causing a selective antiproliferative effect on osteosarcoma cells with lower cytotoxicity than the drug in solution.

Based on our previous data, in the present work we have designed an ordered collagen scaffold loaded with different amounts of VOAsc and evaluated its potential use for regeneration of bone and cartilage. We have determined several physicochemical and biocompatibility properties of the collagen scaffolds. In particular, we have investigated the effects of VOAsc loaded into the membranes on the growth of rat bone marrow progenitor cells (BMPC) and on their phenotypic differentiation into osteoblasts and chondroblasts. We have also evaluated the possible in vitro cytotoxicity of the scaffolds.

2. Materials and Methods

2.1. Membrane Preparation and Characterization. Membranes were prepared using acid-soluble collagen extracted from bovine Achilles tendon. Ordered films were obtained according to standardized procedures of our laboratory [15]. Vanadyl(IV) ascorbate (VOAsc) was synthesized as we have previously described and characterized by UV-vis and infrared spectroscopy [17, 19]. Stock solutions were prepared in distilled water at room temperature and used immediately for the different assays. Membranes of pure collagen were loaded with the VOAsc complex and given an ordered pattern by layering the collagen and vanadium solutions (50, 100, or 200 μg/mL) alternatively on a mold, according to our previously described method [15]. This formulation was selected in order to obtain 10, 20, or 40 μg/mL VOAsc per cm^2 of membrane surface (C1, C2, and C3). An ordered collagen membrane without vanadium was also prepared as a control (basal condition, C0).

Surface characteristics of the membranes were investigated using scanning electron microscopy (SEM; Phillips 505, Netherlands), with an accelerating voltage of 25 kV. The images were analyzed by Soft Imaging System ADDAII. Fourier Transformed Infrared (FTIR) spectra were collected to analyze and compare the material characteristics of all scaffold samples following VOAsc addition. FTIR analysis was carried out using an IRAffinity-1 Spectrum FTIR (Shimadzu). Scaffolds were cut and fixed in a port-cell and spectra were collected between 2000 and 800 cm^{-1}.

Permeability was evaluated by measuring the flux of NaCl across the membranes, with a custom-made glass cell as previously described [15]. In this device the membrane divides the cell in two compartments: one of them (L) was filled with the solution under study and the other (R) with distilled water. The complete cell was kept in a thermostatic bath at 37°C. The two half cells were stirred with a magnetic device to avoid the possible formation of an unstirred layer on the membrane surface. Fluid samples were taken from compartment R at regular intervals for conductivity determination of NaCl concentration with a Radiometer CMD3 conductivity meter. A calibration was previously performed. Permeability is given by

$$P = V\Delta \left(-\frac{\ln\left(1 - 2C\left(t\right)/C\left(0\right)\right)}{2a\Delta t} \right), \tag{1}$$

where V is the cell volume, a the membrane area, $C(0)$ the concentration of the solution at $t = 0$, and $C(t)$ the concentration at time t. A plot of $\ln(1 - 2C(t)/C(0))$ against t was made using the determinations of C. After a least square fit, the slope was taken as the value of P.

2.2. Vanadium Release Kinetic Studies. The release profile of VOAsc from the scaffold was determined by incubating scaffold samples (1 cm^2) loaded with different amounts (10, 20, and 40 μg/mL per cm^2) of VOAsc in 1.0 mL of sterile DMEM without phenol red (pH = 7.4) at 37°C for different periods of time. At appropriate times (every 15 min during the initial hour, then every 30 min until 5 h, and finally every hour for 24 h), the supernatant was removed and replaced by fresh media. The time-dependent release of the drug was followed by monitoring the amount of VOAsc present in the supernatant medium, using a T60 UV-visible spectrophotometer (PG Instrument). A linear calibration curve of VOAsc concentration versus absorbance at 580 nm was obtained using VOAsc standards in the range 0-1 mg/mL. The assay was performed in quadruplicate and results were expressed as the fractional release (M_t/M_∞) versus time of release (t).

2.3. Biocompatibility Studies

2.3.1. Cell Cultures and Incubations. The biocompatibility of Col and Col-V (loaded with different amounts of VOAsc) membranes was evaluated using rat bone marrow progenitor cells (BMPC), investigating possible changes in their morphology, growth, osteoblastic induction, and chondrogenic differentiation when grown on the different scaffold. BMPC were chosen because they represent a better physiologic model of osteogenesis due to its ability of self-renewal and multilineage differentiation compared to cloned established cell lines [20]. Under appropriate conditions, these cells are able to differentiate into osteoblasts, chondroblasts, or adipocytes. Additionally, they could also be used for in vivo studies of bone tissue regeneration. BMPC were isolated from the femora of young male Sprague-Dawley rats and cultured as previously described [21]. Cells were maintained in a basal medium (DMEM-10% FBS) at 37°C until being plated on the different membranes, after which fresh medium was added every 2 days. Prior to their use for cell culture, 1-2 cm^2 scaffold samples were sterilized by immersion in 70% ethanol and irradiation with UV light. After different incubation periods, the C0 and Col-V vanadium-loaded membranes were processed to evaluate cell adhesion, proliferation, or differentiation.

In order to compare the direct effect of VOAsc addition in the cell culture media with the effects of the compound released from the membranes, some experiments were performed with the BMPC plated on standard tissue culture plates. In these cases, cells were incubated in a basal medium with different doses of VOAsc complex in solution as indicated in Figures 1–7. After different incubation periods, cell monolayers were processed to evaluate proliferation and osteogenic differentiation as described below.

2.3.2. Evaluation of Cell Growth.

The cells were washed with phosphate-buffered saline (PBS, pH 7.4), after which adherent cells were fixed with methanol, stained with Giemsa, observed using a TS100 Eclipse Nikon microscope, and photographed with a CCD camera with a 0.7x DXM Nikon lens [16]. Cell adhesion (1 h after seeding) and proliferation (24 h) were evaluated by counting the number of cells/field in 10 representative fields per experimental condition.

2.3.3. BMPC Differentiation.

For BMPC osteoblastic differentiation, cells were incubated for different periods of time in osteogenic media (10% FBS-DMEM supplemented with 5 mM β-glycerol-phosphate and 25 μg/mL ascorbic acid). Osteogenic differentiation was evaluated by alkaline phosphatase specific activity (ALP) and extracellular calcium deposition (mineralization nodules). For ALP determination, cells submitted to a 15-day osteogenic differentiation were washed with PBS and solubilized in 0.5 mL 0.1% Triton X-100. Aliquots of this total cell extract were used for protein determination [22] and for measurement of ALP by spectrophotometric determination of the initial rates (10 min) of hydrolysis of p-nitrophenyl-phosphate to p-nitrophenol at 37°C. Mineralization nodules were measured after 21 days of osteogenic differentiation using Alizarin S red staining. Stained calcium deposits were extracted with 1 mL of 0.1 N sodium hydroxide and the optical density recorded at 548 nm. Additionally, type 1 collagen production was evaluated in the cell monolayer plated into the standard tissue culture plate to assess the effect of VOAsc in solution, by Sirius red staining, as we previously described [17].

Chondrogenic differentiation was assessed after a 21-day culture in a chondrogenic medium [23]. Briefly, 10^7 BMPC/mL were resuspended in 40 μL of serum-free DMEM, and four individual drops (10 μL per drop) were carefully placed on Col or Col-V films included in each well interior of 24-well plates. Cells were allowed to adhere at 37°C for 2 h. Basal medium was then added for an additional 24 hours, after which it was replaced by a chondrogenic medium (serum-free DMEM supplemented with 10 ng/mL of TGF-b3 (PeproTech, USA), 10^{-8} M dexamethasone, and 1x insulin transferring selenium (ITS) supplement (Invitrogen)), in which cells were cultured for 21 days changing the medium every three days. Production of chondroitin sulphate glycosaminoglycan (GAG), a marker of chondrocytic differentiation, was evaluated by Alcian blue staining (pH 3) at the end of the culture period. Briefly, cells were fixed overnight and stained with 0.5% Alcian blue in 0.1 N HCl, rinsed twice with 0.1 N HCl, and once with distilled water. Finally, the dye was extracted with 4 M guanidinium HCl and absorbance was measured at 600 nm.

2.4. Evaluation of Scaffold Cytotoxicity.

RAW 264.7 monocyte-macrophage cells were maintained in DMEM containing 10% FBS, 100 U/mL penicillin, and 100 μg/mL streptomycin at 37°C in a 5% CO_2 atmosphere. This cell line has previously been used in our laboratory to assess cytotoxicity since it represents an adequate and sensitive in vitro model for inflammation [16, 24]. After subculturing with 10% EDTA, cells were seeded on either the Col or Col-V membranes and incubated for 24 h after which the MTT bioassay was performed. Briefly, 5 mg/mL of MTT solution was added and incubated for 3 h [24] until formation of purple formazan crystals, which were dissolved in dimethyl sulfoxide (DMSO) and absorbance measured at 490 nm. In other experiments, morphology of RAW 264.7 cells cultured on either scaffold was analyzed after Giemsa staining, using a TS100 Eclipse Nikon microscope, and photographed with a CCD camera with a 0.7x DXM Nikon lens.

2.5. Statistical Analysis.

Results are expressed as the mean $\pm$ SEM and were obtained from at least three separate experiments. Differences between the groups were assessed by one-way ANOVA using the Tukey post hoc test. For nonnormal distributed data, the nonparametrical Kruskal-Wallis test with Dunn post hoc test was performed, using GraphPad InStat, version 3.00 (GraphPad Software, San Diego, CA, USA). $p < 0.05$ was considered significant for all statistical analyses.

3. Results

3.1. Membrane Characterization.

Inclusion of vanadium in the scaffold was demonstrated by FTIR spectra. The presence of VOAsc in the collagen scaffold was evidenced by the characteristics peaks for ascorbic acid at 1755 cm^{-1} (C=O stretching) and 1670 cm^{-1} (C=C vibration) as has been previously described [19]. In the Col-V membrane spectra, both a characteristic strong band of ν(V=O) appearing at 962 cm^{-1}

(a)

(b)

(c)

(d)

(e)

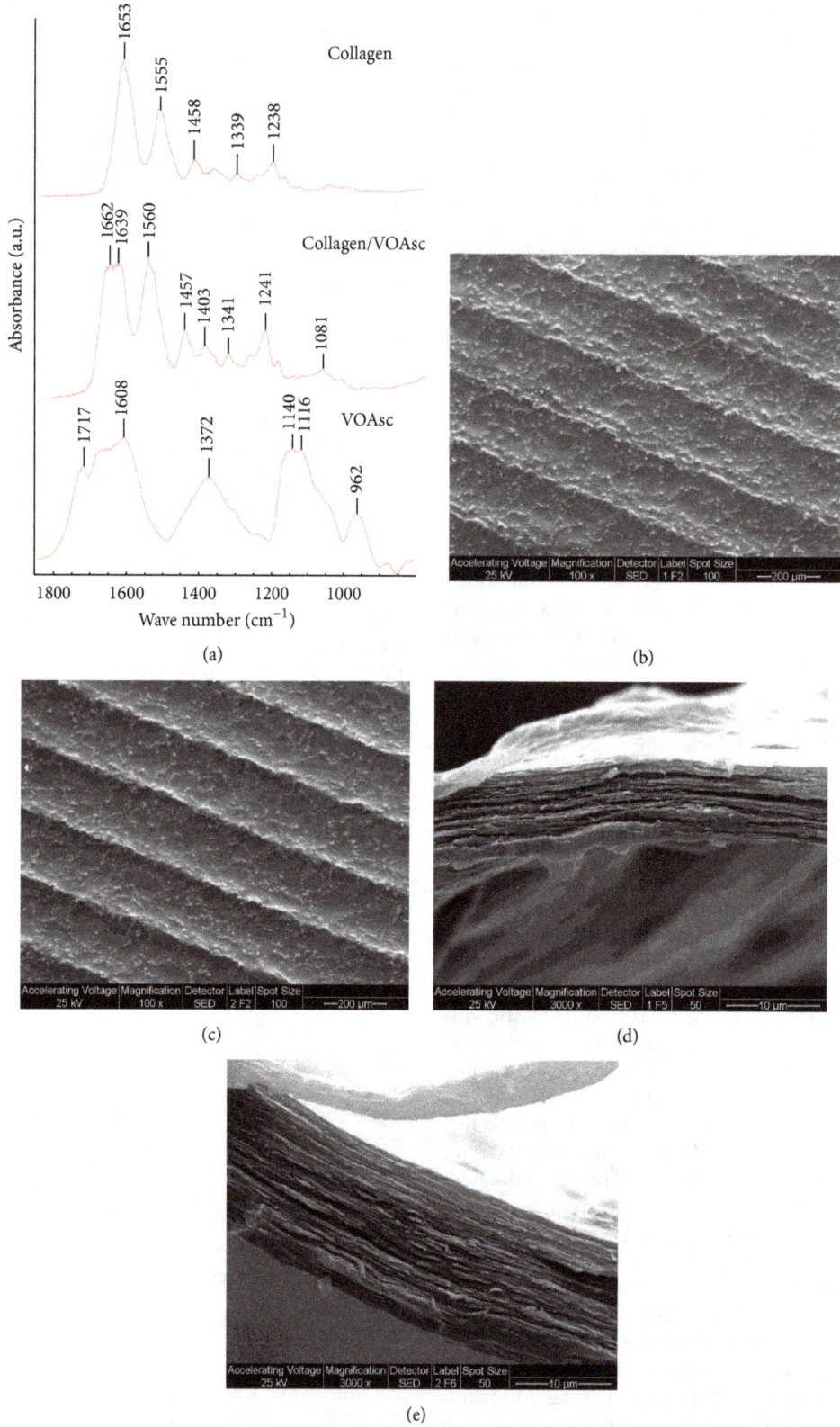

FIGURE 1: Characterization of Col and Col-V scaffolds. FTIR spectra of Col, Col-V (20 μg/mL per cm^2 VOAsc) membrane, and the VOAsc complex (a). SEM images of Col and Col-V surfaces ((b) and (c), resp.) showed an ordered matrix with a grooved structure. Transversal section of Col and Col-V films ((d) and (e), resp.).

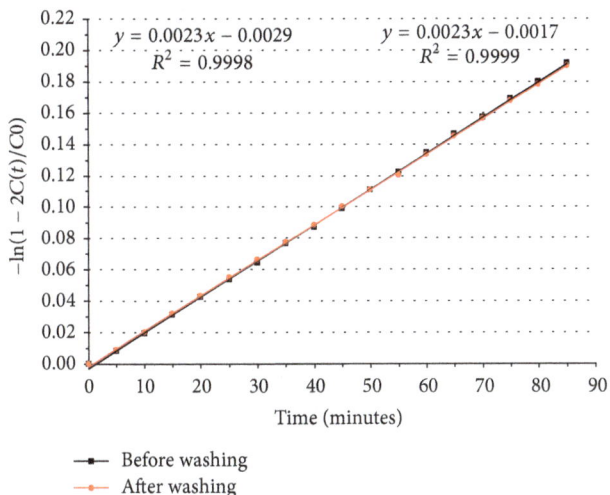

FIGURE 2: NaCl flux experiments across Col-V ($20\,\mu g/mL$ per cm^2 VOAsc) membrane at a temperature of 37°C before (black) and after (red) a 24 h wash of the membrane in distilled water.

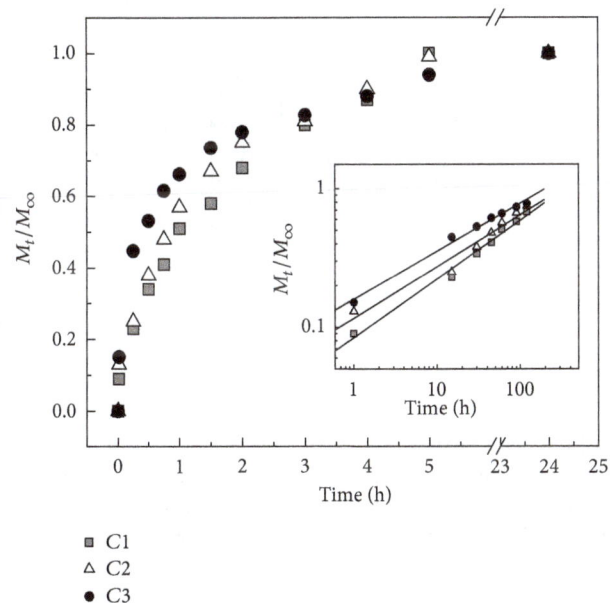

■ C1
△ C2
● C3

FIGURE 3: Fractional VOAsc drug release from the Col-V membranes in vitro. $C1$: $10\,\mu g/mL$ per cm^2 VOAsc; $C2$: $20\,\mu g/mL$ per cm^2 VOAsc; and $C3$: $40\,\mu g/mL$ per cm^2 VOAsc. Release was measured up to 24 h. Insert: plot of $\log M_t/M_\infty$ versus $\log t$ at initial times. Data are presented as mean ($n = 3$).

and a peak at $1372\,cm^{-1}$ corresponding to the (C3-O-)V complex were observed. In addition, characteristic peaks for collagen were evidenced at $1653\,cm^{-1}$ (Amide I), $1550-60\,cm^{-1}$ (Amide II), and $1240\,cm^{-1}$ (Amide III) (Figure 1(a)).

SEM analysis demonstrated a scaffold surface with parallel alignment of collagen fibers and a grooved structure (Figure 1(b)), which was not influenced by the presence of vanadium complex (Figure 1(c)). Superposition of the collagen layers was homogeneous on both Col (Figure 1(d)) and Col-V (Figure 1(e)) scaffolds, revealing no influence of vanadium complex on the morphology of the layers, as

observed by SEM. However, since SEM is not the most sensitive method to evaluate alterations in pore structure of the scaffold we also investigated the permeability properties of the membranes.

Permeability is an important physicochemical characteristic since VOAsc must be delivered from the scaffold in order to act as an osteogenic compound. We measured the flux of NaCl across the membrane as a measure of its permeability at 37°C. We found that the permeability value of Col-V membrane was 2.16×10^{-4} cm/sec, which represents an increase of about 40% compared to our previously described values for Col scaffold (1.47×10^{-4} cm/sec [15]) (Figure 2). We also evaluated the possible effect of VOAsc delivery on scaffold permeability. Thus, permeability of Col-VOAsc scaffolds was determined initially as described above and again using the same membrane after 24 h of washing in distilled water (Figure 2 red plot). We found an overlap for both curves, indicating no differences in permeability coefficient.

3.2. VOAsc Release Kinetics. In order to analyze the kinetics of vanadium release, we used Fick's second law for one-dimensional transport in thin polymeric films with a moderate swelling rate [25]:

$$\frac{M_t}{M_\infty} = kt^n, \qquad (2)$$

where M_t is the cumulative absolute amount of drug released at time t; M_∞ is the absolute cumulative amount of drug profile released at infinite time; k is a constant incorporating structural and geometric characteristics of the device; and n is the release exponent, indicative of the mechanism of drug release. Figure 3 shows the fractional vanadium release from C1 to C3 films (containing different concentrations of the drug) as a function of time. As it can be observed, the VOAsc complex was released to the media in a controlled manner, with a fast and linear kinetic during the first 3 h. At initial rate a slight difference could be observed in the VOAsc release for the different vanadium-loaded membranes. Nevertheless, at the end of the incubation time, a saturation curve was observed for each membrane with the release of vanadium complex sustained for 24 h. The kinetics of such process could be analyzed through the Fick model in order to determine the n exponent. The insert in Figure 3, a plot of $\log M_t/M_\infty$ versus $\log t$, shows the linear regression plots of the fractional vanadium release at short times. The diffusion coefficient values (n) were 0.425 ± 0.02, 0.374 ± 0.03, and 0.347 ± 0.02 for $C1$, $C2$, and $C3$, respectively, although not statistically different. These results suggested that the mechanism of vanadium release occurs by a Fickian diffusion process.

3.3. Effects of VOAsc on BMPC Proliferation and Differentiation. In a first series of experiments, we used BMPC seeded on standard tissue culture plates, in order to investigate the effects of the VOAsc complex added directly into the culture media. Treatment of cells with $2.5-100\,\mu M$ VOAsc for 24 h led to a biphasic effect on cell proliferation. Figure 4(a) shows that $10-50\,\mu M$ VOAsc significantly stimulated BMPC

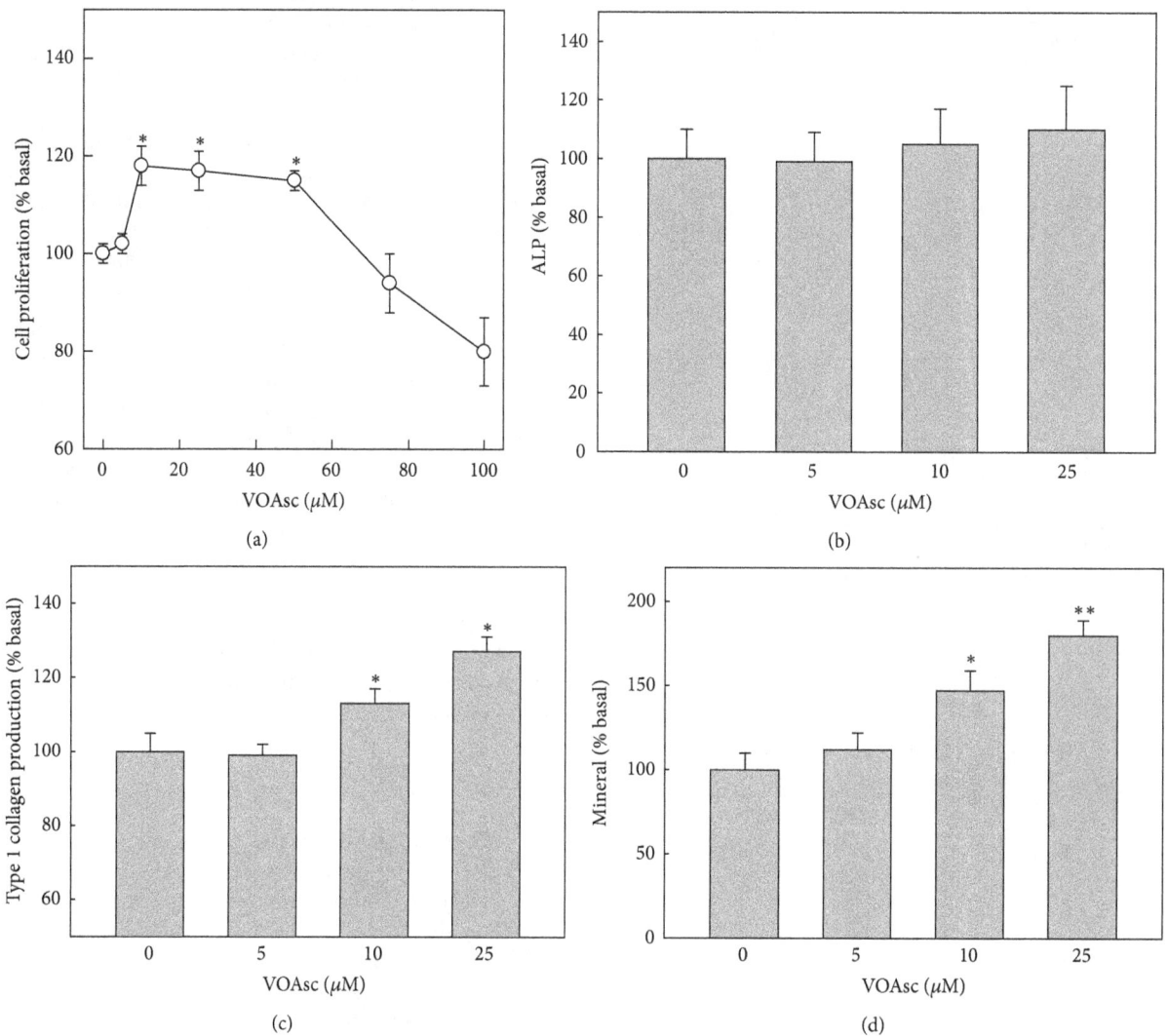

FIGURE 4: Effect of VOAsc on cell proliferation, differentiation, and mineralization. BMPC were incubated with different doses of VOAsc. Cell proliferation was determined after 24 h of culture (a) and osteoblastic differentiation was assessed by ALP (b), type 1 collagen production (c) and matrix mineralization by calcium nodules deposition (d) after 2 weeks of culture in an osteogenic media in the presence of different concetrations of VOAsc. Data represent the mean ± SEM of three independent experiments and are expressed as % basal. $^*p < 0.05$, $^{**}p < 0.01$.

proliferation, while higher doses showed a tendency to inhibit cell growth. The effect of vanadium complex on the BMPC osteogenic potential was assessed after 2 weeks of culture in osteogenic media, by determination of ALP, type 1 collagen production, and nodules of mineral. As can be seen in Figure 4(b), VOAsc did not affect ALP in BMPC. Under similar conditions, VOAsc dose-dependently increases type 1 collagen production and calcium deposition in mineralized nodules. These results indicate that, at low doses, VOAsc is a weak mitogen and that its long term exposure to BMPC results in osteogenic effects.

3.4. Biocompatibility of VOAsc-Collagen Scaffolds on BMPC.
Biocompatibility of the scaffolds was investigated using BMPC. Cells were seeded on C0 and C1 to C3 scaffolds and allowed to adhere (2 h) or growth (24 h). Cells attached to

both C0 and C2 membranes showed a random distribution with homogeneous attachment between crests and valleys (Figures 5(a) and 5(b), resp.). At higher magnification, cells were either rounded or polyhedral with one or more cytoplasmic extensions (Figures 5(c) and 5(d)). Similar results were obtained for the other VOAsc-loaded membranes, C1 and C3 (data not shown).

Then, we investigated the effect of loading different doses of VOAsc on the collagen membranes in BMPC growth and differentiation. As it can be seen in Figures 6(a) and 6(b) there was a doses-dependent increase on both cell adhesion and proliferation of BMPC compared to the ordered collagen membrane (C0). Cell differentiation was also improved in the vanadium-loaded matrices. In those scaffolds, there was a dose-dependent increase on ALP activity (Figure 6(c)) and mineral deposits (Figure 6(d)) compared to collagen ordered

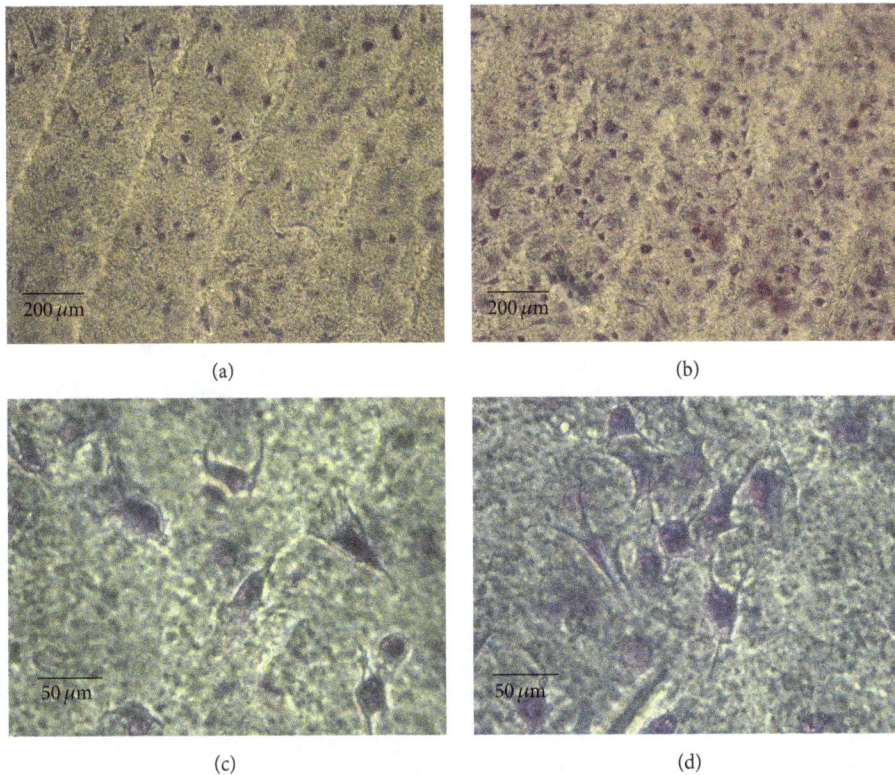

FIGURE 5: BMPC attachment (2 h at 37°C) to C0 ((a) and (c), Obj. ×10.) and C2 (20 μg/mL per cm^2 VOAsc) membranes ((b) and (d), Obj. 40x). Giemsa staining. Morphological images of BMPC are also representatives of the results for C1 and C3.

matrix (C0). In addition, cells differentiated to chondrocytes produced more GAG on C2 (20 μg/mL VOAsc per cm^2) scaffolds than on the C0 membranes (145 ± 11% of Col, Figure 6(e)).

Thus, our observations suggest that the addition of a growth factor such as the vanadium(IV) ascorbate complex can improve the ability of a collagen matrix to support adhesion, growth, and differentiation of BMPC to an osteoblastic and chondrocytic phenotype.

3.5. Cytotoxicity Studies. Although collagen has been previously used for scaffold preparations and it demonstrated being biocompatible, we wondered if our constructs might generate any cytotoxic effect on cultures of RAW 264.7 macrophages. Figure 7(a) shows these cells' growth well on both the C0 and C2 scaffolds, maintaining their round monocytic morphology without any signs of activation (i.e., absence of cytoplasm expansions such as spreading or formation of cell protrusions). MTT evaluation demonstrated the same number of surviving cells growing on C0 and C2 membrane (Figure 7(b)).

4. Discussion

Many strategies for bone tissue repair have focused on the development of biomimetic scaffolds formulating innovative tissue substitutes by the synergistic combination of matrices with cell therapy. An interesting strategy is the combination of synthetic and natural polymers in order to achieve suitable mechanical properties for bone tissue. Additionally, improved osteoconductivity can be accomplished by the incorporation into the scaffold of growth stimulating substances and/or stem cells. Collagen is the principal component of bone extracellular matrix and constitutes an interesting material for bone tissue engineering since it provides the innate biological information required for cell adhesion, proliferation, and orientation and promotes a chemostatic response [26]. On the other hand, collagen can be easily oriented to obtain scaffolds with a micro- or nanometric structure, which have been demonstrated to act as templates for osteoblastic development [15, 16]. In this study, we have extended our previous research to design a collagen-based scaffold that incorporates a vanadium complex (VOAsc) with in vitro osteogenic properties. Inclusion of VOAsc into the collagen matrix did not influence the macrostructure of the scaffold as observed by SEM, although it did increase the permeability of the biomaterial. Interestingly, there was no alteration in the permeability of vanadium-loaded scaffolds before and after 24 h of washing, even when delivery experiments demonstrated that VOAsc was positively released to the media in the same period of time. Thus, we hypothesize that the vanadium complex interacts reversibly with the collagenous matrix creating a more permeable structure but

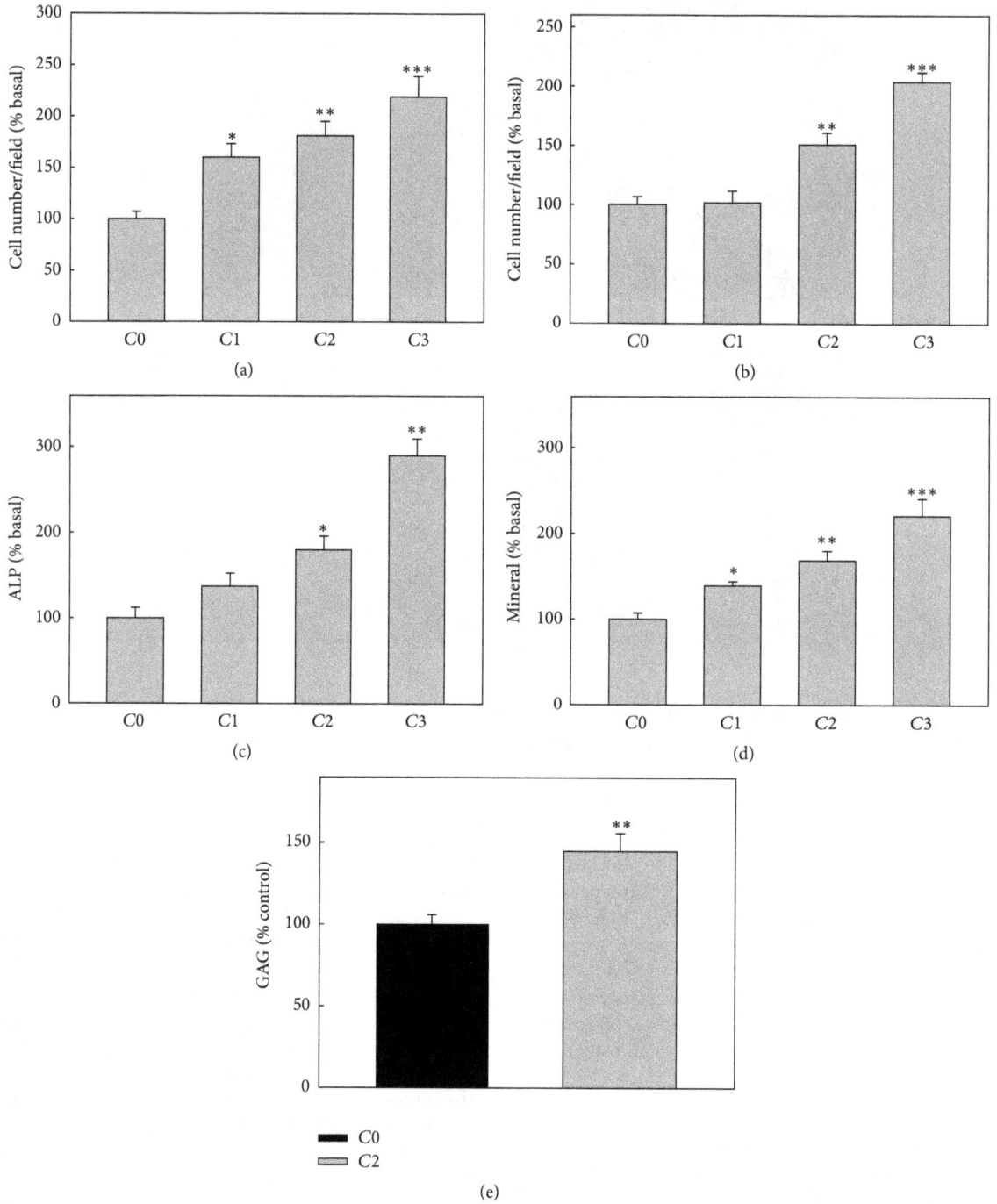

Figure 6: Biocompatibility of BMPC on C0 and VOAsc-loaded scaffolds: C1: 10 μg/mL per cm^2 VOAsc; C2: 20 μg/mL per cm^2 VOAsc; and C3: 40 μg/mL per cm^2 VOAsc. Cells were plated on different collagen membranes and adhesion (1 h (a)) and proliferation (24 h (b)) were evaluated by Giemsa staining. Cells in 10 representative fields per sample were counted and averaged. BMPC growing on different membranes were induced to be differentiated in osteogenic media: ALP (15 days (c)) or mineralization (21 days (d)) or chondrogenic media, GAG production (21 days (e)). Results are expressed as % of basal (cells differentiated in the C0 film) and they are expressed as mean ± SEM. $^*p < 0.05$; $^{**}p < 0.01$; $^{***}p < 0.001$.

does not lodge in the pores of the matrix. Permeability is a physical parameter that describes complicated properties of a material regulating various aspects of molecule transport inside and outside the scaffold. This property is also an important regulator of cell differentiation and scaffold degradation

[27–29]. Higher permeability values appear to be required for osteogenic versus chondrogenic marker expression, an effect correlating with the lower oxygen tension needed for maintenance of chondrocytic phenotype [27]. In our culture conditions, BMPC were able to grow and differentiate into

(a)

(b)

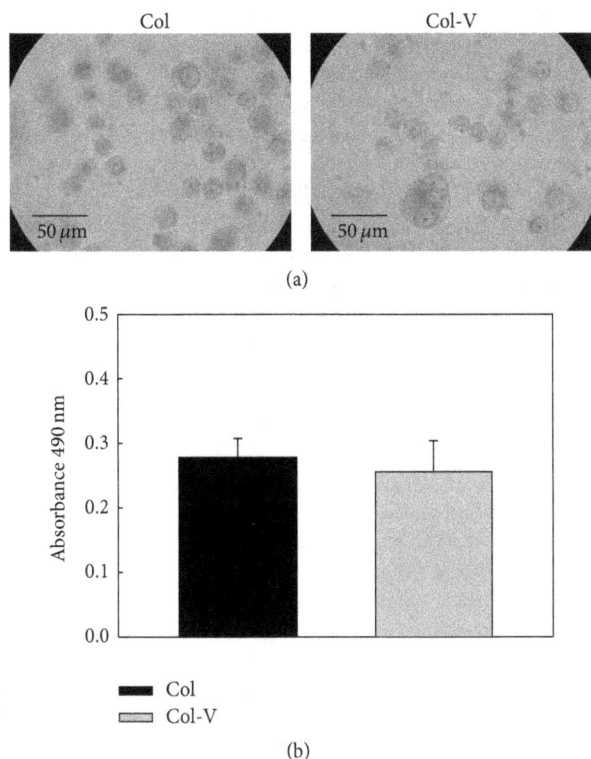

FIGURE 7: Cytotoxicity studies with RAW 264.7 macrophages. (a) Cells were cultured for 24 h on Col or Col-V (20 μg/mL per cm^2 VOAsc) scaffolds, stained with Giemsa and photographed. (b) An MTT assay was performed to evaluate cell survival of BMPC growing on Col or Col-V scaffolds. Results represent the mean $\pm$ SEM.

osteoblasts or chondrocytes on ordered collagen scaffolds, expressing adequate levels of osteogenic or chondrocytic markers when differentiated in lineage-specific conditions. However, since both chondrogenic and osteogenic induction were significantly increased in the more permeable vanadium-loaded scaffold, permeability does not seem to be a primary regulator of cell fate in our system.

To date, several studies have evaluated bone and cartilage regeneration using collagen scaffolds [5, 29, 30]. In these studies, collagen constructs have been found to effectively repair articular cartilage or bone defects. However some unsolved problems remain, such as the need for periosteal scaffold implantation to prevent detachment, articular substitution of hyaline cartilage by fibrocartilage, persisting pain, and inadequate mechanical properties [5, 29–31]. In order to improve the osteoinductivity of our ordered collagen matrix we loaded it with the osteogenic complex VOAsc. We have previously demonstrated that VOAsc in solution stimulates osteoblastic growth and development, increasing collagen secretion and extracellular mineralization in MC3T3E1 cells. These effects correlated with an activation of the extracellular regulated kinase (ERK) pathway [17]. We also shown that VOAsc induces a fast ERK phosphorylation (minutes to hours) and redistribution, suggesting that this complex could regulate osteoblastic growth by MAPK pathways. In our

present study, we confirm previous observation of VOAsc by using rat BMPC. BMPC are characterized by self-renewal capacity and multilineage differentiation under specific culture conditions. Additionally it represents a better model of bone cells, with the advantage that can be used in both in vitro and in vivo studies for bone tissue regeneration. The direct addition of VOAsc to the culture media stimulated proliferation of BMPC growing on standard tissue culture plates in a biphasic manner. Although low doses (10–50 μM) significantly enhance cells growth, higher ones could be inhibitory. In long term experiments, low doses of VOAsc (5–25 μM) significantly enhanced osteogenic potential of BMPC, as indicated by osteoblastic markers type 1 collagen production and matrix mineralization. More importantly, these results suggest nontoxic effect of the vanadium complex on these cells.

In our present system (BMPC grown on collagen scaffolds) it is to be expected that VOAsc must first be released from the scaffold to exert cellular actions. Thus the release kinetics was evaluated, showing a fast VOAsc release within the first three hours of matrix immersion into the culture media, after which the release reached a plateau. This effect was shown not to be dependent on the concentration of VOAsc loaded in the Col scaffold (Figure 3). Besides, the evaluation of diffusion coefficient for the three membranes studied was very similar suggesting that the rate of release kinetic is mainly dependent on the diffusion process. Our results could also be indicating that while an initial greater release of VOAsc might benefit BMPC attachment and growth, lower but sustained release of the vanadium complex thereafter could be acting as an osteogenic or chondrogenic agent. In this context, we found a stimulation of ALP activity on cell growing on Col-V scaffolds in a dose-dependent manner. The direct addition of VOAsc in the media did not modify this marker. Alternatively, the initial interaction of VOAsc present in the membrane with the cells could initiate a signaling to induce specific transcriptor factors associated with the fate of BMPC. Similar suggestions have been proposed by Laurencin's group by the short-term treatment of small osteogenic molecules [32]. Local delivery has emerged as an alternative to systemic delivery as it can avoid adverse drug effects. In this sense other groups have developed controlled-release systems directed to bone tissue, finding both a decrease in drug toxicity and an increase in bone and cartilage matrix deposition [33–35]. Indeed, inclusion of an osteogenic drug in the scaffold would be expected to diminish repair time [12].

Another important matter of concern is the possibility that the implantable scaffolds may cause toxicity, inflammation, or immunogenicity. For this reason we performed experiments to evaluate the response of a macrophage cell line in culture, RAW 264.7 cells, which is an accepted model to study inflammatory response and cytotoxicity of drugs or scaffolds. We have previously demonstrated that some vanadium complexes directly added to culture media can cause toxic effects such as an increase in reactive oxygen and nitrogen species [36, 37]. In addition, oral treatments with some vanadium derivatives have shown gastrointestinal toxicity [38]. In this work, our results demonstrated that there were no morphological changes in cells growing on either collagen or vanadium-loaded collagen scaffolds. Moreover, cells conserved the same proliferation rate on both kinds of membranes. Thus, toxicity was avoided when vanadium was included in the scaffold, a strategy that has been used by other groups for cytotoxic drugs [34, 39, 40]. Although in certain cases toxicity has been associated with scaffold degradation [38–40], in our present system degradation of the collagen scaffold with or without VOAsc, if it exists, does not appear to cause toxicity.

5. Conclusion

In conclusion, we have developed ordered collagen scaffolds for bone and cartilage tissue regeneration. VOAsc inclusion into the scaffolds effectively promoted differentiation of BMPC to osteoblasts and chondrocytes, without toxic effects. Our results indicate that the collagen-VOAsc construct could be of potential use in osteochondral tissue engineering.

Competing Interests

The authors declare no conflict of interests.

Acknowledgments

This work was partially supported by grants from Facultad de Ciencias Exactas, Universidad Nacional de La Plata (UNLP), Comisión de Investigaciones Científicas de la Provincia de Buenos Aires (CICPBA), and Agencia Nacional de Promoción Científica y Técnologica (Prestamo BID-1728/OC-AR, PICT 1083). Ana M. Cortizo is a Member of the Carrera del Investigador of CICPBA, Graciela Ruderman and María Silvina Molinuevo are Members of the Carrera del Investigador of CONICET, Flavia Mazzini is a Fellow of CONICET, and Ines G. Mogilner is supported by CONICET.

References

[1] W. Swieszkowski, B. H. S. Tuan, K. J. Kurzydlowski, and D. W. Hutmacher, "Repair and regeneration of osteochondral defects in the articular joints," *Biomolecular Engineering*, vol. 24, no. 5, pp. 489–495, 2007.

[2] H. Chiang and C.-C. Jiang, "Repair of articular cartilage dafects: review and perspectives," *Journal of the Formosan Medical Association*, vol. 108, no. 2, pp. 87–101, 2009.

[3] R. M. Nerem and A. Sambanis, "Tissue engineering: from biology to biological substitutes," *Tissue Engineering*, vol. 1, no. 1, pp. 3–13, 1995.

[4] A. A. M. Dhollander, V. R. Guevara Sánchez, K. F. Almqvist, R. Verdonk, G. Verbruggen, and P. C. M. Verdonk, "The use of scaffolds in the treatment of osteochondral lesions in the knee: current concepts and future trends," *The Journal of Knee Surgery*, vol. 25, no. 3, pp. 179–186, 2012.

[5] T. Efe, C. Theisen, S. Fuchs-Winkelmann et al., "Cell-free collagen type I matrix for repair of cartilage defects-clinical and magnetic resonance imaging results," *Knee Surgery, Sports Traumatology, Arthroscopy*, vol. 20, no. 10, pp. 1915–1922, 2012.

[6] S. J. Seo, C. Mahapatra, R. K. Singh, J. C. Knowles, and H. W. Kim, "Strategies for osteochondral repair: focus on scaffolds," *Journal of Tissue Engineering*, vol. 5, 2014.

[7] A. Siclari, G. Mascaro, C. Gentili, R. Cancedda, and E. Boux, "A cell-free scaffold-based cartilage repair provides improved function hyaline-like repair at one year," *Clinical Orthopaedics and Related Research*, vol. 470, no. 3, pp. 910–919, 2012.

[8] E. Kon, M. Delcogliano, G. Filardo et al., "A novel nano-composite multi-layered biomaterial for treatment of osteo-chondral lesions: technique note and an early stability pilot clinical trial," *Injury*, vol. 41, no. 7, pp. 693–701, 2010.

[9] X. Li, Y. Li, Y. Zuo et al., "Osteogenesis and chondrogenesis of biomimetic integrated porous PVA/gel/V-n-HA/pa6 scaffolds and BMSCs construct in repair of articular osteochondral defect," *Journal of Biomedical Materials Research Part A*, vol. 103, no. 10, pp. 3226–3236, 2015.

[10] Y. Jiang, B. N. Jahagirdar, R. L. Reinhardt et al., "Pluripotency of mesenchymal stem cells derived from adult marrow," *Nature*, vol. 418, no. 6893, pp. 41–49, 2002.

[11] C. T. Laurencin, K. M. Ashe, N. Henry, H. M. Kan, and K. W.-H. Lo, "Delivery of small molecules for bone regenerative engineering: preclinical studies and potential clinical applications," *Drug Discovery Today*, vol. 19, no. 6, pp. 795–800, 2014.

[12] C. Romagnoli, F. D'Asta, and M. L. Brandi, "Drug delivery using composite scaffolds in the context of bone tissue engineering," *Clinical Cases in Mineral and Bone Metabolism*, vol. 10, no. 3, pp. 155–161, 2013.

[13] J. A. Jansen, J. W. M. Vehof, P. Q. Ruhé et al., "Growth factor-loaded scaffolds for bone engineering," *Journal of Controlled Release*, vol. 101, no. 1–3, pp. 127–136, 2005.

[14] J. F. Mano, G. A. Silva, H. S. Azevedo et al., "Natural origin biodegradable systems in tissue engineering and regenerative medicine: present status and some moving trends," *Journal of the Royal Society Interface*, vol. 4, no. 17, pp. 999–1030, 2007.

[15] G. Ruderman, I. G. Mogilner, E. J. Tolosa, N. Massa, M. Garavaglia, and J. R. Grigera, "Ordered collagen membranes: production and characterization," *Journal of Biomaterials Science, Polymer Edition*, vol. 23, no. 6, pp. 823–832, 2012.

[16] A. M. Cortizo, G. Ruderman, G. Correa, I. G. Mogilner, and E. J. Tolosa, "Effect of surface topography of collagen scaffolds on cytotoxicity and osteoblast differentiation," *Journal of Biomaterials and Tissue Engineering*, vol. 2, no. 2, pp. 125–132, 2012.

[17] A. M. Cortizo, M. S. Molinuevo, D. A. Barrio, and L. Bruzzone, "Osteogenic activity of vanadyl(IV)-ascorbate complex: evaluation of its mechanism of action," *International Journal of Biochemistry and Cell Biology*, vol. 38, no. 7, pp. 1171–1180, 2006.

[18] M. S. Cortizo, J. L. Alessandrini, S. B. Etcheverry, and A. M. Cortizo, "A vanadium/aspirin complex controlled release using a poly(β-propiolactone) film. Effects on osteosarcoma cells," *Journal of Biomaterials Science, Polymer Edition*, vol. 12, no. 9, pp. 945–959, 2001.

[19] E. G. Ferrer, P. A. M. Williams, and E. J. Baran, "Interaction of the vanadyl(IV) cation with l-ascorbic acid and related systems," *Zeitschrift fur Naturforsch*, vol. 53, pp. 256–262, 1998.

[20] H. T. Liao and C. T. Chen, "Osteogenic potential: comparison between bone marrow and adipose-derived mesenchymal stem cells," *World Journal of Stem Cells*, vol. 6, no. 3, pp. 288–295, 2014.

[21] M. S. Molinuevo, L. Schurman, A. D. McCarthy et al., "Effect of metformin on bone marrow progenitor cell differentiation: in

vivo and in vitro studies," *Journal of Bone and Mineral Research*, vol. 25, no. 2, pp. 211–221, 2010.

[22] M. M. Bradford, "A rapid and sensitive method for the quantitation of microgram quantities of protein utilizing the principle of protein-dye binding," *Analytical Biochemistry*, vol. 72, no. 1-2, pp. 248–254, 1976.

[23] S. Bahmanpour and D. F. Paulsen, "Inhibition of chondrogenic differentiation in chick limb-bud mesenchyme microcultures treated with cyclosporine," *Indian Journal of Pharmacology*, vol. 38, no. 1, pp. 43–48, 2006.

[24] J. M. Fernández, M. S. Cortizo, and A. M. Cortizo, "Fumarate/ceramic composite based scaffolds for tissue engineering: evaluation of hydrophylicity, degradability, toxicity and biocompatibility," *Journal of Biomaterials and Tissue Engineering*, vol. 4, no. 3, pp. 227–234, 2014.

[25] P. L. Ritger and N. A. Peppas, "A simple equation for description of solute release I. Fickian and non-fickian release from non-swellable devices in the form of slabs, spheres, cylinders or discs," *Journal of Controlled Release*, vol. 5, no. 1, pp. 23–36, 1987.

[26] A. M. Ferreira, P. Gentile, V. Chiono, and G. Ciardelli, "Collagen for bone tissue regeneration," *Acta Biomaterialia*, vol. 8, no. 9, pp. 3191–3200, 2012.

[27] J. M. Kemppainen and S. J. Hollister, "Differential effects of designed scaffold permeability on chondrogenesis by chondrocytes and bone marrow stromal cells," *Biomaterials*, vol. 31, no. 2, pp. 279–287, 2010.

[28] A. G. Mitsak, J. M. Kemppainen, M. T. Harris, and S. J. Hollister, "Effect of polycaprolactone scaffold permeability on bone regeneration in vivo," *Tissue Engineering Part A*, vol. 17, no. 13-14, pp. 1831–1839, 2011.

[29] D. Enea, S. Cecconi, S. Calcagno, A. Busilacchi, S. Manzotti, and A. Gigante, "One-step cartilage repair in the knee: collagen-covered microfracture and autologous bone marrow concentrate. A pilot study," *Knee*, vol. 22, no. 1, pp. 30–35, 2015.

[30] J. Iwasa, L. Engebretsen, Y. Shima, and M. Ochi, "Clinical application of scaffolds for cartilage tissue engineering," *Knee Surgery, Sports Traumatology, Arthroscopy*, vol. 17, no. 6, pp. 561–577, 2009.

[31] J. Gille, E. Schuseil, J. Wimmer, J. Gellissen, A. P. Schulz, and P. Behrens, "Mid-term results of Autologous Matrix-Induced Chondrogenesis for treatment of focal cartilage defects in the knee," *Knee Surgery, Sports Traumatology, Arthroscopy*, vol. 18, no. 11, pp. 1456–1464, 2010.

[32] K. W.-H. Lo, H. M. Kan, and C. T. Laurencin, "Short-term administration of small molecule phenamil induced a protracted osteogenic effect on osteoblast-like MC3T3-E1 cells," *Journal of Tissue Engineering and Regenerative Medicine*, 2013.

[33] K. W.-H. Lo, K. M. Ashe, H. M. Kan, and C. T. Laurencin, "The role of small molecules in musculoskeletal regeneration," *Regenerative Medicine*, vol. 7, no. 4, pp. 535–549, 2012.

[34] C. M. Murphy, A. Schindeler, J. P. Gleeson et al., "A collagen-hydroxyapatite scaffold allows for binding and co-delivery of recombinant bone morphogenetic proteins and bisphosphonates," *Acta Biomaterialia*, vol. 10, no. 5, pp. 2250–2258, 2014.

[35] L. Nie, G. Zhang, R. Hou, H. Xu, Y. Li, and J. Fu, "Controllable promotion of chondrocyte adhesion and growth on PVA hydrogels by controlled release of TGF-β1 from porous PLGA microspheres," *Colloids and Surfaces B: Biointerfaces*, vol. 125, pp. 51–57, 2015.

[36] A. M. Cortizo, L. Bruzzone, S. Molinuevo, and S. B. Etcheverry, "A possible role of oxidative stress in the vanadium-induced

cytotoxicity in the MC3T3E1 osteoblast and UMR106 osteosarcoma cell lines," *Toxicology*, vol. 147, no. 2, pp. 89–99, 2000.

[37] M. S. Molinuevo, S. B. Etcheverry, and A. M. Cortizo, "Macrophage activation by a vanadyl-aspirin complex is dependent on L-type calcium channel and the generation of nitric oxide," *Toxicology*, vol. 210, no. 2-3, pp. 205–212, 2005.

[38] A. K. Srivastava, "Anti-diabetic and toxic effects of vanadium compounds," *Molecular and Cellular Biochemistry*, vol. 206, no. 1-2, pp. 177–182, 2000.

[39] O. Belfrage, G. Flivik, M. Sundberg, U. Kesteris, and M. Tägil, "Local treatment of cancellous bone grafts with BMP-7 and zoledronate increases both the bone formation rate and bone density: a bone chamber study in rats," *Acta Orthopaedica*, vol. 82, no. 2, pp. 228–233, 2011.

[40] C. Jeppsson, J. Åstrand, M. Tägil, and P. Aspenberg, "A combination of bisphosphonate and BMP additives in impacted bone allografts," *Acta Orthopaedica Scandinavica*, vol. 74, no. 4, pp. 483–489, 2003.

Preparation Methods for Improving PEEK's Bioactivity for Orthopedic and Dental Application

Davood Almasi,[1] **Nida Iqbal,**[2] **Maliheh Sadeghi,**[3] **Izman Sudin,**[1] **Mohammed Rafiq Abdul Kadir,**[2] **and Tunku Kamarul**[4]

[1]*Department of Materials, Manufacturing and Industrial Engineering, Faculty of Mechanical Engineering, Universiti Teknologi Malaysia, 81310 Skudai, Johor, Malaysia*
[2]*Medical Implant Technology Group (MEDITEG), Faculty of Bioscience and Medical Engineering, Universiti Teknologi Malaysia, 81310 Skudai, Johor, Malaysia*
[3]*Faculty of Chemical and Energy Engineering, Universiti Teknologi Malaysia, 81310 Skudai, Johor, Malaysia*
[4]*Department of Orthopaedic Surgery, NOCERAL, Faculty of Medicine, University of Malaya, 50603 Kuala Lumpur, Malaysia*

Correspondence should be addressed to Izman Sudin; izman@utm.my

Academic Editor: Rosalind Labow

There is an increased interest in the use of polyether ether ketone (PEEK) for orthopedic and dental implant applications due to its elastic modulus close to that of bone, biocompatibility, and its radiolucent properties. However, PEEK is still categorized as bioinert due to its low integration with surrounding tissues. Many studies have reported on methods to increase the bioactivity of PEEK, but there is still one-preparation method for preparing bioactive PEEK implant where the produced implant with desirable mechanical and bioactivity properties is required. The aim of this review is to present the progress of the preparation methods for improvement of the bioactivity of PEEK and to discuss the strengths and weaknesses of the existing methods.

1. Introduction

PEEK with high chemical resistance, radiolucency, mechanical characteristics compared to those of human bones [1–6], and local inflammation and stress shielding problem of the metallic implant [7, 8] has become a very interesting biomaterial for scientists and a promising good alternative for metallic implants. The radiolucency of PEEK is vital especially for the postoperative radiotherapy follows the surgical removal of the tumor. The presence of metallic implants can change the local dose distribution [9, 10]. In addition, it can be repeatedly sterilized and shaped by machining and heat contouring to fit the contour of bones [11]. PEEK has been used for load bearing orthopedic applications such as spinal cage, dental implant, and screws [12, 13]. Despite these excellent properties, PEEK is still categorized as bioinert due to its very low reaction with the surrounding tissue, which limits its potential applications [1]. For overcoming this problem, several methods have been proposed which can broadly be divided into two main categories: incorporation of bioactive materials such as hydroxyapatite (HA) and titanium dioxide (TiO_2) into PEEK composite and surface treatment techniques such as laser surface modification, coating with the bioactive material, and wet chemical treatment [14–22].

A review of presently available methods to improve the bioactivity of PEEK was conducted with the aim of providing sufficient information regarding known preparation techniques and to compare the pros and cons of each of these methods. It is hoped that this will lead to a better understanding of the methods available and a clear reason as to why a method should be ultimately chosen by a researcher or an implant manufacturer.

2. PEEK's Bioactivity

One of the important factors that lead to successful implantation is the biological response to the implant, which very

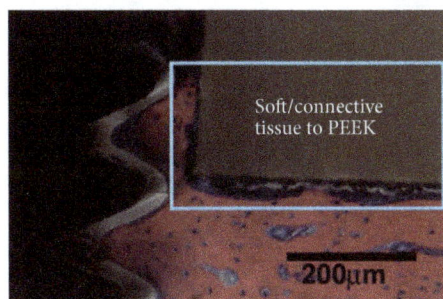

FIGURE 1: PEEK bioinert properties and growth of soft tissue around it [30].

much depends on the bioactivity of the implant. A material is considered bioactive if it obtains a particular biological answer to the interface of the element, which ends in the formation of a bond between the tissue and the substance [23].

When an implant is placed in the body, molecules of water are one of the first molecules to reach the implant surface. The absorption of proteins on the surface is influenced by the initially adsorbed water molecules and is affected by surface structure, chemistry, charge, and wettability [24]. Subsequently, these adsorbed proteins influence cellular interactions and eventually tissue growth [25, 26]. Surfaces with moderate hydrophilicity properties showed the best interactions with cells and surrounding tissues [27].

An essential problem with most polymers, including PEEK, is their low-surface energy. This hydrophobic property of the surface can reduce cellular adhesion. The lack of response from the biological environment caused PEEK to be categorized as bioinert [28, 29]. As explained above one of the most important applications of PEEK is for orthopedic area. The bioinert properties of PEEK in the orthopedic area mean the growth of soft tissues around the PEEK implant instead of bone growth (Figure 1) [30].

By changing the surface energy of the polymer, the reactions of the surrounded tissue to the polymeric implant can be improved, which could broaden its applications in the medical field, where direct bone interaction is important. Many methods have been used to alter the surface energy, and these methods can be broadly divided into two groups: compounding PEEK with a bioactive material and producing a composite and through surface modification. Figure 2 shows the general categorization of the existing methods for improving the bioactivity of PEEK.

3. Surface Modification of PEEK's Implant

Surface modification is a series of approaches which alter the properties of the surface of the material but do not affect the bulk properties of the material. Surface modification methods can be broadly divided into two broad categories: direct surface modification and deposition methods.

3.1. Direct Surface Modification. Direct surface modification methods are techniques that changed the surface properties of the material without depositing any layer of new material on the surface. These techniques consist of the following.

3.1.1. Wet Chemical Treatments. This is a method which is based on changing the surface chemistry of the implant and affects the bioactivity of the surface. Several studies reported that the bioactivity of PEEK could be increased by wet chemical treatment. Various chemical treatments modifying PEEK surface chemistry to PEEK-ONa, PEEK-OH, PEEK-F, and PEEK-OH (CFCl$_3$) showed a decrease in water contact angle of the implant and, therefore, increase the bioactivity of PEEK [31]. Another study showed that the amine and carboxyl functional group on the surface of PEEK could improve cellular adhesion and growth [28].

In vitro study on Fibronectin (FN) adsorption for probing the bioactivity of PEEK-OH, PEEK-NH$_2$, and PEEK-NCO produced by wet chemical treatment showed protein can merely be adsorbed onto PEEK-NCO that Fibronectin covalently grafted to PEEK-NCO [32]. The performances of the FN-grafted substrate improved adhesion and spreading of Caco-2 cells in the absence of serum in comparison with PEEK substrates, which were simply coated with FN [33]. In another study, wet chemical treatment was used as a pretreatment for enhancement of apatite formation via immersion in SBF. The effect of NaOH pretreatments on apatite formation of PEEK in SBF showed the growth of apatite coating layer was enhanced with NaOH pretreatment [34].

A recent study probed the effect of sulphonation and the production of 3D porous and nanostructured network on *in vitro* cellular behavior and *in vivo* osseointegration and apatite formation. Two types of sulphonated PEEK (SPEEK) samples, SPEEK-W (sulphonated PEEK with just subsequent water immersion) and SPEEK-WA (SPEEK-W with additional acetone immersion) were probed. They showed new bone can grow and penetrate the porous sulphonated layer. The SPEEK-WA samples showed better cytocompatibility, bioactivity, osseointegration, and bone-implant bonding strength [22]. Table 1 presents the summary of the existing functional groups which have been deposited on PEEK via wet chemical deposition to enhance the bioactivity of it.

3.1.2. Plasma Surface Treatment. Plasma is often known as the fourth state of matter in which the gases are ionized and electrons are separated from their atoms. There are two types of plasma, hot plasma and cold plasma. In hot plasma using very high temperature, the gas is ionized. In cold plasma, the gas is ionized using low pressure in ambient temperature. The plasma can be used for altering the surface chemistry of the material. Plasma treatment of PEEK in oxygen, air, nitrogen, ammonia, and argon showed increasing of the wettability [35, 36].

In vitro study with osteoblast cells and wettability study carried out on plasma treated PEEK in N$_2$/O$_2$ showed the plasma treatment of PEEK reduced the water contact angle. *In vitro* study with osteoblast cells showed the plasma treatment does not have disadvantages on cell viability [37]. Plasma treated PEEK in NH$_3$ showed lower water contact angle and

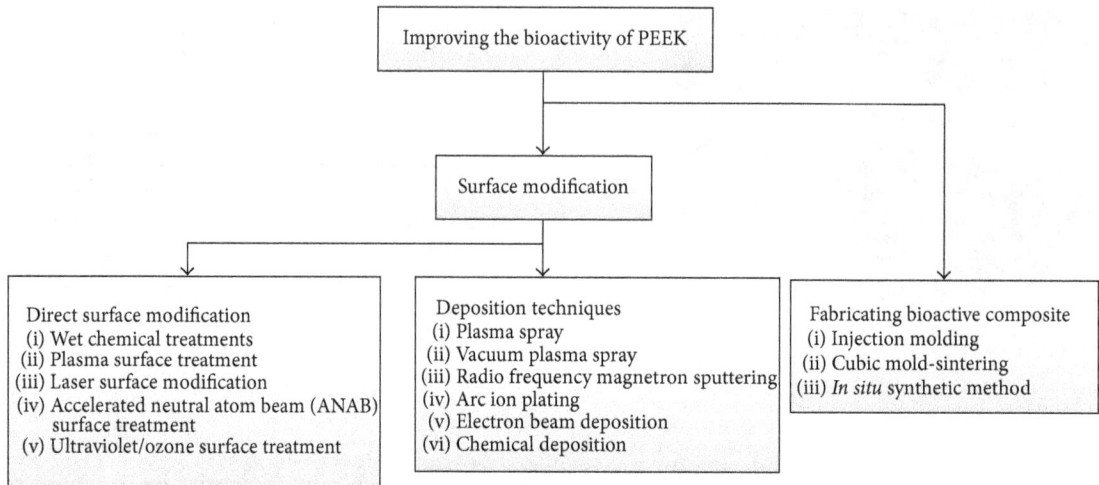

FIGURE 2: General categorization of the techniques for improving PEEK's bioactivity.

TABLE 1: Deposited functional groups on PEEK via wet chemical deposition.

Functional group	Results
–ONa	Enhancement of wettability [31]. Enhancement of apatite formation [34].
–OH	Enhancement of wettability [31]. Disable to graft to Fibronectin [32].
–F	Enhancement of wettability [31].
–OH(CFCl$_3$)	Enhancement of wettability [31].
Amine	Improvement of cellular adhesion and growth [28].
Carboxyl	Improvement of cellular adhesion and growth [28].
–NH$_2$	Disable to graft to Fibronectin [32].
–NCO	Fibronectin covalently grafted to PEEK-NCO [32].
Fibronectin grafting	Enhancement of adhesion and spreading of Caco-2 cells in the absence of serum in comparison with PEEK substrates, which were simply coated with FN [33].
–SO$_3$H	Producing 3D nanostructured treated layer. *In vitro* (cell culture and apatite formation) and *in vivo* study showed enhancement of bioactivity [22].

increased cell growth [38]. Osteoblast biocompatibility test showed required biocompatibility for plasma treated PEEK in ammonia/argon and hydrogen/argon. Higher rate of cell proliferation and lower contact angle were demonstrated for plasma treated PEEK in comparison with untreated PEEK [39]. Plasma treatment of PEEK in chamber of CH$_4$/O$_2$ gas mixture showed better cell adhesion and lower water contact angle [40]. *In vivo* study of oxygen plasma, modified PEEK in cortical and cancellous bone of the sheep showed an increase in push-out force test and the percentage of the bone-implant contact area in comparison of untreated PEEK [41]. *In vitro* study via osteoblast precursor cells MC3T3-E1 and rat bone mesenchymal stem cells on plasma immersion ion implantation treatment with a gas mixture of water vapor as a plasma resource and argon as an ionization assistant of PEEK showed improvement of osteoblast adhesion, spreading, proliferation, and early osteogenic differentiation [42]. Also tuned PEEK by argon plasma treatment showed increasing of the surface roughness in comparison with pristine PEEK. As a consequence due to higher surface roughness and changing the surface chemistry of the treated PEEK, significant enhancements in terms of cell adhesion, proliferation, and metabolic activity were observed when compared to pristine PEEK [43]. Probing the effect of plasma treatment of PEEK by O$_2$/Ar or NH$_4$ on adhesion, proliferation, and osteogenic differentiation of adipose tissue-derived mesenchymal stem cells (adMSC) showed an improvement of bioactivity of plasma treated samples in comparison with nontreated samples [44]. Table 2 summarizes the ionization assistants which have been used for enhancement of the bioactivity of PEEK via plasma surface treatment method.

3.1.3. Laser Surface Modification. Laser is a high energy photon source which can alter the surface roughness and wettability of the polymers. Laser treatments are used due to their low cost, high resolution, high-operating speed, and the fact that lasers do not change the bulk properties of implant. For these reasons, lasers become very interesting for scientists in order to improve the surface energy of the implants [45, 46]. This surface treatment technique can modify the surface chemistry of PEEK [47, 48]. Investigation into the effect of laser wavelengths on the wettability of PEEK showed the capability of this method in increasing the wettability of the PEEK for biomedical applications [49].

3.1.4. Accelerated Neutral Atom Beam (ANAB) Surface Treatment. This technique is a method which is used to enhance the bioactivity of PEEK and improve the bone-implant

TABLE 2: Different ionization assistants which have been used for improving the bioactivity of PEEK via plasma treatment.

Ionization assistant	Results
Oxygen	Enhancement of wettability [35, 36]. Increase of push-out force and bone-implant contact area [41].
Air	Enhancement of wettability [35, 36].
Nitrogen	Enhancement of wettability [35, 36].
Ammonia	Enhancement of wettability [35, 36].
Argon	Enhancement of wettability [35, 36]. Using vapor as a plasma resource showed improvement of osteoblast adhesion, spreading, proliferation, and early osteogenic differentiation [42]. Increasing surface roughness, enhancement of cell adhesion, proliferation, and metabolic activity [43].
N_2/O_2	In vitro study via osteoblast cells showed no disadvantages on cell viability [37].
NH_3	Enhancement of wettability and increasing cell growth [38].
Ammonia/argon	Enhancement of cell proliferation rate and enhancement of wettability [39].
Hydrogen/argon	Enhancement of cell proliferation rate and enhancement of wettability [39].
CH_4/O_2	Enhancement of cell adhesion and enhancement of wettability [40].
O_2/Ar	Enhancement of cell adhesion, proliferation, and osteogenic differentiation of adMSC [40].
NH_4	Enhancement of cell adhesion, proliferation, and osteogenic differentiation of adMSC [40].

integrity. In this technique a powerful beam of cluster-like packets of accelerated unbonded neutral argon (Ar) gas atoms is used to modify the surface of PEEK. The results showed that ANAB treatment of PEEK modified the surface in the nanometer scale, increased surface wettability, and improved human osteoblast cell proliferation to a level comparable with titanium. The in vivo study shows the bone tissue formation on the ANAB treated PEEK while no growth of bone tissue on the untreated PEEK was observed [50]. The atomic force microscope examination showed the effect of ANAB technique in producing nanoscale texturing on the surface. In vitro study of ANAB treated PEEK showed better osteoblast cell adhesion in comparison with untreated PEEK [51].

3.1.5. Ultraviolet/Ozone Surface Treatment.
Polymers can be degraded by exposure to sunlight because of the chemical reaction activation due to short wavelengths of ultraviolet (UV) of sunlight and photon-activation cross-linking or fragmentation of the polymer. UV/ozone treatment method for PEEK was used to change the surface energy of PEEK. The results showed increasing of the surface wettability of treated PEEK by UV/ozone [52].

3.2. Deposition Techniques.
Several methods exist for depositing bioactive material on PEEK such as plasma spraying, vacuum deposition, sol gel, dip coating, and immersion in SBF method [53]. In this section, the trend of progress of PEEK's coating is described based on the coated materials.

Hydroxyapatite is one of the most important materials which have been used widely for coating of biomaterials. HA coating on carbon fiber reinforced PEEK (CF/PEEK) via plasma spray method showed low adhesion of the coating layer to the substrate [54]. The authors explained that the high temperature used in plasma spray method caused the evaporation of the PEEK substrate preventing close contact between coating layer and substrate. In the next study, they coated titanium intermediate layer via vacuum-plasma-sprayed and after that coated hydroxyapatite layer on CF/PEEK for increasing the adhesion between the coating layer and the substrate. The cross section study showed very good interlocking between the PEEK substrate and the intermediate Ti layer [55]. To prevent damage to the PEEK substrate due to the high temperature during the coating process and damage to the PEEK substrate during the sintering, intermediate coating layer of yttria-stabilized zirconia (YSZ) was first deposited onto PEEK and after that the HA coating layer was deposited via radio frequency magnetron sputtering method. For increasing the adhesion between the substrate and coating layer, preplasma treatment was used for substrate. Microwave was used for sintering and forming crystalline HA coating layer. The authors showed the crystalline YSZ layer encouraged the HA layer during the sintering procedure by providing nucleation site for HA grain formation [56]. Hydroxyapatite coating via plasma spraying method on different PEEK (unfilled and carbon fiber reinforced composite) specimens was studied and chemical, crystallographic compositions, adhesions, and microstructures of HA coating via plasma spraying method on different PEEK (unfilled and CF/PEEK) specimens and comparison with HA coating on Ti-6Al-4V showed almost the same structure of HA coatings for PEEK and Ti-6Al-4V substrate. Mechanical tests showed the plasma spraying method does not have a negative effect on mechanical properties of PEEK implant [57]. In vitro study with human bone marrow mesenchymal stem cells of HA coated PEEK via cold spray method showed early cell adhesion, viability improvement, and increased cell differentiation and proliferation. In vivo study on rabbits showed promotion of bone growth and integrity with the implant after coating [58]. HA coating on medical-grade PEEK via aerosol deposition showed dense microstructure with no pores and cracks with high-adhesion strength of HA coating layer without damaging the PEEK substrate. In vitro and in vivo study in terms of cell proliferation, differentiation, adhesion morphology, and bone-implant contact ratio showed enhancement for HA coated sample in comparison to uncoated PEEK [59]. In vivo osseointegration (histomorphometry) study of surface modified PEEK implants showed the nano-HA coated implants have more bone area and more bone-to-implant contact in comparison to uncoated PEEK [60]. In our recent study the HA crystalline particles were chemically deposited on the PEEK's surface whereby crystallization process and high temperature for deposition of the HA were eliminated. For depositing the HA particles, the surface of the PEEK was sulphonated first to establish the –SO$_3$H functional group, and then the polarity property of the HA particles was used to attach the particles to the functional group. The surface treatment was able to decrease the water contact angle from 72 to 36.4 degrees [61]. *In vitro* study comprising apatite formation via SBF immersion and mesenchymal stem cell proliferation confirmed enhancement of bioactivity of treating PEEK via this method [62].

In vitro osteoblast study of PEEK substrate coated with TiO$_2$ via arc ion plating method showed a significant improvement in cell adhesion, proliferation, and differentiation compared with an uncoated PEEK substrate [21]. The anatase-rich titanium dioxide (A-TiO$_2$) and especially rutile-rich titanium dioxide (R-TiO$_2$) intermediate layer onto the PEEK substrate showed enhancement of produced HA layer after immersion in SBF in comparison with pure PEEK. The authors explained that the intermediate layer, by providing nucleation site for growing HA, improves the produced HA layer. Osteocompatibility evaluation showed the produced HA layer improves osteocompatibility, in which R-TiO$_2$ achieves the best result [63]. In another study the bone morphogenetic protein-2 (BMP-2) was immobilized on porous TiO$_2$ coating layer on PEEK. The bone-to-implant contact ratio study showed better interaction of TiO$_2$/BMP-2 coating layer in comparison with TiO$_2$, and BMP-2 coating layer and pure PEEK [64].

In vivo study on sheep was performed on titanium plasma spray coating on the PEEK screw. Histological investigation showed higher bone-to-implant contact and lower soft tissue around coated samples in comparison with pure PEEK [65]. Electron beam deposition of Ti on PEEK produced a dense coating layer at low temperature. *In vitro* study in terms of proliferation and differentiation of MC3T3-E1 cells showed more than double improvement after Ti coating in comparison with pure PEEK. *In vivo* study showed that the bone-to-implant contact ratio increased with coating Ti on the PEEK substrate [6]. In another study the vacuum-plasma-sprayed Ti coating layer on CF/PEEK substrate was treated by sodium hydroxide (NaOH) solution for improving its bioactivity. *In vitro* study via SBF showed apatite formation on the coated samples while no apatite was formed on the untreated PEEK samples [66]. *In vivo* comparative study for probing the effect of two different methods of PVD and VPS for deposition of the Ti on CF/PEEK screws showed no significant difference between these two methods in terms of bioactivity. The coated screws by these two methods showed better bone deposition and higher removal torque in comparison with uncoated screws [67].

An *in vivo* study of Ti-coated CF/PEEK for dental implant application via plasma vapor deposition was carried out to evaluate the bioactivity of Ti-coated CF/PEEK. The results showed direct growth of new bone for both coated and uncoated PEEK samples, but the coated samples showed better bone growth around the coated implant. However, the push-out test revealed almost the same interface strength between the coated and uncoated samples by new bone growth [68]. In another study, electron beam deposition method was used to deposit pure titanium on PEEK. The

Ti coating layer showed superb adhesion properties to the PEEK substrate. Contact angle analysis showed the Ti coating enhances the wettability of PEEK. *In vitro* study by MC3T3-E1 cells for methoxyphenyl tetrazolium salt (MTS) assay to measure the proliferation of the cells shows enhancement of more than double for coated samples. Alkaline phosphatase (ALP) assay showed double differentiation level of cells for Ti-coated samples. Furthermore, an *in vivo* animal study showed much higher bone-in-contact (BIC) ratio for Ti-coated PEEK samples in comparison with the pure PEEK samples [6].

Zirconium and titanium tetra(tert-butoxides) are another bioactive material which was deposited on the surface PEEK at room temperature via vapor deposition to enhance the bioactivity of PEEK. The deposited metal layer reacted with the phosphonic acid for attachment of monolayer phosphonates. *In vitro* study showed significant enhancement of osteoblast cell growth as compared to the untreated surface [69]. Diamond-like carbon (DLC) is another material which was used to coat PEEK implant for increasing bioactivity. *In vitro* study via osteoblast showed better attachment, proliferation, and differentiation on DLC-coated PEEK compared to uncoated PEEK [70]. Table 3 presents the summary of the existing deposition methods/materials which have been used for enhancement of PEEK bioactivity.

4. Bioactive PEEK Composites

As explained before compounding with bioactive material is one strategy to increase the bioactivity of the PEEK implants. Different bioactive material such as HA, strontium-containing hydroxyapatite, TiO$_2$, βTCP, and bioactive glass was compounded with PEEK for increasing the bioactivity of PEEK's implant. PEEK composites were produced for different applications. The most important application is load bearing implant application [71], but several other studies were carried out to show the feasibility of producing three-dimensional porous scaffold PEEK/HA for tissue engineering application [72–74] and cervical spinal fusion cages [75]. One of the most significant disadvantages of the PEEK composites is the low mechanical properties in comparison with PEEK [76–78]. Thus previous studies focused on probing the effect of different parameters on two important aspects of mechanical properties and bioactivity. In this part, previous studies of the PEEK composites were first broadly categorized as bioactivity and mechanical properties study, and in each category the trend of progress of PEEK's composites is described based on the compound material.

4.1. In Vitro and In Vivo Bioactivity Study of PEEK Composite. Several studies have been conducted to probe the effect of compounding PEEK with bioactive materials on *in vitro* and *in vivo* bioactivity of the produced composite. PEEK/HA composites with different volume fraction of HA up to 40 vol% via injection molding method were evaluated *in vivo*. Preliminary histological *in vivo* study of composite with 20 vol% of HA showed the enhancement of the presence of fibroblast cells which stimulate vascularization. Osteoblastic

TABLE 3: Summary of the existing deposition methods/materials for improving PEEK bioactivity.

Deposited material	Deposition method	Area of studies	Findings
HA	Plasma spray	—	Low adhesion of the coating layer to the substrate [54].
	Vacuum-plasma-sprayed	Using titanium intermediate coating layer	Good interlocking between PEEK substrate and intermediate Ti layer and preventing damage of the substrate [55].
	Radio frequency magnetron sputtering	Crystalline YSZ layer was deposited as an intermediate layer	Enhancement crystallinity of HA deposited layer during sintering [56].
	Plasma spraying	Crystallographic compositions, adhesions, and microstructures of HA coating via plasma spraying method on different PEEK (unfilled and CF/PEEK) specimens were studied and compared with HA coating on Ti-6Al-4V	Almost the same structure of HA coatings for PEEK and Ti-6Al-4V substrate. Plasma spraying method does not have a negative effect on mechanical properties of PEEK [57].
	Vacuum-plasma-sprayed	*In vitro* study with human bone marrow mesenchymal stem cells and *in vivo* study	Viability improvement and enhancement of cell differentiation and proliferation. Promoting of bone growth [58].
	Aerosol deposition	Microstructure, *in vivo*, *in vitro* study	Dense microstructure with no pores and cracks. Enhancement of bioactivity in terms of cell proliferation, differentiation, adhesion morphology, and bone-implant contact ratio [59].
	Spin coating	*In vivo* osseointegration (histomorphometry) study	Improvement of bone-to-implant contact area [60].
	Chemical deposition	$-SO_3H$ functional group was created via sulphonation and HA crystalline particles were chemically deposited	The proposed method did not use high temperature and improved the wettability [61].
A-TiO_2 and R-TiO_2	Arc ion plating	*In vitro* SBF immersion and osteocompatibility study	Enhancement of apatite formation and improvement of osteocompatibility, in which R-TiO_2 achieves the best result [63].
TiO_2	Arc ion plating	*In vitro* osteoblast study	Improvement in cell adhesion, proliferation, and differentiation [21].
TiO_2/BMP-2	Immobilization	*In vivo* study	Enhancement of bone-to-implant contact ratio in comparison with TiO_2 and BMP-2 coating layer and bare PEEK [64].
Ti	Plasma spray	*In vivo* study	Enhancement bone-to-implant contact ratio [65].
	Electron beam deposition	*In vitro* study in terms of proliferation and differentiation of MC3T3-E1 cells and *in vivo* study	Enhancement of *in vitro* bioactivity and bone-to-implant contact ratio [6].
	VPS	Probing the effect of pretreatment of the substrate with NaOH solution on bioactivity via *in vitro* SBF immersion study	Improvement bioactivity in terms of apatite formation [66].
	PVD and VPS	*In vivo* comparative study for probing the effect of PVD and VPS methods on the Ti deposited on CF/PEEK substrate	No significant difference between these two methods in terms of bioactivity [67].
	PVD	*In vivo* study of Ti-coated CF/PEEK for dental implant application	Coated samples showed better bone growth around the coated implant but the same push-out force for coated and uncoated samples by new bone growth [68].
	Electron beam deposition	Wettability, *in vitro* study via MC3T3-E1 cell and *in vivo* study	Enhancement of *in vitro* bioactivity and bone-in-contact ratio [6].
Zirconium and titanium tetra	PVD	*In vitro* study via osteoblast	Enhancement of osteoblast cell growth [69].
DLC	Plasma immersion ion implantation and deposition	*In vitro* study via osteoblast	Enhancement of attachment, proliferation, and differentiation of osteoblast [70].

activities study showed the formation of osteoid and osteocytes within lamellar bone in developing mature bone at longer implantation periods [15]. The SBF bioactivity test on HA/PEEK composites with different volume fraction up to 40% which were prepared by mixing of HA and PEEK powders, compaction, and sintering showed the higher rate of HA growth for the composite with higher volume fraction percentage of HA [14]. Biological study of HA/PEEK composites which were prepared by mixing and sintering the material powders using simple cubic mold shows the capability of this technique to replace the injection molding which is a high-cost method. *In vitro* study via SBF and cell seeding tests confirmed the bioactivity of the composite [79]. For better dispersion of HA particles in HA/PEEK composite nanosized HA (nHA)/PEEK with different nHA contents (15.1, 21.6, 29.2, and 38.2 vol%) was fabricated by Li et al. [80]. *In vitro* study via SBF immersion, cell adhesion, and proliferation showed nanocomposite with 29.2 vol% of nHA content has the best bioactivity in comparison with other samples. For the improvement, the bonding between HA and PEEK of the HA/PEEK composite was fabricated via *in situ* synthetic method [81–83]. The biocompatibility study of *in situ* synthetic method for fabrication of composite showed the fabricated composites are nontoxic, and the bioactivity study showed the produced composites are bioactive.

Study of the bioactivity of βTCP-PEEK composite via injection molding method showed lower rates of osteoblast growth on the βTCP-PEEK compared to pure PEEK [84]. *In vitro* study with osteoblast cells confirmed the nontoxicity of laser sintering method for producing βTCP/PEEK composite but showed no advantage of adding βTCP as fillers on cell growth [85, 86]. However, *in vivo* study of the laser sintered PEEK/βTCP implant revealed the PEEK/βTCP implants showed better interaction with surrounding bone and direct connection to the surrounding bone in comparison with pure PEEK [87].

In vitro study with osteoblast cells confirms the nontoxicity of laser sintering method for producing carbon black/PEEK composite but showed no advantage of adding carbon black as fillers on cell growth [85]. *In vitro* study of HA/PEEK composite via selective laser sintering method showed improvement in bioactivity of the composite in comparison with pure PEEK. The results showed higher content of HA exhibited enhancement in cell proliferation and osteogenic differentiation [88].

In vitro osteoblast cell proliferation and viability study from PEEK, PEEK/carbon, PEEK/carbon/βTCP, and PEEK/carbon/bioglass 4s5S5 composites via laser sintering method revealed that all samples were nontoxic. However, the cell culture test did not show any advantageous effect of βTCP in the PEEK composite on the bioactivity properties of the samples. High-proliferation rates of osteoblasts on PEEK/carbon/bioglass composite showed the significant effect of bioglass on improving the bioactivity of the composite [86]. *In vitro* study via MG-63 cells on glass fiber/PEEK composite showed a higher rate of cell proliferation on the surface of the composite compared to pure PEEK [89, 90].

Nano-TiO$_2$ is another additive which is used for improvement in the bioactivity of PEEK composite. *In vitro* and *vivo* studies confirmed the positive effect of nano-TiO$_2$ on improvement of bioactivity of PEEK. *In vitro* study demonstrated that compounding PEEK with nano-TiO$_2$ was able to increase cell attachment and enhanced osteoblast cell spreading. In *in vivo* studies, the enhancement of the bone regeneration around the nano-TiO$_2$/PEEK composite implant was observed by higher bone volume/tissue volume in comparison with the PEEK implant [20].

In another study of increasing the bioactivity of PEEK, strontium-containing hydroxyapatite/polyether ketone (Sr-HA/PEEK) composites were fabricated by compression molding technique. *In vitro* study involving apatite formation in SBF and MG-63-mediated mineralization confirmed higher bioactivity in comparison to HA/PEEK composite [16]. Also, calcium oxide and silicon dioxide (CS) were used as bioactive additives to PEEK composite. *In vitro* bioactivity study via SBF showed that by increasing the volume fraction of CS the bioactivity of the composite increased [91]. Table 4 summarizes the effect of the compound materials on the enhancement of the bioactivity of the PEEK composites.

4.2. Mechanical Properties of PEEK Composite. PEEK exhibits superb mechanical properties appropriate for load bearing orthopedic applications. However, as mentioned before the low mechanical properties of bioactive PEEK composites in comparison to PEEK are one of the biggest concerns of scientists and a lot of works in this field have been done. In this part, the present works based on the additives are described.

Studies showed that increasing the volume fraction of HA in the HA/PEEK composite increased Young's modulus and microhardness of the composite, though strength and strain at the fracture point decreased [76]. However, cyclic load on the PEEK/HA composite with different content of HA showed the HA/PEEK composite is a promising fatigue-resistant material for biomedical applications [92]. For improving the mechanical properties of the HA/PEEK composites the composites were prepared via *in situ* process. The composite showed strong physical bonding between HA and PEEK matrix due to improvement of mechanical properties of the composite in comparison with previously prepared HA/PEEK composites by other methods [81–83].

The mechanical properties of PEEK/HA nanoparticle composite showed the initial increase of tensile strength by increasing the content of HA nanoparticles to 5 vol% and after that decreasing the tensile strength. The authors described the first increase in tensile strength that was due to the "strong interactivity of nanoparticles and PEEK chains," and they explained the agglomeration of HA nanoparticles for the contents of over 10 vol% which was due to decreased binding between nanoparticles and PEEK and reduction in the tensile strength of the composite [77, 78].

PEEK/HA whiskers composite via compression molding method showed the additive HA whiskers were oriented in the direction of viscous flow due to the production of composites with anisotropy mechanical properties. The results

TABLE 4: Effect of the compound materials on the bioactivity of the PEEK composite.

Compound material	Studied areas	Results
HA	Probing the effect of HA volume fraction on bioactivity via *in vivo* study.	Enhancement of the presence of fibroblast cells, formation of osteoid and osteocytes within lamellar bone [15].
	Probing the effect of HA volume fraction on bioactivity via SBF immersion test.	Higher rate of HA growth for the composite with higher volume fraction of HA [14].
	In vitro study of the new method of simple cubic molding and sintering.	Confirmed improvement of bioactivity of the composite [79].
	Biocompatibility and bioactivity study of the produced composite via *in situ* synthetic method.	Produced composite showed nontoxic and the bioactive properties [81–83].
	In vitro bioactivity study of HA/PEEK composite produced by selective laser sintering method.	Improvement in bioactivity of the composite and higher content of HA exhibited higher bioactivity rate [88].
nHA	Probing the effect of nHA volume fraction on bioactivity via *in vitro* study by SBF immersion, cell adhesion, and proliferation.	Nanocomposite with 29.2 vol% of nHA content showed the best bioactivity in comparison with other samples [80].
βTCP	*In vitro* bioactivity study via osteoblast cells.	Lower rates of osteoblast growth on the βTCP-PEEK compared to pure PEEK [84].
	Biocompatibility study of laser sintering method for producing βTCP/PEEK via *in vitro* study by osteoblast cells.	Confirmed nontoxicity of laser sintering method for producing βTCP/PEEK composite but showed no advantage of adding βTCP as an additive on cell growth [85, 86].
	In vivo bioactivity study of the laser sintered PEEK/βTCP composite.	Better interaction with surrounding bone and direct connection to the surrounding bone [87].
Carbon black	Biocompatibility study of laser sintering method for producing carbon black/PEEK composite via *in vitro* study by osteoblast cells.	Confirmed nontoxicity of laser sintering method for producing carbon black/PEEK composite but showed no advantage of adding carbon black as an additive on cell growth [85].
Carbon, carbon/βTCP, and carbon/bioglass 4s5S5	Biocompatibility and bioactivity study of produced composites via laser sintering method.	Produced composite via laser sintering method was nontoxic. PEEK/carbon/bioglass composite showed improvement in the bioactivity property [86].
Glass fiber	*In vitro* study via MG-63 cells.	Higher rate of cell proliferation [89, 90].
Nano-TiO$_2$	*In vitro* and *in vivo* study.	Increasing in cell attachment and enhanced osteoblast cell spreading. Enhancement of the bone regeneration around the nano-TiO$_2$/PEEK composite [20].
Sr-HA	*In vitro* study contains apatite formation in SBF and MG-63-mediated mineralization.	Enhancement of bioactivity [16].
CS	Probing the effect of CS volume fraction on bioactivity via *in vitro* bioactivity study by SBF immersion.	By increasing the volume fraction of CS the bioactivity of the composite increased [91].

of mechanical properties showed an increase in the volume fraction of HA whisker reinforcement due to increased elastic modulus of the composite but caused a decrease in the ultimate tensile strength/strain at the failure point [18].

Polyether ketone (PEKK) reinforced with 0, 20, and 40 vol% HA whiskers specimens by compression molding method and subsequent annealing showed a decrease of the fatigue life with the increase in the volume fraction of the HA whiskers [93]. Effect of HA contents and mold temperature on the mechanical properties of PEKK/HA whiskers scaffolds was studied. The elastic modulus of the scaffold increased from 0 to 20 vol% HA with the increase of HA value from 20 to 40 vol%, while the yield strength and strain at

the fracture point were decreased with increasing volume fraction of HA. Elastic modulus, yield strength, and yield strain were also increased by increasing the mold temperature [94].

The bending modulus of strontium-containing hydroxyapatite/polyether ketone (Sr-HA/PEEK) increased with increasing the volume fraction of Sr-HA. The elastic modulus of 25 vol% and 30 vol% Sr-HA reinforcement showed 113% and 136% increase, respectively, in comparison with pure PEEK. The bending strengths of 25 vol% and 30 vol% Sr-HA reinforcement showed 25% and 29% decrease, respectively, in comparison with pure PEEK [16]. Table 5 presents the summary of the effect of different compounds on the mechanical properties of the PEEK composite.

TABLE 5: Effect of the compound materials on the mechanical properties of the PEEK composites.

Compound material	Studied mechanical properties	Results
HA	E, microhardness, ultimate tensile strength/strain	Young's modulus and microhardness of composite increased, ultimate tensile strength and strain at the fracture point decreased [76].
	Fatigue-resistant	Showing enough fatigue-resistant property for biomedical applications [92].
	Ultimate tensile strength	Prepared composite via *in situ* process showed strong physical bonding between HA and PEEK matrix and enhanced ultimate tensile strength [81–83].
HAnp	Ultimate tensile strength	Initial increase of tensile strength by increasing HAnp content to 5 vol% and after that decreasing the tensile strength [77, 78].
Whiskers HA	E, isotropy property, ultimate tensile strength/strain	Anisotropy mechanical properties, increasing of E and decreasing in the ultimate tensile strength/strain by increasing of the volume fraction of HA whisker reinforcement [18].
	Fatigue life	Decreasing of the fatigue life with increase in the volume fraction of the HA whiskers in PEKK [93].
	E, ultimate strength and strain	Elastic modulus increased, while the ultimate tensile strength and strain decreased with increasing volume fraction of HA. Elastic modulus, yield strength, and yield strain were increased by increasing the mold temperature [94].
Sr-HA	E, bending strength	The bending modulus, elastic modulus increased with the volume fraction ratio of Sr-HA. The elastic modulus of 25 vol% and 30 vol% Sr-HA reinforcement showed 113% and 136% increase, respectively, in comparison with pure PEEK. The bending strengths of 25 vol% and 30 vol% Sr-HA reinforcement showed 25% and 29% decrease, respectively, in comparison with pure PEEK [16].

5. Summary and Conclusion

For long term load bearing implant applications, PEEK is the only commercial material that offers characteristics with good chemical resistance, radiolucency, and mechanical properties similar to those of human bones. However, bioactivity of PEEK is the biggest hindrance which causes reduction in the acceleration of worldwide spreading. We have summarized the previous study of bioactivation of PEEK and categorized them broadly to the bioactive PEEK composites and surface modified PEEK. The biggest concern about the PEEK composite is its mechanical properties. Thus, the PEEK bioactive composites were subcategorized to probe the previous studies from the bioactivity and mechanical aspects. Although different bioactive additives such as HA, Ti, TiO_2, β-tricalcium phosphate, and bioactive glass improve the bioactivity of PEEK's composite, the low mechanical properties of PEEK's composite are still its most important weakness. The surface modification of PEEK for biomedical application was subcategorized based on the techniques which were used for modifying the surface of the PEEK's implants. Between these methods the deposition of HA via plasma spraying method is the only method which qualified for commercial usage. However, there are still some concerns with this method such as damaging the surface chemistry of PEEK substrate and therefore in-depth research is needed. The trend of research in the bioactivity of PEEK shows a very encouraging result which has potential to overcome the existing problems in the current techniques and production of bioactive PEEK implant and spreading its application as bioactive material in orthopedic and dental implant areas.

Competing Interests

The authors declare that they have no competing interests.

Acknowledgments

The authors would like to acknowledge Universiti Teknologi Malaysia (UTM) for providing research facilities and financial support under grant of Potential Academic Staff with Grant no. Q.J130000.2745.01K62 and FRGS no. PY/2015/05371.

References

[1] S. M. Kurtz and J. N. Devine, "PEEK biomaterials in trauma, orthopedic, and spinal implants," *Biomaterials*, vol. 28, no. 32, pp. 4845–4869, 2007.

[2] D. Williams, "Polyetheretherketone for long-term implantable devices," *Medical Device Technology*, vol. 19, no. 1, pp. 8–11, 2008.

[3] J. M. Toth, M. Wang, B. T. Estes, J. L. Scifert, H. B. Seim III, and A. S. Turner, "Polyetheretherketone as a biomaterial for spinal applications," *Biomaterials*, vol. 27, no. 3, pp. 324–334, 2006.

[4] P. Xing, G. P. Robertson, M. D. Guiver, S. D. Mikhailenko, K. Wang, and S. Kaliaguine, "Synthesis and characterization of sulfonated poly(ether ether ketone) for proton exchange membranes," *Journal of Membrane Science*, vol. 229, no. 1-2, pp. 95–106, 2004.

[5] M. C. Sobieraj, S. M. Kurtz, and C. M. Rimnac, "Notch sensitivity of PEEK in monotonic tension," *Biomaterials*, vol. 30, no. 33, pp. 6485–6494, 2009.

[6] C.-M. Han, E.-J. Lee, H.-E. Kim et al., "The electron beam deposition of titanium on polyetheretherketone (PEEK) and the resulting enhanced biological properties," *Biomaterials*, vol. 31, no. 13, pp. 3465–3470, 2010.

[7] V. A. Stadelmann, A. Terrier, and D. P. Pioletti, "Microstimulation at the bone-implant interface upregulates osteoclast activation pathways," *Bone*, vol. 42, no. 2, pp. 358–364, 2008.

[8] S. Ramakrishna, J. Mayer, E. Wintermantel, and K. W. Leong, "Biomedical applications of polymer-composite materials: a review," *Composites Science and Technology*, vol. 61, no. 9, pp. 1189–1224, 2001.

[9] J. Tams, F. R. Rozema, R. R. M. Bos, J. L. N. Roodenburg, P. G. J. Nikkels, and A. Vermey, "Poly(L-lactide) bone plates and screws for internal fixation of mandibular swing osteotomies," *International Journal of Oral and Maxillofacial Surgery*, vol. 25, no. 1, pp. 20–24, 1996.

[10] N. Scher, D. Poe, F. Kuchnir, C. Reft, R. Weichselbaum, and W. R. Panje, "Radiotherapy of the resected mandible following stainless steel plate fixation," *The Laryngoscope*, vol. 98, no. 5, pp. 561–563, 1988.

[11] A. J. Barton, R. D. Sagers, and W. G. Pitt, "Bacterial adhesion to orthopedic implant polymers," *Journal of Biomedical Materials Research*, vol. 30, no. 3, pp. 403–410, 1996.

[12] D. F. Williams, A. McNamara, and R. M. Turner, "Potential of polyetheretherketone (PEEK) and carbon-fibre-reinforced PEEK in medical applications," *Journal of Materials Science Letters*, vol. 6, no. 2, pp. 188–190, 1987.

[13] A. Schwitalla and W.-D. Müller, "PEEK dental implants: a review of the literature," *Journal of Oral Implantology*, vol. 39, no. 6, pp. 743–749, 2013.

[14] S. Yu, K. P. Hariram, R. Kumar, P. Cheang, and K. K. Aik, "In vitro apatite formation and its growth kinetics on hydroxyapatite/polyetheretherketone biocomposites," *Biomaterials*, vol. 26, no. 15, pp. 2343–2352, 2005.

[15] M. S. Abu Bakar, M. H. W. Cheng, S. M. Tang et al., "Tensile properties, tension-tension fatigue and biological response of polyetheretherketone-hydroxyapatite composites for load-bearing orthopedic implants," *Biomaterials*, vol. 24, no. 13, pp. 2245–2250, 2003.

[16] K. L. Wong, C. T. Wong, W. C. Liu et al., "Mechanical properties and in vitro response of strontium-containing hydroxyapatite/polyetheretherketone composites," *Biomaterials*, vol. 30, no. 23-24, pp. 3810–3817, 2009.

[17] G. L. Converse, T. L. Conrad, C. H. Merrill, and R. K. Roeder, "Hydroxyapatite whisker-reinforced polyetherketoneketone bone ingrowth scaffolds," *Acta Biomaterialia*, vol. 6, no. 3, pp. 856–863, 2010.

[18] G. L. Converse, W. Yue, and R. K. Roeder, "Processing and tensile properties of hydroxyapatite-whisker-reinforced polyetheretherketone," *Biomaterials*, vol. 28, no. 6, pp. 927–935, 2007.

[19] J. P. Fan, C. P. Tsui, C. Y. Tang, and C. L. Chow, "Influence of interphase layer on the overall elasto-plastic behaviors of HA/PEEK biocomposite," *Biomaterials*, vol. 25, no. 23, pp. 5363–5373, 2004.

[20] X. Wu, X. Liu, J. Wei, J. Ma, F. Deng, and S. Wei, "Nano-TiO$_2$/PEEK bioactive composite as a bone substitute material: in vitro and in vivo studies," *International Journal of Nanomedicine*, vol. 7, pp. 1215–1225, 2012.

[21] H.-K. Tsou, P.-Y. Hsieh, C.-J. Chung, C.-H. Tang, T.-W. Shyr, and J.-L. He, "Low-temperature deposition of anatase TiO2 on

medical grade polyetheretherketone to assist osseous integration," *Surface and Coatings Technology*, vol. 204, no. 6-7, pp. 1121–1125, 2009.

[22] Y. Zhao, H. M. Wong, W. Wang et al., "Cytocompatibility, osseointegration, and bioactivity of three-dimensional porous and nanostructured network on polyetheretherketone," *Biomaterials*, vol. 34, no. 37, pp. 9264–9277, 2013.

[23] L. L. Hench, R. J. Splinter, W. C. Allen, and T. K. Greenlee, "Bonding mechanisms at the interface of ceramic prosthetic materials," *Journal of Biomedical Materials Research*, vol. 5, no. 6, pp. 117–141, 1972.

[24] B. Kasemo, "Biological surface science," *Surface Science*, vol. 500, no. 1–3, pp. 656–677, 2002.

[25] B. Kasemo and J. Lausmaa, "Material-tissue interfaces: the role of surface properties and processes," *Environmental Health Perspectives*, vol. 102, no. 5, pp. 41–45, 1994.

[26] M. Mrksich, "A surface chemistry approach to studying cell adhesion," *Chemical Society Reviews*, vol. 29, no. 4, pp. 267–273, 2000.

[27] G. Altankov and T. Groth, "Reorganization of substratum-bound fibronectin on hydrophilic and hydrophobic materials is related to biocompatibility," *Journal of Materials Science: Materials in Medicine*, vol. 5, no. 9-10, pp. 732–737, 1994.

[28] O. Noiset, Y.-J. Schneider, and J. Marchand-Brynaert, "Surface modification of poly(aryl ether ether ketone) (PEEK) film by covalent coupling of amines and amino acids through a spacer arm," *Journal of Polymer Science Part A: Polymer Chemistry*, vol. 35, no. 17, pp. 3779–3790, 1997.

[29] A. Katzer, H. Marquardt, J. Westendorf, J. V. Wening, and G. von Foerster, "Polyetheretherketone—cytotoxicity and mutagenicity in vitro," *Biomaterials*, vol. 23, no. 8, pp. 1749–1759, 2002.

[30] A. H. C. Poulsson and R. G. Richards, "Chapter 10—surface modification techniques of polyetheretherketone, including plasma surface treatment," in *PEEK Biomaterials Handbook*, S. M. Kurtz, Ed., chapter 10, pp. 145–161, William Andrew, Oxford, UK, 2012.

[31] J. Marchand-Brynaert, G. Pantano, and O. Noiset, "Surface fluorination of PEEK film by selective wet-chemistry," *Polymer*, vol. 38, no. 6, pp. 1387–1394, 1997.

[32] O. Noiset, Y. J. Schneider, and J. Marchand-Brynaert, "Fibronectin adsorption or/and covalent grafting on chemically modified PEEK film surfaces," *Journal of Biomaterials Science, Polymer Edition*, vol. 10, no. 6, pp. 657–677, 1999.

[33] O. Noiset, Y.-J. Schneider, and J. Marchand-Brynaert, "Adhesion and growth of CaCo2 cells on surface-modified PEEK substrata," *Journal of Biomaterials Science, Polymer Edition*, vol. 11, no. 7, pp. 767–786, 2000.

[34] M. Pino, N. Stingelin, and K. E. Tanner, "Nucleation and growth of apatite on NaOH-treated PEEK, HDPE and UHMWPE for artificial cornea materials," *Acta Biomaterialia*, vol. 4, no. 6, pp. 1827–1836, 2008.

[35] E. Occhiello, M. Morra, G. L. Guerrini, and F. Garbassi, "Adhesion properties of plasma-treated carbon/PEEK composites," *Composites*, vol. 23, no. 3, pp. 193–200, 1992.

[36] J. Comyn, L. Mascia, G. Xiao, and B. M. Parker, "Plasma-treatment of polyetheretherketone (PEEK) for adhesive bonding," *International Journal of Adhesion and Adhesives*, vol. 16, no. 2, pp. 97–104, 1996.

[37] S.-W. Ha, M. Kirch, F. Birchler et al., "Surface activation of polyetheretherketone (PEEK) and formation of calcium phosphate coatings by precipitation," *Journal of Materials Science: Materials in Medicine*, vol. 8, no. 11, pp. 683–690, 1997.

[38] K. Schröder, A. Meyer-Plath, D. Keller, and A. Ohl, "On the applicability of plasma assisted chemical micropatterning to different polymeric biomaterials," *Plasmas and Polymers*, vol. 7, no. 2, pp. 103–125, 2002.

[39] D. Briem, S. Strametz, K. Schröoder et al., "Response of primary fibroblasts and osteoblasts to plasma treated polyetheretherketone (PEEK) surfaces," *Journal of Materials Science: Materials in Medicine*, vol. 16, no. 7, pp. 671–677, 2005.

[40] F. Awaja, D. V. Bax, S. Zhang, N. James, and D. R. McKenzie, "Cell adhesion to PEEK treated by plasma immersion ion implantation and deposition for active medical implants," *Plasma Processes and Polymers*, vol. 9, no. 4, pp. 355–362, 2012.

[41] A. H. C. Poulsson, D. Eglin, S. Zeiter et al., "Osseointegration of machined, injection moulded and oxygen plasma modified PEEK implants in a sheep model," *Biomaterials*, vol. 35, no. 12, pp. 3717–3728, 2014.

[42] H. Wang, T. Lu, F. Meng, H. Zhu, and X. Liu, "Enhanced osteoblast responses to poly ether ether ketone surface modified by water plasma immersion ion implantation," *Colloids and Surfaces B: Biointerfaces*, vol. 117, pp. 89–97, 2014.

[43] Z. Novotna, A. Reznickova, S. Rimpelova, M. Vesely, Z. Kolska, and V. Svorcik, "Tailoring of PEEK bioactivity for improved cell interaction: plasma treatment in action," *RSC Advances*, vol. 5, no. 52, pp. 41428–41436, 2015.

[44] J. Waser-Althaus, A. Salamon, M. Waser et al., "Differentiation of human mesenchymal stem cells on plasma-treated polyetheretherketone," *Journal of Materials Science: Materials in Medicine*, vol. 25, no. 2, pp. 515–525, 2014.

[45] R. Comesaña, F. Quintero, F. Lusquiños et al., "Laser cladding of bioactive glass coatings," *Acta Biomaterialia*, vol. 6, no. 3, pp. 953–961, 2010.

[46] P. Laurens, B. Sadras, F. Décobert, F. Aréfi-Khonsari, and J. Amouroux, "Laser-induced surface modifications of poly(ether ether ketone): influence of the excimer laser wavelength," *Journal of Adhesion Science and Technology*, vol. 13, no. 9, pp. 983–997, 1999.

[47] P. Laurens, M. Ould Bouali, F. Meducin, and B. Sadras, "Characterization of modifications of polymer surfaces after excimer laser treatments below the ablation threshold," *Applied Surface Science*, vol. 154-155, pp. 211–216, 2000.

[48] P. Laurens, B. Sadras, F. Decobert, F. Arefi-Khonsari, and J. Amouroux, "Enhancement of the adhesive bonding properties of PEEK by excimer laser treatment," *International Journal of Adhesion and Adhesives*, vol. 18, no. 1, pp. 19–27, 1998.

[49] A. Riveiro, R. Soto, R. Comesaña et al., "Laser surface modification of PEEK," *Applied Surface Science*, vol. 258, no. 23, pp. 9437–9442, 2012.

[50] J. Khoury, S. R. Kirkpatrick, M. Maxwell, R. E. Cherian, A. Kirkpatrick, and R. C. Svrluga, "Neutral atom beam technique enhances bioactivity of PEEK," *Nuclear Instruments and Methods in Physics Research Section B: Beam Interactions with Materials and Atoms*, vol. 307, pp. 630–634, 2013.

[51] A. Kirkpatrick, S. Kirkpatrick, M. Walsh et al., "Investigation of accelerated neutral atom beams created from gas cluster ion beams," *Nuclear Instruments and Methods in Physics Research, Section B: Beam Interactions with Materials and Atoms*, vol. 307, pp. 281–289, 2013.

[52] I. Mathieson and R. H. Bradley, "Improved adhesion to polymers by UV/ozone surface oxidation," *International Journal of Adhesion and Adhesives*, vol. 16, no. 1, pp. 29–31, 1996.

[53] H. Garg, G. Bedi, and A. Garg, "Implant surface modifications: a review," *Journal of Clinical and Diagnostic Research*, vol. 6, no. 2, pp. 319–324, 2012.

[54] S.-W. Ha, J. Mayer, B. Koch, and E. Wintermantel, "Plasma-sprayed hydroxylapatite coating on carbon fibre reinforced thermoplastic composite materials," *Journal of Materials Science: Materials in Medicine*, vol. 5, no. 6-7, pp. 481–484, 1994.

[55] S.-W. Ha, A. Gisep, J. Mayer, E. Wintermantel, H. Gruner, and M. Wieland, "Topographical characterization and microstructural interface analysis of vacuum-plasma-sprayed titanium and hydroxyapatite coatings on carbon fibre-reinforced poly(etheretherketone)," *Journal of Materials Science: Materials in Medicine*, vol. 8, no. 12, pp. 891–896, 1997.

[56] A. Rabiei and S. Sandukas, "Processing and evaluation of bioactive coatings on polymeric implants," *Journal of Biomedical Materials Research—Part A*, vol. 101, no. 9, pp. 2621–2629, 2013.

[57] S. Beauvais and O. Decaux, "Plasma sprayed biocompatible coatings on PEEK implants," in *Proceedings of the International Thermal Spray Conference*, B. R. Marple, Ed., Beijing, China, May 2007.

[58] J. H. Lee, H. L. Jang, K. M. Lee et al., "In vitro and in vivo evaluation of the bioactivity of hydroxyapatite-coated polyetheretherketone biocomposites created by cold spray technology," *Acta Biomaterialia*, vol. 9, no. 4, pp. 6177–6187, 2013.

[59] B.-D. Hahn, D.-S. Park, J.-J. Choi et al., "Osteoconductive hydroxyapatite coated PEEK for spinal fusion surgery," *Applied Surface Science*, vol. 283, pp. 6–11, 2013.

[60] S. Barkarmo, A. Wennerberg, M. Hoffman et al., "Nano-hydroxyapatite-coated PEEK implants: a pilot study in rabbit bone," *Journal of Biomedical Materials Research Part A*, vol. 101, no. 2, pp. 465–471, 2013.

[61] D. Almasi, S. Izman, M. Assadian, M. Ghanbari, and M. R. Abdul Kadir, "Crystalline ha coating on peek via chemical deposition," *Applied Surface Science*, vol. 314, pp. 1034–1040, 2014.

[62] D. Almasi, S. Izman, M. Sadeghi et al., "In vitro evaluation of bioactivity of chemically deposited hydroxyapatite on polyether ether ketone," *International Journal of Biomaterials*, vol. 2015, Article ID 475435, 5 pages, 2015.

[63] M.-H. Chi, H.-K. Tsou, C.-J. Chung, and J.-L. He, "Biomimetic hydroxyapatite grown on biomedical polymer coated with titanium dioxide interlayer to assist osteocompatible performance," *Thin Solid Films*, vol. 549, pp. 98–102, 2013.

[64] C.-M. Han, T.-S. Jang, H.-E. Kim, and Y.-H. Koh, "Creation of nanoporous TiO_2 surface onto polyetheretherketone for effective immobilization and delivery of bone morphogenetic protein," *Journal of Biomedical Materials Research A*, vol. 102, no. 3, pp. 793–800, 2014.

[65] A. G. R. Wieling, "Osteointegrative surfaces for CF/PEEK implants," *European Cells and Materials*, vol. 17, supplement 1, p. 10, 2009.

[66] S.-W. Ha, K.-L. Eckert, E. Wintermantel, H. Gruner, M. Guecheva, and H. Vonmont, "NaOH treatment of vacuum-plasma-sprayed titanium on carbon fibre-reinforced poly(etheretherketone)," *Journal of Materials Science: Materials in Medicine*, vol. 8, no. 12, pp. 881–886, 1997.

[67] D. M. Devine, J. Hahn, R. G. Richards, H. Gruner, R. Wieling, and S. G. Pearce, "Coating of carbon fiber-reinforced polyetheretherketone implants with titanium to improve bone apposition," *Journal of Biomedical Materials Research B: Applied Biomaterials*, vol. 101, no. 4, pp. 591–598, 2013.

[68] S. D. Cook and A. M. Rust-Dawicki, "Preliminary evaluation of titanium-coated PEEK dental implants," *The Journal of Oral Implantology*, vol. 21, no. 3, pp. 176–181, 1995.

[69] T. Joseph Dennes and J. Schwartz, "A nanoscale adhesion layer to promote cell attachment on PEEK," *Journal of the American Chemical Society*, vol. 131, no. 10, pp. 3456–3457, 2009.

[70] H. Wang, M. Xu, W. Zhang et al., "Mechanical and biological characteristics of diamond-like carbon coated poly aryl-ether-ether-ketone," *Biomaterials*, vol. 31, no. 32, pp. 8181–8187, 2010.

[71] M. S. Abu Bakar, P. Cheang, and K. A. Khor, "Thermal processing of hydroxyapatite reinforced polyetheretherketone composites," *Journal of Materials Processing Technology*, vol. 89-90, pp. 462–466, 1999.

[72] K. H. Tan, C. K. Chua, K. F. Leong et al., "Scaffold development using selective laser sintering of polyetheretherketone-hydroxyapatite biocomposite blends," *Biomaterials*, vol. 24, no. 18, pp. 3115–3123, 2003.

[73] K. H. Tan, C. K. Chua, K. F. Leong et al., "Selective laser sintering of biocompatible polymers for applications in tissue engineering," *Bio-Medical Materials and Engineering*, vol. 15, no. 1-2, pp. 113–124, 2005.

[74] K. H. Tan, C. K. Chua, K. F. Leong, M. W. Naing, and C. M. Cheah, "Fabrication and characterization of three-dimensional poly(ether-ether-ketone)/-hydroxyapatite biocomposite scaffolds using laser sintering," *Proceedings of the Institution of Mechanical Engineers Part H: Journal of Engineering in Medicine*, vol. 219, no. 3, pp. 183–194, 2005.

[75] R. K. Roeder, S. M. Smith, T. L. Conrad, N. J. Yanchak, C. H. Merrill, and G. L. Converse, "Porous and bioactive PEEK implants for interbody spinal fusion," *Journal of Advanced Materials and Processing*, vol. 167, no. 10, pp. 46–48, 2009.

[76] M. S. Abu Bakar, P. Cheang, and K. A. Khor, "Tensile properties and microstructural analysis of spheroidized hydroxyapatite-poly (etheretherketone) biocomposites," *Materials Science and Engineering A*, vol. 345, no. 1-2, pp. 55–63, 2003.

[77] L. Wang, L. Weng, S. Song, and Q. Sun, "Mechanical properties and microstructure of polyetheretherketone-hydroxyapatite nanocomposite materials," *Materials Letters*, vol. 64, no. 20, pp. 2201–2204, 2010.

[78] L. Wang, L. Weng, S. Song, Z. Zhang, S. Tian, and R. Ma, "Characterization of polyetheretherketone-hydroxyapatite nanocomposite materials," *Materials Science and Engineering A*, vol. 528, no. 10-11, pp. 3689–3696, 2011.

[79] C. Hengky, B. Kelsen, and P. Saraswati, "Mechanical and biological characterization of pressureless sintered hydroxapatite-polyetheretherketone biocomposite," in *13th International Conference on Biomedical Engineering: ICBME 2008 3–6 December 2008 Singapore*, vol. 23 of *IFMBE Proceedings*, pp. 261–264, Springer, Berlin, Germany, 2009.

[80] K. Li, C. Y. Yeung, K. W. K. Yeung, and S. C. Tjong, "Sintered hydroxyapatite/polyetheretherketone nanocomposites: mechanical behavior and biocompatibility," *Advanced Engineering Materials*, vol. 14, no. 4, pp. B155–B165, 2012.

[81] R. Ma, L. Weng, X. Bao, S. Song, and Y. Zhang, "In vivo biocompatibility and bioactivity of in situ synthesized hydroxyapatite/polyetheretherketone composite materials," *Journal of Applied Polymer Science*, vol. 127, no. 4, pp. 2581–2587, 2013.

[82] R. Ma, L. Weng, X. Bao, Z. Ni, S. Song, and W. Cai, "Characterization of in situ synthesized hydroxyapatite/polyetheretherketone composite materials," *Materials Letters*, vol. 71, pp. 117–119, 2012.

[83] R. Ma, L. Weng, L. Fang, Z. Luo, and S. Song, "Structure and mechanical performance of in situ synthesized hydroxyapatite/polyetheretherketone nanocomposite materials," *Journal of Sol-Gel Science and Technology*, vol. 62, no. 1, pp. 52–56, 2012.

[84] L. Petrovic, D. Pohle, H. Münstedt, T. Rechtenwald, K. A. Schlegel, and S. Rupprecht, "Effect of βTCP filled polyetheretherketone on osteoblast cell proliferation in vitro," *Journal of Biomedical Science*, vol. 13, no. 1, pp. 41–46, 2006.

[85] D. Pohle, S. Ponader, T. Rechtenwald et al., "Processing of three-dimensional laser sintered polyetheretherketone composites and testing of osteoblast proliferation in vitro," *Macromolecular Symposia*, vol. 253, pp. 65–70, 2007.

[86] C. Von Wilmowsky, E. Vairaktaris, D. Pohle et al., "Effects of bioactive glass and β-TCP containing three-dimensional laser sintered polyetheretherketone composites on osteoblasts in vitro," *Journal of Biomedical Materials Research A*, vol. 87, no. 4, pp. 896–902, 2008.

[87] C. Von Wilmonsky, R. Lutz, U. Meisel et al., "In Vivo evaluation of β-TCP containing 3D laser sintered poly(ether ether ketone) composites in pigs," *Journal of Bioactive and Compatible Polymers*, vol. 24, no. 2, pp. 169–184, 2009.

[88] Y. Zhang, L. Hao, M. M. Savalani, R. A. Harris, L. Di Silvio, and K. E. Tanner, "In vitro biocompatibility of hydroxyapatite-reinforced polymeric composites manufactured by selective laser sintering," *Journal of Biomedical Materials Research A*, vol. 91, no. 4, pp. 1018–1027, 2009.

[89] T. W. Lin, A. A. Corvelli, C. G. Frondoza, J. C. Roberts, and D. S. Hungerford, "Glass peek composite promotes proliferation and osteocalcin production of human osteoblastic cells," *Journal of Biomedical Materials Research*, vol. 36, no. 2, pp. 137–144, 1997.

[90] A. A. Corvelli, J. C. Roberts, P. J. Biermann, and J. H. Cranmer, "Characterization of a peek composite segmental bone replacement implant," *Journal of Materials Science*, vol. 34, no. 10, pp. 2421–2431, 1999.

[91] I. Y. Kim, A. Sugino, K. Kikuta, C. Ohtsuki, and S. B. Cho, "Bioactive composites consisting of PEEK and calcium silicate powders," *Journal of Biomaterials Applications*, vol. 24, no. 2, pp. 105–118, 2009.

[92] S. M. Tang, P. Cheang, M. S. AbuBakar, K. A. Khor, and K. Liao, "Tension-tension fatigue behavior of hydroxyapatite reinforced polyetheretherketone composites," *International Journal of Fatigue*, vol. 26, no. 1, pp. 49–57, 2004.

[93] G. L. Converse, T. L. Conrad, and R. K. Roeder, "Fatigue life of hydroxyapatite whisker reinforced polyetherketoneketone," *Transactions of the Society for Biomaterials*, vol. 32, p. 584, 2009.

[94] G. L. Converse, T. L. Conrad, and R. K. Roeder, "Mechanical properties of hydroxyapatite whisker reinforced polyetherketoneketone composite scaffolds," *Journal of the Mechanical Behavior of Biomedical Materials*, vol. 2, no. 6, pp. 627–635, 2009.

Intrinsic Osteoinductivity of Porous Titanium Scaffold for Bone Tissue Engineering

Maryam Tamaddon,[1] **Sorousheh Samizadeh,**[1] **Ling Wang,**[2]
Gordon Blunn,[1] **and Chaozong Liu**[1]

[1]*Institute of Orthopaedic & Musculoskeletal Science, University College London, Royal National Orthopaedic Hospital, Stanmore HA7 4LP, UK*
[2]*State Key Laboratory for Manufacturing System Engineering, Xi'an Jiaotong University, Xi'an, Shanxi Province 710049, China*

Correspondence should be addressed to Chaozong Liu; Chaozong.liu@ucl.ac.uk

Academic Editor: Kheng-Lim Goh

Large bone defects and nonunions are serious complications that are caused by extensive trauma or tumour. As traditional therapies fail to repair these critical-sized defects, tissue engineering scaffolds can be used to regenerate the damaged tissue. Highly porous titanium scaffolds, produced by selective laser sintering with mechanical properties in range of trabecular bone (compressive strength 35 MPa and modulus 73 MPa), can be used in these orthopaedic applications, if a stable mechanical fixation is provided. Hydroxyapatite coatings are generally considered essential and/or beneficial for bone formation; however, debonding of the coatings is one of the main concerns. We hypothesised that the titanium scaffolds have an intrinsic potential to induce bone formation without the need for a hydroxyapatite coating. In this paper, titanium scaffolds coated with hydroxyapatite using electrochemical method were fabricated and osteoinductivity of coated and noncoated scaffolds was compared in vitro. Alizarin Red quantification confirmed osteogenesis independent of coating. Bone formation and ingrowth into the titanium scaffolds were evaluated in sheep stifle joints. The examinations after 3 months revealed 70% bone ingrowth into the scaffold confirming its osteoinductive capacity. It is shown that the developed titanium scaffold has an intrinsic capacity for bone formation and is a suitable scaffold for bone tissue engineering.

1. Introduction

Massive traumatic injuries or tumour resections are among the factors which can contribute to substantial bone loss [1, 2]. Thanks to a spontaneous capacity for regeneration, most bone lesions, such as fractures, can be repaired with conventional therapies. The process of fracture healing is a sequence that begins with hematoma formation and then moves to inflammation, destruction of nonvital debris, granulation tissue proliferation, callus formation, conversion of woven bone to lamellar bone, and, finally, remodelling of the healed bone [3]. However, in cases of large defects and osseous congenital deformities, bone grafts (e.g., xeno-, allo-, and autografts) or substitutes are needed to aid healing [4]. The current gold standard for repair of large bone defects [1] is autograft where host bone is removed from another non-load-bearing site

to fill the defect. However, the complication rate is as high as 30% due to donor site morbidity, pain, hematoma, and inflammation. In many cases, this has been proven a challenging treatment for critical-sized defects [1].

Tissue engineering (TE) approaches, which use body's natural ability to repair injured bone with new bone tissue and to remodel newly produced bone in response to the local stresses, are being explored as alternatives for large bone defect repairs [5]. There are three key ingredients necessary in TE: a scaffold, which may be either natural or synthetic, cells [6], and inductive signals (i.e., growth factors or proteins) [7].

Studies have suggested that cells might be unable to establish themselves properly within a defect without matrix guidance [8]. Therefore, a scaffold must be developed to provide a three-dimensional structure to support the cells, aid their proliferation, and help them be differentiated, while its architecture defines the ultimate shape of the new bone [9, 10].

In addition to general requirements for TE scaffolds such as biocompatibility and ability to be sterilised, the key requirements for the development of an orthopaedic scaffold include the following [1]:

(1) Mechanical stability to be retained in the affected area

(2) Interconnected porous architecture (porosity exceeding 90%) [4, 11] to allow for vascularization and bone ingrowth and to act as a channel for delivery of nutrients and gases to the cells deep inside the scaffold and, at the same time, removal of the metabolic waste from cells

(3) Supporting and promoting osteogenic differentiation of undifferentiated cells (osteoinduction) and growth of differentiated bone cells (osteoconduction) [12]

(4) Enhancing cellular activity towards scaffold-host tissue integration (osseointegration).

Mechanical properties are especially important in scaffolds for hard and ductile tissues such as bone [13, 14] because the scaffolds must also interact with their physiological surroundings to transmit mechanical signals to cells and regulate cell behaviour (i.e., differentiation, motility, and contractility). The stiffness of scaffold can have effects at a transcriptional level, determining whether stem cells make the decision to become cells as functionally diverse as osteoblasts [15].

Biomaterials used in tissue engineering of bone are usually categorized into four major groups: natural polymers, synthetic polymers, metallic materials, and inorganic materials such as ceramics and bioactive glasses. Multicomponent systems can be designed to generate composites of enhanced performance [16].

Naturally derived polymers have the advantage of native biological function [17, 18] but their low mechanical strength makes them less attractive as an option for bone tissue repair. In synthetic polymers, on the other hand, it is possible to precisely control the mechanical properties; however, they exhibit poor cell adhesion [18]. Bioceramics are known to enhance and promote biomineralization [14, 16], but their brittleness and low fracture toughness means that they are mostly suitable only in combination with other materials and in form of composites.

Metallic scaffolds are promising alternatives for hard load-bearing tissue repairs. These biomaterials in their solid form have been widely used for fabrication of the implants replacing hard human tissues for many years [19], and therefore in their porous form they can be possible candidates for TE approaches. Titanium and its alloys are of great interest in biomedical applications due to their excellent combinations of mechanical properties, biocompatibility, and chemical stability [20] and one of their drawbacks, which is the mismatch of mechanical properties between bulk titanium and natural bone which leads to stress shielding, and eventual implant loosening, that can be rectified by producing a low modulus porous network [21, 22]. In fact, Ti meshes have been successfully used in spine fusion surgeries

and for oral and maxillofacial structures [23]. Introduction of porosity and pore interconnectivity improves mechanical fixation and osteointegration by allowing extensive body fluid transport through the porous implant. This can provoke bone tissue ingrowth, consequently leading to the development of a stable interface between the scaffold and host tissue [19].

The porous Ti matrices can be produced using techniques such as powder metallurgy (PM) [19, 22] or additive manufacturing technique (AM) [20, 24]. The drawbacks of the PM including limited control over the size, shape, and distribution of the porosity [25] can be resolved using AM techniques. Selective laser melting (SLM) and Selective Laser Sintering (SLS) are two of the AM processes that are able to produce complex structures layer by layer with high precision. Where SLS uses a very precise nanolaser beam to sinter the powder material to build up the structure [24], SLM fully melts the powder to form a solid mass. Direct metal laser sintering (DMLS) is essentially similar to SLS in terms of method, as it involves sintering rather than melting, but where SLS is used for polymers, ceramics as well as metals, DMLS is used exclusively for metals. With DMLS it is possible to control the porosity of each layer, pore interconnectivity, size, shape, and distribution, and consequently the 3D architecture of the implant, by changing the processing parameters, such as laser power, laser spot diameter, and layer thickness, or by modifying the size of the original titanium particles [26, 27].

Titanium is considered a bioinert material, which does not possess osteoconductivity or osteoinductivity by itself; however, the surface can be modified to induce osteoconductivity [28] or osteoinductivity [29]; hydroxyapatite (HAp) is a very good biomaterial which has excellent osteoinductivity and has been widely used in bone defect repairs. HAp coatings can be used to prompt osteogenesis without the need for additional osteogenic cells or bone morphogenic proteins (BMP) [28]. Conventional HAp coating of solid titanium surfaces involves plasma spraying which is a line of sight technique and fails to coat inner surfaces of porous structures. HAp coatings can be produced using techniques such as electrochemical deposition and biomimetic method. Both methods are based on precipitation from aqueous solutions, take place at low temperature, are economical, and allow coating of complex shapes. The biomimetic method uses simulated body fluids (SBF) that mimic physiological ionic strength and pH. In a typical electrochemical deposition, a precursor (brushite) is first formed that is converted into hydroxyapatite (HA) through an ageing process. Thus this method offers a control over crystallinity [30]. However, debonding and loss of HAp coatings in vivo are one of the main concerns in using these coatings [31, 32].

In this study, we have produced a porous titanium scaffold for bone tissue engineering using SLS technique. We have used both electrochemical and biomimetic processes to coat the three-dimensional Ti matrix with hydroxyapatite and compared the osteoinductivity of coated and noncoated scaffolds in vitro. Electrochemical deposition leads to a more uniform coating and was used as the main coating method for any further analysis. Our results showed that interestingly HAp coating did not significantly increase osteogenecity (Alizarin Red production) in vitro and noncoated Ti scaffolds

TABLE 1: Composition of SBF coating solutions.

	NaCl	NaHCO$_3$	K$_2$HPO$_4$·3H$_2$O	MgCL$_2$·6H$_2$O	CaCl$_2$·2H$_2$O
Solution A	680 mM	21 mM	5 mM	7.6 mM	9.9 mM
Solution B	680 mM	10 mM	5 mM	1.5 mM	9.9 mM

TABLE 2: Scaffold formulations in terms of coating technique and conditions.

Coating method	Condition		Acronym
	Solution A (hrs)	Solution B (hrs)	
Biomimetic method	24	48	BM24
	48	48	BM48
Electrochemical method	Current density (mA/cm^2)		
	6	10	EM6 and EM10

were also osteoinductive. These scaffolds were then implanted in sheep femoral condyle to investigate bone formation and ingrowth. Extensive osteoinduction and osteointegration (70% bone ingrowth) were observed in vivo, confirming the intrinsic capacity of the produced porous Ti scaffolds for bone regeneration.

2. Materials and Methods

2.1. Fabrication of Ti Matrix. Titanium lattices were fabricated from commercially pure titanium powder (cp-Ti) using a DMLS system (EO SINTM270). Cp-Ti and titanium alloys (typically Ti6Al4V) are used extensively as dental and orthopaedic implants, respectively. Alloying improves the mechanical properties of titanium for use in high load-bearing applications; however, some concerns related to the toxicity of various alloying elements do exist [33]. Since no significant differences were observed in terms of osseointegration, biomechanical anchorage, and bacterial interaction between cp-Ti and Ti alloys [33], we selected cp-Ti (grade 4, 99% purity, density 4.51 g/cm^3, and Young's modulus of 105 GPa) in this study to avoid this concern.

A 200 W Yb fiber laser was used to sinter Ti powder. The scaffold was built at a speed of 4 mm^2/s with layer thickness of 40 μm, resulting in a cylindrical scaffold (10 mm × 8 mm) with strut thickness of 1.5 mm, pitch size of 0.75 mm, and porosity of 72%.

Prior to coating, the scaffolds were ultrasonically cleaned in 10% Decon®90 (Decon Laboratory Limited, UK), distilled water, and ethanol for 15 mins each, and dried in air.

2.2. HAp Coating Deposition and Characterisation. Two methods for HAp coating of porous Ti matrix were employed: biomimetic coating process and electrochemical deposition. In the biomimetic coating procedure, saturated simulated body fluid (5x SBF) was used according to a previously published method [30]. Briefly, coating solutions A and B were prepared (Table 1) by dissolving reagent grade salts (Sigma-Aldrich, UK) at 37°C with a constant 5% CO$_2$ supply and stirring in distilled water. Samples were firstly soaked in solution A, for 24 or 48 hrs at 37°C and then in solution B for 48 hrs at 40°C with constant stirring.

In the electrochemical method, titanium lattices were immersed in the CaP solution (0.13 M, Ca(H$_2$PO$_4$)$_2$, Sigma-Aldrich, UK) and attached to the negative terminal of a DC Dual Power Supply pack (Peak Tech, Telonic Instruments Ltd, UK). Two different electrical current densities of 10 and 6 mA/cm^2 were applied between the two electrodes for 10 mins. The samples were then soaked in 0.1 M NaOH solution for 72 h, cleaned in distilled water, and air-dried. Scaffold formulations are summarized in Table 2.

Morphology of deposited HAp was observed by scanning electron microscopy (SEM, JEOL JSM 5500 LV, at 10 kV) and elemental analysis was performed by energy dispersive X-ray spectroscopy (EDAX, EDAX Inc., USA).

2.3. Evaluation of Mechanical Properties and In Vitro Mechanical Stability. The mechanical properties of the structures were determined in a universal material testing machine (Instron® 5565) under uniaxial compressive load. Ti scaffolds (n = 3) were placed between two hard metal compression inserts. The force and deformation were recorded during the strain-controlled compression phase with a constant strain of 1 mm/min at room temperature.

Mechanical stability of scaffolds in a defect site was examined using mechanical push-in and push-out tests (n = 3) in dry state in Sawbones©. Tapered cylindrical defects matching the dimensions of the scaffolds were made using appropriate drill bits in Sawbones© polyurethane foam (Sawbones Europe AB, Malmö, Sweden). We used Sawbones© foam with a density of 160 kg/m^3 and a compressive modulus of 66 MPa comparable to that of cancellous bone [34]. For push-in/push-out tests, two types of experiments were designed: blind-hole experiment to determine the push-in depth and strength, and a through-hole experiment to observe the push-out and interfacial strengths. Samples were inserted into the created defects in Sawbones©, and the load required to push them in/out (a ramp compressive extension of 1 mm/min) was monitored. The schematic illustration of each setup is depicted in Figure 1. Based on the maximum load achieved (from the load-displacement curves) and the area of scaffolds in contact with Sawbones© (lateral surface area of a truncated cone × 28% scaffold density) the interface strength between the scaffolds and their surroundings was calculated.

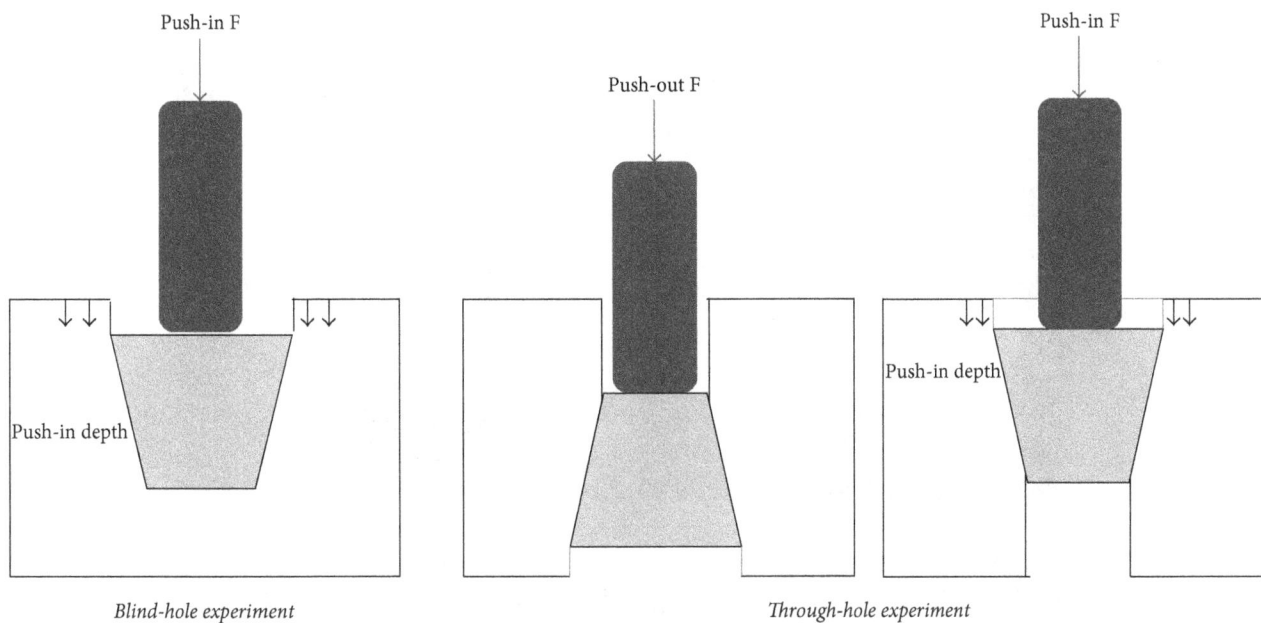

FIGURE 1: Schematics of blind and through holes experiments.

2.4. In Vitro Evaluation of Osteoinductivity. Viability and osteoinductivity of sheep bone marrow mesenchymal stem cells (BMMSCs) were evaluated on HAP coated and non-coated Ti scaffolds. BMMSCs were isolated from sheep bone marrow aspirate, expanded, and maintained in tissue culture flasks containing Dulbecco's modified eagles medium (DMEM, Sigma-Aldrich, UK) supplemented with 10% fetal calf serum (FCS, First Link, UK) and 100 Units/mL of Penicillin and Streptomycin (P/S, Gibco, UK). Flasks were kept at $37°C$ with 5% CO_2 and passaged when 80% of confluency was reached. BMMSCs were characterised by demonstrating their multipotency by differentiating them down the adipogenic, chondrogenic, and osteogenic lineages.

Three sample groups were tested: coated Ti scaffold (EM10), noncoated Ti scaffold, and Thermanox™ discs (Nalge Nunc International, USA) as control ($n = 3$) in osteogenic media.

Samples were sterilised by autoclaving and were then seeded with 10,000 BMMSCs (passage 3) in a total volume of 50 μL basal cell culture media (DMEM, 10% FCS, 1% P/S). After incubation for 1 hr at $37°C$ with 5% CO_2, 2-3 mL of osteogenic (basal media with 0.1 lM Dexamethasone, 500 lM Ascorbic Acid, and 10 mM b-glycerophosphate; all from Sigma-Aldrich, UK) cell culture media was added to each well. Media were changed every 3–5 days. Cell adhesion and morphology (by DAPI and phalloidin stainings at days 7 and 21), proliferation (by AlamarBlue© activity at days 1, 14, and 28), and differentiation into the osteogenic lineage (Alizarin Red staining at day 28) and cell colonisation and morphology (SEM at day 21) were studied for all groups and controls.

To assess cell adhesion and morphology, scaffolds were fixed in 4% paraformaldehyde (PFA) for 30 min, washed twice with PBS, and permeabilised with 0.25% (v/v) Triton X-100 in PBS for 30 min. Samples were blocked with 3% (w/v) bovine serum albumin (BSA) in PBS for 30 min and the actin cytoskeleton was stained with Alexa Fluor 568 phalloidin (Invitrogen; 1 : 200) for 1 h. Nuclei were stained with DAPI (4,6-Diamidino-2-phenylindole, dihydrochloride, Sigma; Fluoroshield) for 1 min. Samples were mounted on glass slides and cells were observed under a ZEISS ApoTome.2 Fluorescent Microscope (ZEISS, Germany).

AlamarBlue© was used to examine the proliferation of cells on the samples. AlamarBlue© (AbD Serotec, UK) was diluted in phenol free DMEM (Sigma, UK) to make a 10% working solution. Samples were washed with PBS and incubated with 1 mL of the working solution at $37°C$ and 5% CO_2. After 4 h, 200 μL from each sample was loaded in triplicate into a FluoroNunc™ white 96-well plate and fluorescence was measured at 530–560 nm excitation and 590 nm emission using a microplate reader (Infinite® 200 PRO, Tecan, Switzerland). Results were compared to those of an empty well loaded with 1 mL of the working solution at the beginning of the assay.

To assess late stage of osteogenesis, Alizarin Red staining was performed quantitatively. At day 28, samples were fixed with 4% (w/v) PFA for 30 mins and then washed with PBS. Samples were then incubated at room temperature with Alizarin Red solution for 30 mins, after which they were washed with PBS and incubated with 200 μl 10% CPC in 10 mM sodium phosphate (pH 7) for 15 mins. Duplicates of 100 μl of supernatant were transferred to a Nunc® 96-well plate, and the absorbance was measured with a microplate reader (Infinite 200 PRO, Tecan, Switzerland) at 570 nm.

2.5. In Vivo Evaluations of Mechanical and Biological Fixation. The bone tissue reaction to the porous Ti scaffold was examined by animal tests using sheep condyle. Three sheep (77–82 kg) were sedated by intravenous administration of ketamine and midazolam and sedation was maintained using gaseous anaesthesia with 2.5% isoflurane. The sheep were

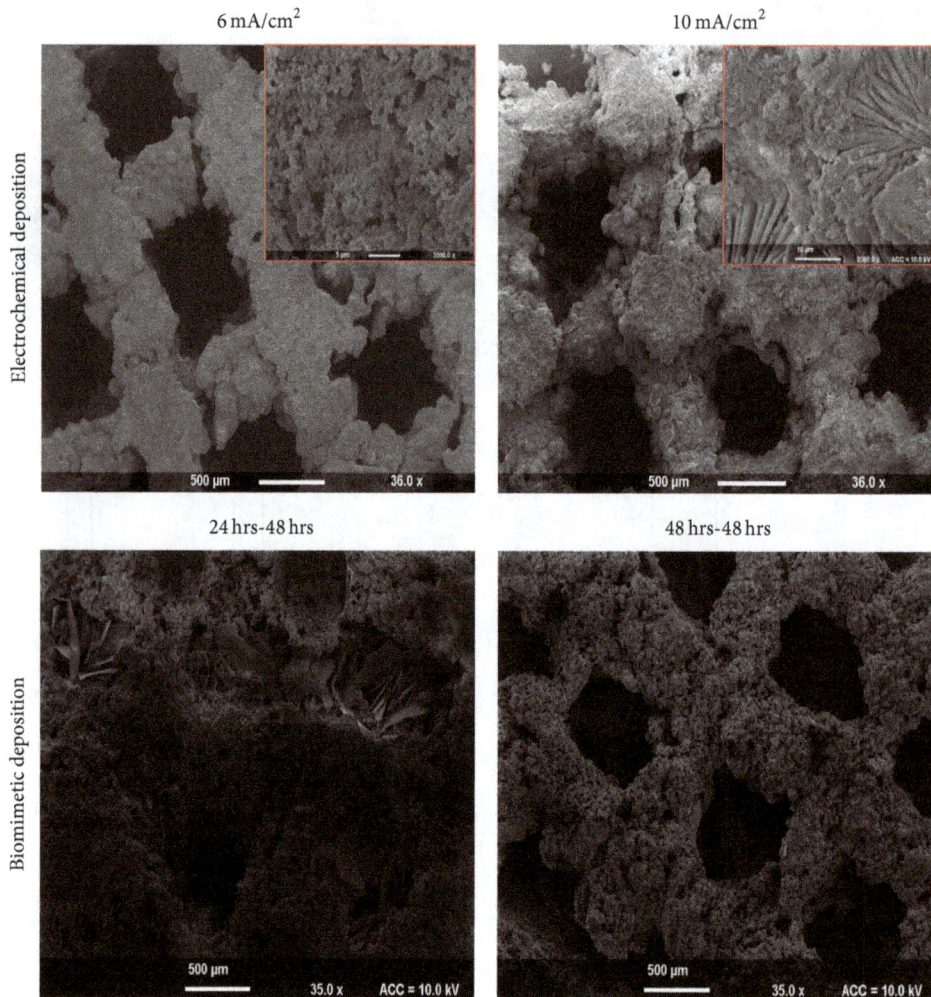

FIGURE 2: SEM micrographs of HAp coated Ti scaffolds using biomimetic and electrochemical methods.

given Ceporex injections (active ingredient cephalexin), an antibiotic, on days 0, 1, 2, and 3. Each sheep also had fentanyl patches on days 1 and 3.

A collagen type-I–HAp scaffold was fabricated to act as the control using a freeze-drying method. Briefly, lyophilised collagen powder (Sigma-Aldrich, UK) was dispersed in distilled water (pH 3.2 with acetic acid) using IKA blender while HAp powder (Sigma-Aldrich, UK) was added to the mix. The suspension was then casted into 3D printed resin moulds, frozen, and freeze-dried. The noncoated Ti scaffolds, as well as collagen-HAp scaffolds, were implanted into 10 mm (depth and upper diameter) bone defects in the left medial condyle of sheep stifle joints and fixed by press-fit only. The limbs were scanned radiologically and implanted titanium scaffold and surrounding tissue were retrieved 12 weeks after operation. Micro-computed tomography (micro-CT) analysis was performed on samples using a Nikon XT H 225 with 110 kVP X-ray source and 112 mA (resolution of ~18–22 μm) in order to assess the new bone formed within the titanium matrix. Three-dimensional reconstructions were performed using Nikon CT Agent. Subsequent visualization

and analysis were performed in Bruker Software CTVOX and CTAN. Subchondral bone repair was expressed as percentage bone volume over the total volume (% BV/TV), while SEM (JEOL JSM 5500 LV, at 10 kV) observations were used to assess bone-scaffold interface.

2.6. Statistical Analysis. Statistical analyses were performed using OriginPro 2015 (OriginLab). The data are presented as means with standard deviation. One-way ANOVA and subsequent post hoc Tukey tests were used to analyze differences among the groups at significant level of 0.05.

3. Results and Discussion

3.1. Characterisation of HAp Coating. Porous Ti matrices were coated with HAp using both a biomimetic and an electrochemical method in order to select the most suitable coating technique in terms of homogeneity. The morphology of the HAp coatings, examined by SEM, is shown in Figure 2 and the compositions of the coatings, determined by EDX analysis, are listed in Table 3.

TABLE 3: Composition (Ca/P ratio) of different coatings obtained by EDX; * is used for further cell analyses.

Sample	Ca/P ratio
EM6	1.84 ± 0.54
EM10*	1.85 ± 0.25
BM24	1.79 ± 0.21
BM48	1.85 ± 0.08

TABLE 4: Mechanical properties of porous Ti scaffold and its mechanical fixation in Sawbones. Data are expressed as mean ± standard deviation; * shows significant difference at $p = 0.004$.

Mechanical characteristics	
Compressive strength	35 MPa
Compressive modulus	73 MPa
Young's modulus	0.55 MPa
Mechanical stability in vitro	
Push-in depth	5.8 mm
Push-in strength	5.67 MPa (±1.54)*
Push-out strength	1.44 MPa (±1.14)*
Interfacial shear strength	0.81 MPa (±0.36)

From the EDX analysis, the molar ratio of calcium to phosphorous (Ca/P) of HAp was 1.79–1.85 for both coating methods showing a slightly calcium rich HAp compared to stoichiometric HAp (Ca/P = 1.67). While the Ca/P ratio was similar in all groups, it became apparent from SEM images that the coating was more uniform in electrochemical deposition method, whereas the biomimetic procedure usually led to nonuniform coating and blocking of the matrix pores.

Based on the uniformity criteria of the coatings, electrochemical method with 10 mA/cm² setting (sample EM10) was chosen for tests on cell adhesion proliferation and mineralisation.

3.2. Mechanical Properties and In Vitro Fixation.
The mechanical properties of Ti scaffold were determined in compression mode and are summarized in Table 4. It was observed that the Ti scaffolds exhibited a compressive strength of 35 MPa and Young's modulus of 0.55 GPa, which is within the range of trabecular bone (e.g., 1–100 MPa for compressive strength) [35, 36], and is comparable to values (24 MPa) achieved in a 75% porosity PM processed Ti matrix [19]. In terms of bone, mid-range values for the modulus of trabecular bone are 90–400 MPa. The compressive modulus of our Ti scaffold reached 73 MPa, which is slightly lower than the lower range of native bone modulus. However, it must be noted that the values of native bone vary considerably across different locations and patients. An example is the compression modulus of human cancellous bone obtained by Martens et al. (1983), where superior-anterior femoral head showed a modulus of 900 ± 714 MPa, while the anterior-posterior showed a modulus of only 12 ± 6 MPa and medial-lateral a modulus of 63 ± 7 MPa [37].

To evaluate the mechanical stability of scaffolds when placed in a defect, the push-in, push-out, and interfacial strengths as well as the push-in depth were assessed, and these are reported in Table 4. Mechanical stability can be influenced by interlocking of the newly formed bone, the macroscopic design of the implant, and its stiffness and interface stress, as well as the interface friction and the bone-implant gap size [38]. Because there is no contribution from ingrown bone, the stability of scaffold in vitro solely relies on the macroscopic design of scaffold, friction coefficient, and gap size between bone and scaffold.

In fact, friction influences the mechanical stability of an implant and its relative migration in the bone and affect seating of the implant in the bone [39]. The friction coefficient values obtained in Sawbones® in dry state are usually lower than those in the lubricated human bone resembling the actual implant condition.

We observed that the achieved push-in and push-out strengths were in fact greatly affected by the degree to which the scaffolds were fitted in the defects. Even slightest mismatch between geometries could lead to a significant decrease in the interface strength.

The push-in strength was determined to be significantly higher than the push-out strength ($p = 0.004$). It is believed that the tapered design of the scaffold contributed the higher values of push-in strength compared to push-out strength.

In a study on bone-implant interface strength and osseointegration in rats, the ex vivo interface shear strength of a Ti alloy implant after 4 weeks was reported to be in the range of 0.57–1.50 MPa [40]. The ultimate interface shear strength of porous Ti scaffold in our study is 0.81 MPa, demonstrating that the scaffold mechanical fixation in vitro is comparable to that of a Ti alloy implant in vivo and would fulfil one of the key requirements, that is, mechanical stability to be retained in the affected area, for bone scaffolds.

3.3. Evaluation of Osteoinductivity In Vitro.
All scaffolds (HAp coated and noncoated) allowed for cell attachment and viability throughout 28-day culture period. Presence of metabolically active cells and their proliferation, as determined by AlamarBlue® analysis, indicated an increase in the number of cells attached on the scaffolds during the culture period (Figure 3). AlamarBlue activity was higher on day 1 on noncoated scaffolds compared to HAp coated scaffolds, showing higher cell attachment to noncoated samples. Metabolic activity peaked at day 14, again significantly higher on noncoated scaffolds, which may indicate higher cell division on these samples. The decrease in metabolic activity after day 14 may demonstrate stem cell differentiation towards an osteogenic lineage, which is confirmed by Alizarin Red quantification results on day 28.

Immunofluorescence staining for actin cytoskeleton showed a well-developed cytoskeleton suggesting cells could adopt a flattened and spread morphology on noncoated scaffolds, while DAPI staining indicated the presence of cells as evidenced by staining of their nuclei covering the entire scaffold surface (Figure 4). Scanning electron microscopy further confirmed the attachment and proliferation of the seeded cells on both coated and noncoated scaffolds. Spindle-like cells, with cell –cell contact were observed attached to the surface of the HAp coated and noncoated titanium scaffolds

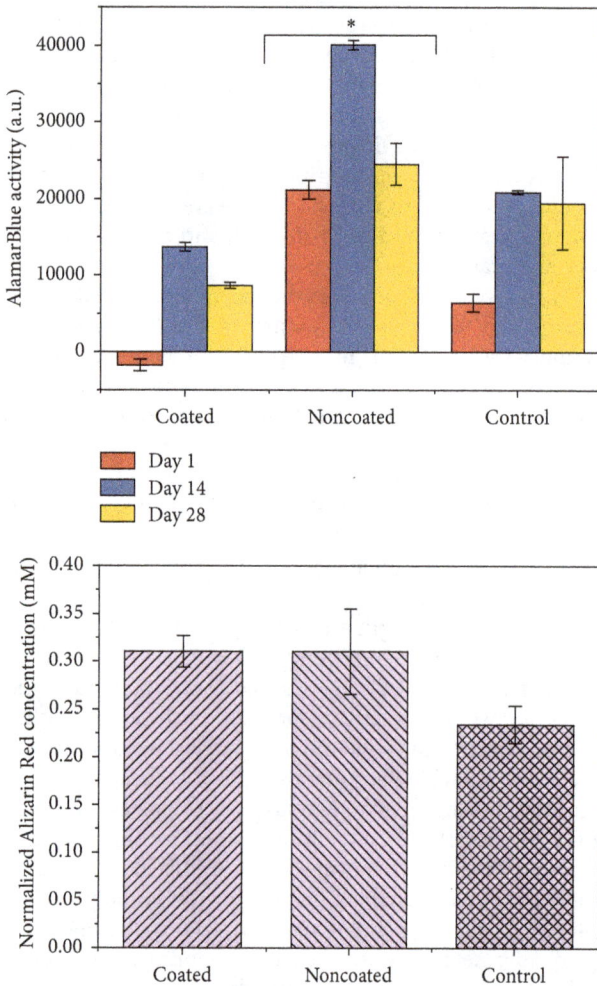

FIGURE 3: Biocompatibility and osteoinductivity of porous Ti scaffolds. AlamarBlue absorbance of BMMSCs seeded on HAp coated noncoated scaffolds over 28 days; Alizarin Red absorbance normalized to noncoated samples at day 28. * shows significantly higher value compared to other samples ($p < 0.05$).

(Figure 5). Osteoblastic like cells and mineral nodules were observed on and within the core of the scaffolds by day 21 of the culture.

Alizarin Red staining was used to determine late-stage osteoblastic differentiation. Alizarin Red quantification over 21 days of culture period confirmed osteoblastic differentiation of the seeded cells on both coated and HAp noncoated scaffolds (Figure 3). Noncoated samples indicated higher Alizarin Red absorbance; however this may be associated with the poor release of the stain from the HA coating during the quantification process. By normalizing the concentrations to that of noncoated samples both HAp coated and noncoated scaffolds indicated similar levels of differentiation (Figure 3).

3.4. Bone Formation and Ingrowth. For the clinical determination of the bone ingrowth inside the scaffold recent advances in the μCT imaging have shown sufficient resolution for the accurate identification of the bone ingrowth within the metallic porous structure [23]. A 3D volume

of bone-scaffold was reconstructed from μCT images and sagittal, coronal, and transverse sections are presented in Figure 6, showing bone trabeculae inside the scaffold porous structure. We observed areas of incomplete bone regeneration below the scaffold; the void is speculated to be generated during the surgery with the drill, which has been unable to heal. The amount of bone ingrowth in scaffold and control groups was quantified using the principles for bone histomorphometry and is reported in Figure 6. We observed that the BV/TV was significantly higher ($p = 0.01$) in the scaffold group compared to the control group, which consisted of a collagen-HAp matrix. The bone-implant contact was calculated to be 70%, and the interface between the Ti struts and the regenerated bone, demonstrating a very close contact between the two, is shown by SEM images in Figure 7, which further confirms integration of Ti scaffold and the newly formed bone.

We speculate that the significant increase in bone regeneration in the defects treated with porous titanium scaffolds compared to collagen-HAp scaffold may be related to the scaffold structure and its mechanical properties, as it possesses an interconnected porous structure and mechanical properties in range of trabecular bone. The mechanism of osteoinduction in porous biomaterials and its biological effects are still largely unknown. However, it has been argued that biomaterials must meet very specific requirements in terms of macrostructure (e.g., geometry and porosity), microstructure (e.g., microporosity and surface roughness), and chemical composition (calcification ability) in order to be osteoinductive [29, 41]. For example, it has been previously shown that porous Ti containing no calcium phosphate can become osteoinductive when it has a complex interconnecting porous structure and bioactive surfaces activated by simple chemical and thermal treatments [28, 29]. However, in this study, we have shown that surface modification and/or HAp coating might not be required for osteoinduction of porous titanium. The developed 3D printed Ti scaffold possess a rough surface (microstructure), an interconnected structure of pores with over 70% porosity, and mechanical properties in range of trabecular bone (macro- and microstructure), but they do not contain any source of calcium phosphate. In materials containing calcium phosphate, the Ca^{2+}, PO_4^{3-}, and HPO_4^{2-} are liberated from the surface into the surrounding which may increase the local supersaturation of the biologic fluid, causing precipitation of carbonated apatite that incorporates calcium, phosphate, and other ions [42]. The dissolution part of this process is missing in the materials that initially do not contain calcium phosphate; however, their physicochemical properties are such that they provide nucleation sites for the deposition of a biological apatite layer [41]. It is plausible that similar events occurred in the developed Ti scaffold providing nucleation sites for calcification and bone formation.

4. Conclusions

We have developed a highly porous (72%) Ti scaffold for bone tissue engineering using additive manufacturing (selective laser sintering) technique. The mechanical properties,

Day 7 Day 21

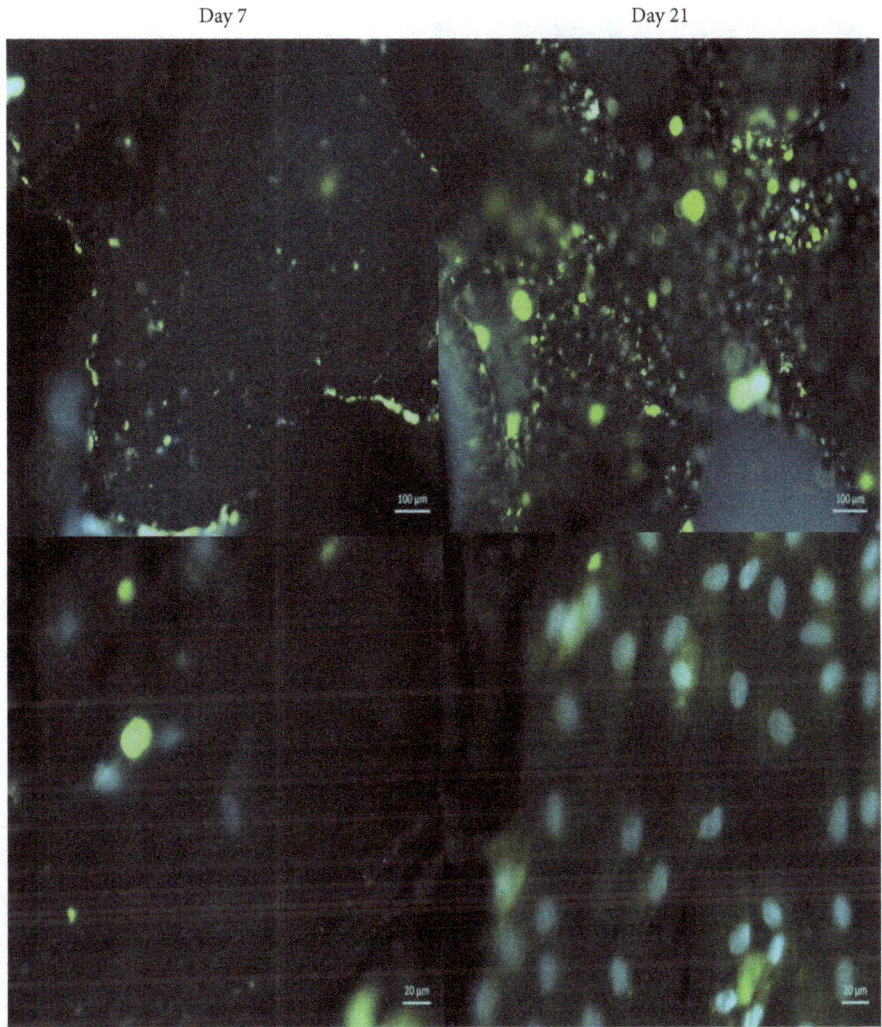

FIGURE 4: Cell-scaffold interaction; immunostaining of cells on porous noncoated Ti scaffold shows cell nuclei (DAPI, blue) and cytoskeleton (phalloidin, green) at days 7 and 21.

Noncoated Ti HAp coated Ti

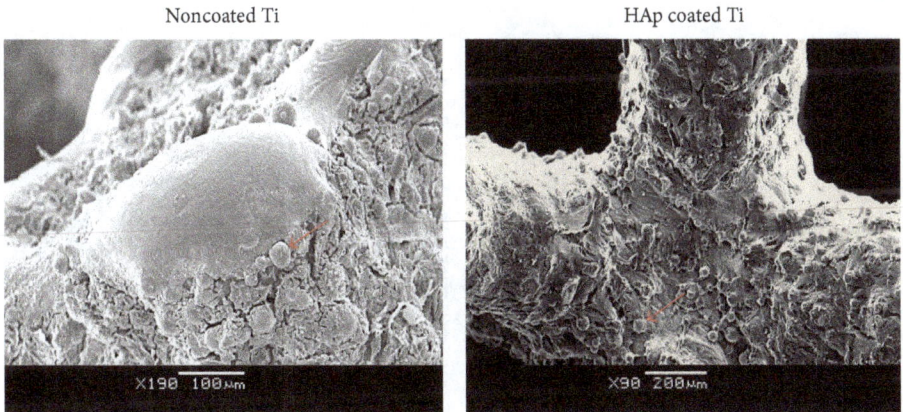

FIGURE 5: Cell morphology and osteogenesis on coated and noncoated Ti scaffolds; noncoated: flat cuboidal osteoblastic cells and mineral nodules (red arrow) on the scaffold peaks; coated: spindle-like cells and round mineral nodules (red arrow) attached to HAp coated scaffold surface.

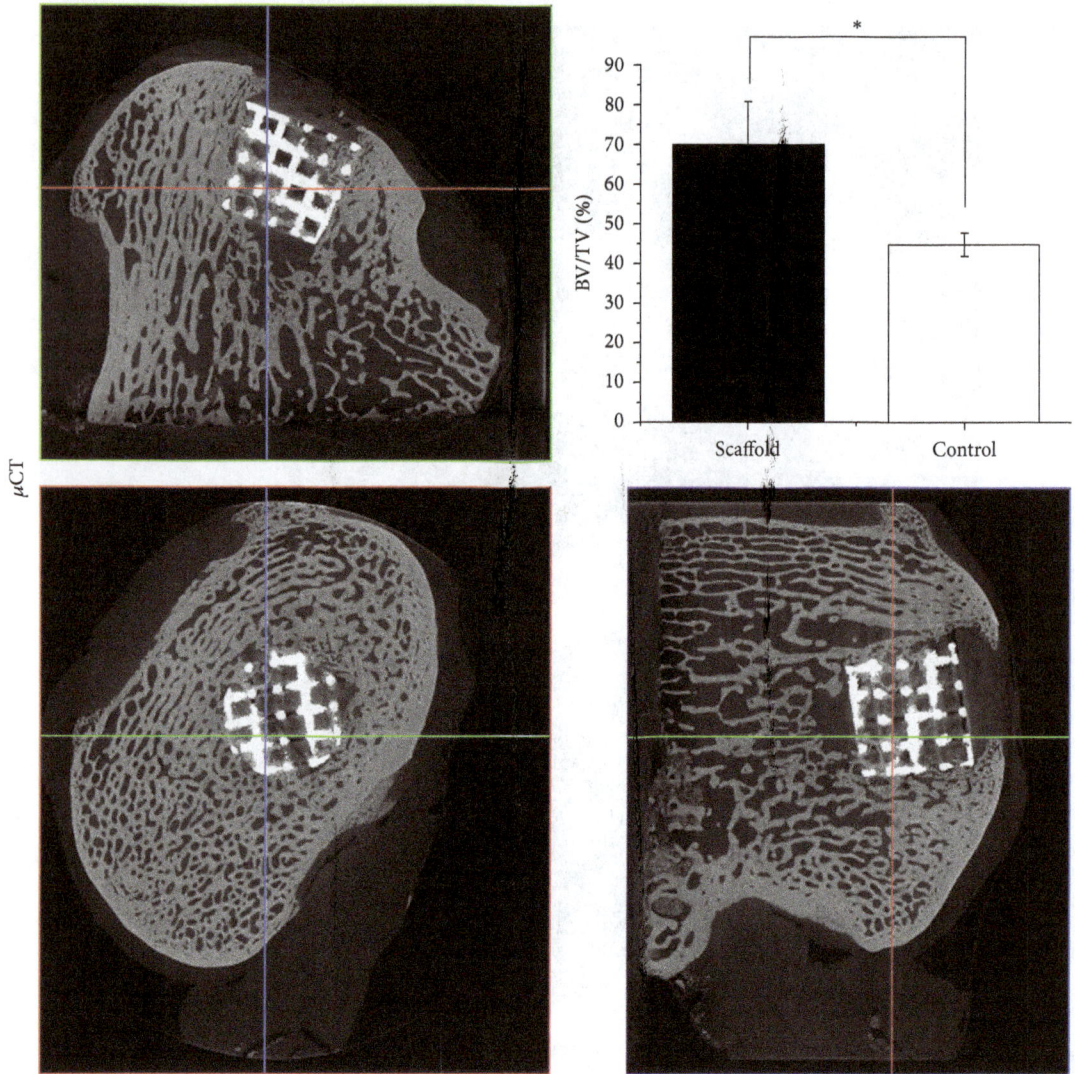

FIGURE 6: Bone ingrowth into Ti matrix. μCT images show bone formation within the Ti scaffold; * shows significantly higher bone volume formation in the scaffold compared to control ($p = 0.01$).

FIGURE 7: Bone-scaffold interface; SEM images show bone (B) in contact with the metal (M).

including compressive strength and stiffness of the produced scaffold, were in range of human trabecular bone, and they showed a stable mechanical fixation in vitro, comparable to fixations observed after 4 weeks of bone ingrowth. We showed that both HAp coated and noncoated scaffolds promote osteogenesis in vitro and HAp coating did not produce a significant increase in late osteogenesis. Bone formation and ingrowth in sheep stifle joints confirmed that the Ti scaffolds—as produced and without any coating—exhibited an intrinsic capacity for bone formation and osteoinduction. Therefore, these scaffolds have the potential to be used for tissue engineering of large bone defects and nonunions.

Conflicts of Interest

The authors declare that there are no conflicts of interest regarding the publication of this article.

Acknowledgments

This work was financially supported by Arthritis Research UK (Grant no. 21160) and Rosetrees Trust (Project no. A1184). The authors thank Dr. Xuekun Lu (Chemical Engineering Department, UCL) for help with micro-CT measurements.

References

[1] J. R. Porter, T. T. Ruckh, and K. C. Popat, "Bone tissue engineering: a review in bone biomimetics and drug delivery strategies," *Biotechnology Progress*, vol. 25, no. 6, pp. 1539–1560, 2009.

[2] P. Lichte, H. C. Pape, T. Pufe, P. Kobbe, and H. Fischer, "Scaffolds for bone healing: concepts, materials and evidence," *Injury*, vol. 42, no. 6, pp. 569–573, 2011.

[3] V. Karageorgiou and D. Kaplan, "Porosity of 3D biomaterial scaffolds and osteogenesis," *Biomaterials*, vol. 26, no. 27, pp. 5474–5491, 2005.

[4] P. Leng, Y.-Z. Wang, and H.-N. Zhang, "Repair of large osteochondral defects with mix-mosaicplasty in a goat model," *Orthopedics*, vol. 36, no. 3, pp. e331–e336, 2013.

[5] V. I. Sikavitsas, J. S. Temenoff, and A. G. Mikos, "Biomaterials and bone mechanotransduction," *Biomaterials*, vol. 22, no. 19, pp. 2581–2593, 2001.

[6] H. Petite, K. Vandamme, L. Monfoulet, and D. Logeart-Avramoglou, "Strategies for improving the efficacy of bioengineered bone constructs: A perspective," *Osteoporosis International*, vol. 22, no. 6, pp. 2017–2021, 2011.

[7] A. K. Lynn, R. A. Brooks, W. Bonfield, and N. Rushton, "Repair of defects in articular joints. Prospects for material-based solutions in tissue engineering," *Journal of Bone and Joint Surgery - Series B*, vol. 86, no. 8, pp. 1093–1099, 2004.

[8] T. Aigner and J. Stöve, "Collagens - Major component of the physiological cartilage matrix, major target of cartilage degeneration, major tool in cartilage repair," *Advanced Drug Delivery Reviews*, vol. 55, no. 12, pp. 1569–1593, 2003.

[9] E. B. Hunziker, "Articular cartilage repair: basic science and clinical progress. A review of the current status and prospects," *Osteoarthritis and Cartilage*, vol. 10, no. 6, pp. 432–463, 2002.

[10] S. Martino, F. D'Angelo, I. Armentano, J. M. Kenny, and A. Orlacchio, "Stem cell-biomaterial interactions for regenerative medicine," *Biotechnology Advances*, vol. 30, no. 1, pp. 338–351, 2012.

[11] T. L. Livingston, S. Gordon, and M. Archambault, *Mesenchymal stem cells combined with biphasic calcium phosphate ceramics promote bone regeneration*, Chapman Hall, UK, 2003.

[12] T. Albrektsson and C. Johansson, "Osteoinduction, osteoconduction and osseointegration," *European Spine Journal*, vol. 10, no. 2, pp. S96–S101, 2001.

[13] J. S. Temenoff and A. G. Mikos, "Review: Tissue engineering for regeneration of articular cartilage," *Biomaterials*, vol. 21, no. 5, pp. 431–440, 2000.

[14] D. Hutmacher, "Scaffolds in tissue engineering bone and cartilage," *Biomaterials*, vol. 21, no. 24, pp. 2529–2543, 2000.

[15] E. S. Place, N. D. Evans, and M. M. Stevens, "Complexity in biomaterials for tissue engineering," *Nature Materials*, vol. 8, no. 6, pp. 457–470, 2009.

[16] A. Yousefi, M. E. Hoque, R. G. Prasad, and N. Uth, "Current strategies in multiphasic scaffold design for osteochondral tissue engineering: a review," *Journal of Biomedical Materials Research Part A*, vol. 103, no. 7, pp. 2460–2481, 2014.

[17] P. B. Malafaya, G. A. Silva, and R. L. Reis, "Natural-origin polymers as carriers and scaffolds for biomolecules and cell delivery in tissue engineering applications," *Advanced Drug Delivery Reviews*, vol. 59, no. 4-5, pp. 207–233, 2007.

[18] M. Chen, D. Q. S. Le, A. Baatrup et al., "Self-assembled composite matrix in a hierarchical 3-D scaffold for bone tissue engineering," *Acta Biomaterialia*, vol. 7, no. 5, pp. 2244–2255, 2011.

[19] B. Dabrowski, W. Swieszkowski, D. Godlinski, and K. J. Kurzydlowski, "Highly porous titanium scaffolds for orthopaedic applications," *Journal of Biomedical Materials Research Part B Applied Biomaterials*, vol. 95, no. 1, pp. 53–61, 2010.

[20] Y. Wang, Y. Shen, Z. Wang, J. Yang, N. Liu, and W. Huang, "Development of highly porous titanium scaffolds by selective laser melting," *Materials Letters*, vol. 64, no. 6, pp. 674–676, 2010.

[21] M. A. Lopez-Heredia, J. Sohier, C. Gaillard, S. Quillard, M. Dorget, and P. Layrolle, "Rapid prototyped porous titanium coated with calcium phosphate as a scaffold for bone tissue engineering," *Biomaterials*, vol. 29, no. 17, pp. 2608–2615, 2008.

[22] G. E. Ryan, A. S. Pandit, and D. P. Apatsidis, "Porous titanium scaffolds fabricated using a rapid prototyping and powder metallurgy technique," *Biomaterials*, vol. 29, no. 27, pp. 3625–3635, 2008.

[23] K. Alvarez and H. Nakajima, "Metallic scaffolds for bone regeneration," *Materials*, vol. 2, no. 3, pp. 790–832, 2009.

[24] P. H. Warnke, T. Douglas, P. Wollny et al., "Rapid prototyping: Porous titanium alloy scaffolds produced by selective laser melting for bone tissue engineering," *Tissue Engineering - Part C: Methods*, vol. 15, no. 2, pp. 115–124, 2009.

[25] W. Xue, B. V. Krishna, A. Bandyopadhyay, and S. Bose, "Processing and biocompatibility evaluation of laser processed porous titanium," *Acta Biomaterialia*, vol. 3, no. 6, pp. 1007–1018, 2007.

[26] L. Mullen, R. C. Stamp, W. K. Brooks, E. Jones, and C. J. Sutcliffe, "Selective laser melting: a regular unit cell approach for the manufacture of porous, titanium, bone in-growth constructs, suitable for orthopedic applications," *Journal of Biomedical Materials Research Part B Applied Biomaterials*, vol. 89, no. 2, pp. 325–334, 2009.

[27] F. Mangano, L. Chambrone, R. van Noort, C. Miller, P. Hatton, and C. Mangano, "Direct metal laser sintering titanium dental

implants: a review of the current literature," *International Journal of Biomaterials*, vol. 2014, Article ID 461534, 11 pages, 2014.

[28] S. Fujibayashi, M. Neo, H.-M. Kim, T. Kokubo, and T. Nakamura, "Osteoinduction of porous bioactive titanium metal," *Biomaterials*, vol. 25, no. 3, pp. 443–450, 2004.

[29] A. Fukuda, M. Takemoto, and T. Saito, "Osteoinduction of porous Ti implants with a channel structure fabricated by selective laser melting," *Acta Biomaterialia*, vol. 7, no. 5, pp. 2327–2336, 2011.

[30] E. García-Gareta, J. Hua, J. C. Knowles, and G. W. Blunn, "Comparison of mesenchymal stem cell proliferation and differentiation between biomimetic and electrochemical coatings on different topographic surfaces," *Journal of Materials Science: Materials in Medicine*, vol. 24, no. 1, pp. 199–210, 2013.

[31] J. P. Collier, V. A. Surprenant, M. B. Mayor, M. Wrona, R. E. Jensen, and H. P. Surprenant, "Loss of hydroxyapatite coating on retrieved, total hip components," *The Journal of Arthroplasty*, vol. 8, no. 4, pp. 389–393, 1993.

[32] O. Reikerås and R. B. Gunderson, "Failure of HA coating on a gritblasted acetabular cup 155 patients followed for 7-10 years," *Acta Orthopaedica Scandinavica*, vol. 73, no. 1, pp. 104–108, 2002.

[33] F. A. Shah, M. Trobos, P. Thomsen, and A. Palmquist, "Commercially pure titanium (cp-Ti) versus titanium alloy (Ti6Al4V) materials as bone anchored implants - Is one truly better than the other?" *Materials Science and Engineering C*, vol. 62, pp. 960–966, 2016.

[34] L. A. Crawford, E. S. Powell, and I. A. Trail, "The fixation strength of scaphoid bone screws: An in vitro investigation using polyurethane foam," *Journal of Hand Surgery*, vol. 37, no. 2, pp. 255–260, 2012.

[35] W. Suchanek and M. Yoshimura, "Processing and properties of hydroxyapatite-based biomaterials for use as hard tissue replacement implants," *Journal of Materials Research*, vol. 13, pp. 94–117, 1998.

[36] L. J. Gibson, "The mechanical behaviour of cancellous bone," *Journal of Biomechanics*, vol. 18, no. 5, pp. 317–328, 1985.

[37] M. Martens, R. van Audekercke, P. Delport, P. de Meester, and J. C. Mulier, "The mechanical characteristics of cancellous bone at the upper femoral region," *Journal of Biomechanics*, vol. 16, no. 12, pp. 971–983, 1983.

[38] B. Johnston, "High-strength fully porous biomaterials for bone replacement and their application to a total hip replacement," in *Department of Mechanical Engineering*, McGill, Montreal, Canada, 2016.

[39] J. A. Grant, N. E. Bishop, N. Götzen, C. Sprecher, M. Honl, and M. M. Morlock, "Artificial composite bone as a model of human trabecular bone: The implant-bone interface," *Journal of Biomechanics*, vol. 40, no. 5, pp. 1158–1164, 2007.

[40] C. Castellani, R. A. Lindtner, P. Hausbrandt et al., "Bone-implant interface strength and osseointegration: biodegradable magnesium alloy versus standard titanium control," *Acta Biomaterialia*, vol. 7, no. 1, pp. 432–440, 2011.

[41] A. Barradas, H. Yuan, C. van Blitterswijk, and P. Habibovic, "Osteoinductive biomaterials: current knowledge of properties, experimental models and biological mechanisms," *European Cells and Materials*, vol. 21, pp. 407–429, 2011.

[42] P. Habibovic, *Properties and clinical relevance of osteoinductive biomaterials*, University of Twente, 2005.

9

Strain and Vibration in Mesenchymal Stem Cells

Brooke McClarren ⓘ **and Ronke Olabisi** ⓘ

Department of Biomedical Engineering, Rutgers University, 599 Taylor Rd, Piscataway, NJ 08854, USA

Correspondence should be addressed to Ronke Olabisi; ronke.olabisi@rutgers.edu

Academic Editor: Junling Guo

Mesenchymal stem cells (MSCs) are multipotent cells capable of differentiating into any mesenchymal tissue, including bone, cartilage, muscle, and fat. MSC differentiation can be influenced by a variety of stimuli, including environmental and mechanical stimulation, scaffold physical properties, or applied loads. Numerous studies have evaluated the effects of vibration or cyclic tensile strain on MSCs towards developing a mechanically based method of differentiation, but there is no consensus between studies and each investigation uses different culture conditions, which also influence MSC fate. Here we present an overview of the response of MSCs to vibration and cyclic tension, focusing on the effect of various culture conditions and strain or vibration parameters. Our review reveals that scaffold type (e.g., natural versus synthetic; 2D versus 3D) can influence cell response to vibration and strain to the same degree as loading parameters. Hence, in the efforts to use mechanical loading as a reliable method to differentiate cells, scaffold selection is as important as method of loading.

1. Introduction

In tissue engineering and regenerative medicine, mesenchymal stem cells (MSCs) are often preferable to fully differentiated cells, which are limited in supply and do not multiply as rapidly or to as great an extent [1]. MSCs can proliferate for numerous passages. The MSC response to tensile strain and vibration has been researched using various scaffolds and stimulation parameters. Typical MSC responses to various mechanical loading include differentiation into osteocytes and chondrocytes, often guided by the presence of growth factors and calcium. Cell responses have also been guided by the microenvironment, whether cells are in their native environment, a transplanted *in vivo* environment, or cultured using tissue culture plastic, 2D scaffolds, or 3D scaffolds. Even the choice of scaffold material has an impact, that is, whether the scaffold is derived from natural or synthetic material. Although it is well known that MSCs respond to mechanical loading, it is not known how to best load these cells to achieve the desired differentiation. Identifying which combination of scaffold and loading protocol are associated with which MSC fate may permit researchers to reliably control differentiation without using differentiation media. Bending, tension, mechanical compression, hydrostatic compression, fluid shear, and vibration are all experienced by MSCs *in vivo*, as such their effects on MSCs have been explored extensively *in vitro*. When examining the aforementioned loading conditions, tensile strain of tissues is perhaps the easiest to measure *in vivo*, while vibration is the easiest to apply *in vivo*. Thus, for tensile strain and vibration, it is possible to compare their effects when applied *in vivo* versus *in vitro*. Therefore, this review focuses on the effect of tensile strain and vibration on the fate of MSCs in a variety of culture environments.

2. Common Methods to Differentiate MSCs

When maintained *in vitro*, MSCs can be chemically and mechanically differentiated into a variety of tissues such as bone, cartilage, tendon, and ligament [2]. The biochemical factors that promote specific cell responses are well understood and thus enable researchers to successfully guide cell differentiation. Adding chemical factors to cell culture media such as ascorbate, dexamethasone (dex), and bone morphogenetic proteins (BMPs) promotes osteogenesis; serum-free medium and transforming growth factor-β_1 (TGF-β_1) promote chondrogenesis; growth and differentiation factor

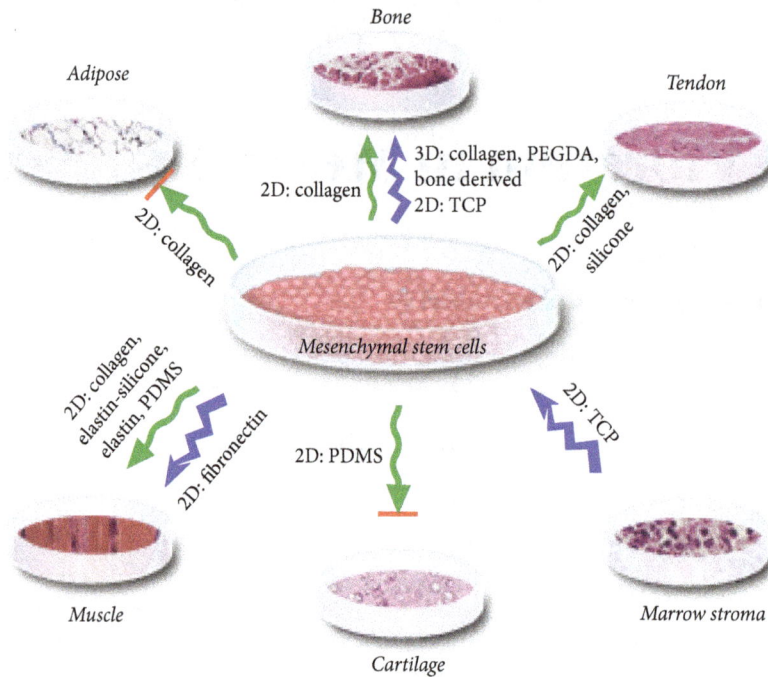

FIGURE 1: Diagram representing the effects of vibration (blue jagged arrow) and cyclic tensile strain (green squiggly arrow) on MSCs. The arrows depict the loading type. The italics detail the *in vitro* culture conditions in which the differentiation into the indicated lineage was observed. Red lines indicate inhibition of the downstream lineage. Tissue images from Tuan et al. [4], CC BY 4.0.

5 (GDF-5) promotes tenogenesis; platelet derived growth factor (PDGF) promotes myogenesis; and dex, insulin, and 3-isobutyl-1-methylxanthine (IBMX) promote adipogenesis (Figure 1) [3–5]. In addition to chemical factors, mechanical properties such as scaffold stiffness can be used to guide MSC differentiation [1, 6]. For instance, dex is widely used to promote MSC osteodifferentiation and alternately scaffolds with elastic moduli comparable to bone promote MSC osteodifferentiation [3, 6].

In their native environment tissues are subjected to a variety of biochemical signals in addition to a multitude of loading conditions that influence their development. In the absence of appropriate biochemical factors or scaffold mechanical properties, appropriate loading can drive MSC differentiation towards a desired fate [7]. Conversely, inappropriate loading can inhibit a desired fate [8]. MSC response to loading is dependent on stress/strain magnitude, duration, loading type, and force propagation through cytoskeletal configuration and attachment site geometry (for a recent review, see Delaine-Smith and Reilly) [3]. Loading types include such parameters as tension, compression, shear, bending, torsion, electromagnetic inputs, and vibration. Furthermore, loading can be separated into static or cyclic loading. All these loading types are experienced in situ by cells and often in combination. For example, in bone marrow, tension, compression, and fluid-induced shear may all be present but the effects of these forces on the stem cells within the bone marrow are not well understood [3, 9]. A challenge of tissue engineering is identifying the appropriate combination of chemical and mechanical parameters that will differentiate harvested MSCs into specific cell types *in vitro*. Although chemical inducers alone can drive differentiation *in vitro*, after scaffold implantation any biofactors it contains will eventually dissipate; thus the success of the scaffold will be maximized if its mechanical properties continue to influence cells.

3. The Effect of Vibration on Mesenchymal Stem Cells

Although vibration is not necessarily a loading condition experienced in nature, an extensive number of *in vivo* studies have been conducted with whole body vibration [33–39]. Whole body vibration studies have been used to model the cyclic tensile strain imparted on muscle or bone during physical actives such as walking, stair climbing, or weight lifting exercises [34]. The vibration stimulates the skeleton in a manner similar to walking or running and has been found to increase bone mass and bone strength [33–35, 40]. Whole body vibration stimulates osteogenesis of MSCs through mechanotransduction, resulting in a bone mass increase [35, 39]. Whole body vibration may also elicit a response from differentiated cells, which influence MSC differentiation [35]. Investigations exploring the effect of whole body vibration on MSCs may be confounded by the concurrent effect of vibration on differentiated cells. The mechanism of mechanotransduction during vibration in MSCs and subsequent response is not fully understood [36–38]. Additionally, the in situ response of cells to an external load cannot be separated from the systemic response of the whole organism.

Thus, researchers also explore *in vitro* parallels to whole body vibration to tease out the response of MSCs to vibration (summarized in Table 1). As such, the cell environment and

TABLE 1: Effect of vibration on MSCs.

Environment		Cell	Acceleration	Frequency [Hz]	Results	Ref.
	TCP	hMSC	0.3 g	30, 400, 800	Osteogenic	[10]
		mMSC	0.15 g	90	Osteogenic	[11]
		rMSC	0.3 g	60	Osteogenesis inhibited*	[12]
		hMSC	0.1–0.6 g	10, 20, 30, 40	Cell proliferation	[13]
2D	Gelatin	hMSC	0.02 g	15, 30, 45, 60	Inconclusive*	[14]
	Collagen	mMSC	10 μstrain	90	Adipogenesis inhibited	[15]
	Fibronectin	hMSC	Acoustic	200	Myogenic	[16]
3D	Collagen Sponge	hMSC	0.3 g	30	Osteogenic	[13]
	Bone derived	rMSC	0.3 g	40	Osteogenic	[17]
	PEGDA	hMSC	0.3, 3, 6 g	100	Osteogenic	[18]

∗ indicates the addition of differentiation media.

loading factors can be controlled. When subjecting MSCs to vibration, investigators specifically select loading parameters, biochemical additives, and cellular environment to observe or induce differentiation of cells.

3.1. Tissue Culture Plastic. Most in vitro vibration studies are performed with a cell monolayer cultured on tissue culture plastic (TCP) [10–13, 41]. The MSC response to vibration depends on frequency, acceleration, and duration of stimulation.

Chen et al. investigated the effects of 0.3 g acoustic vibration at 30, 400, and 800 Hz on human MSCs (hMSCs) in cell culture plates [10]. Cells were stimulated 30 min/day for 7 days. The investigators selected 30 Hz because they noted that osteogenesis was promoted following whole body vibration under 100 Hz [35] and they selected higher frequencies because higher frequencies are more suited for localized body vibrations [10]. The authors found that cell proliferation, calcium deposition, and collagen 1 (Col I) gene expression increased the most following 800 Hz vibrations at 0.3 g (Figure 2). At 800 Hz, adipogenic gene expression and lipid accumulation were decreased while at 30 Hz adipogenesis was promoted. Though acoustic vibration differs from direct mechanical vibration, both methods impart physical vibration to the cells.

Demiray and Özçivici cultured mouse MSCs on glass cover slides within 6 well plates [11]. Plates were stimulated with low magnitude (<1 g) and high frequency (20–90 Hz; LMHF) vibrations of 90 Hz at 0.15 g over 7 days for 15 min/day. The authors hypothesized that low intensity vibrations would induce MSC differentiation into osteogenic cells. Following vibration, cells were tested for osteogenic markers Runx2 and osteocalcin (OC) to identify osteogenic differentiation. While gene expression of vibrated and control cells was similar, the vibrated cells exhibited increased proliferation and morphological changes. Vibrated cells also displayed increased cellular height and increased molecular expression of focal adhesion kinase.

Lau et al. studied the effects of LMHF vibration on rat MSCs cultured on TCP while using osteogenic media [12]. Cells were stimulated with 60 Hz vibrations at 0.3 g for six 1-hour bouts. The authors hypothesized that the vibration

would promote osteogenesis based on prior animal and human studies [35, 42]. Following vibration, cells were tested for osteoblast-specific transcription factor Osterix (Osx) to detect osteoblastic differentiation. The MSCs displayed decreased Osx levels and inhibited mineralization, indicating that LMHF vibration did not enhance osteogenic differentiation but seemed to inhibit it. Further, LMHF vibration did not affect proliferation rate. As both the control and test groups contained osteogenic media, rather than a true investigation of the effects of LMHF, the study was more an investigation of the combined effect of LMHF and osteogenic media compared to osteogenic media alone.

Kim et al. tested hMSCs with a wide array of vibration frequencies and accelerations [13]. MSCs were seeded on TCP or a collagen sponge. The collagen sponge was prepared from a cross reaction of chondroitin-6-sulfate and type I collagen [43]. The cells were seeded within the pores of the sponge, creating a multidimensional scaffold. Cells were subjected to varying accelerations of vertical vibration for 10 min/day for 5 days using a custom platform on a shaker. Accelerations varied between 0.1, 0.2, 0.3, 0.4, 0.5, and 0.6 g and frequencies varied between 10, 20, 30, and 40 Hz. Vibration on TCP resulted in a minor increase of proliferation. At 0.2 g and 0.3 g accelerations, proliferation rates increased as frequency increased. For all frequencies, proliferation was significantly higher at 0.3 g compared to other accelerations. The highest proliferation was observed at 0.3 g for both 30 Hz and 40 Hz. Thus, their subsequent experiments were performed with 30 Hz vibrations delivering 0.3 g accelerations. In their differentiation assays, the authors found that the osteoblastic differentiation markers, alkaline phosphatase (ALP) and osteopontin (OPN), were upregulated in vibrated cells while the osteoblastic markers, OC and bone sialoprotein (BSP), were unaffected. Alizarin red staining was increased in MSC monolayers receiving vibration and osteogenic media compared to control, though staining was not increased for vibrated cells that did not receive osteogenic media. MSCs behaved differently on scaffolds compared to TCP. Specifically, while OPG, Col I, and VEGF expression showed significant increases in vibrated groups compared to nonvibrated groups, this effect was observed only in MSCs cultured on scaffolds. These differing results

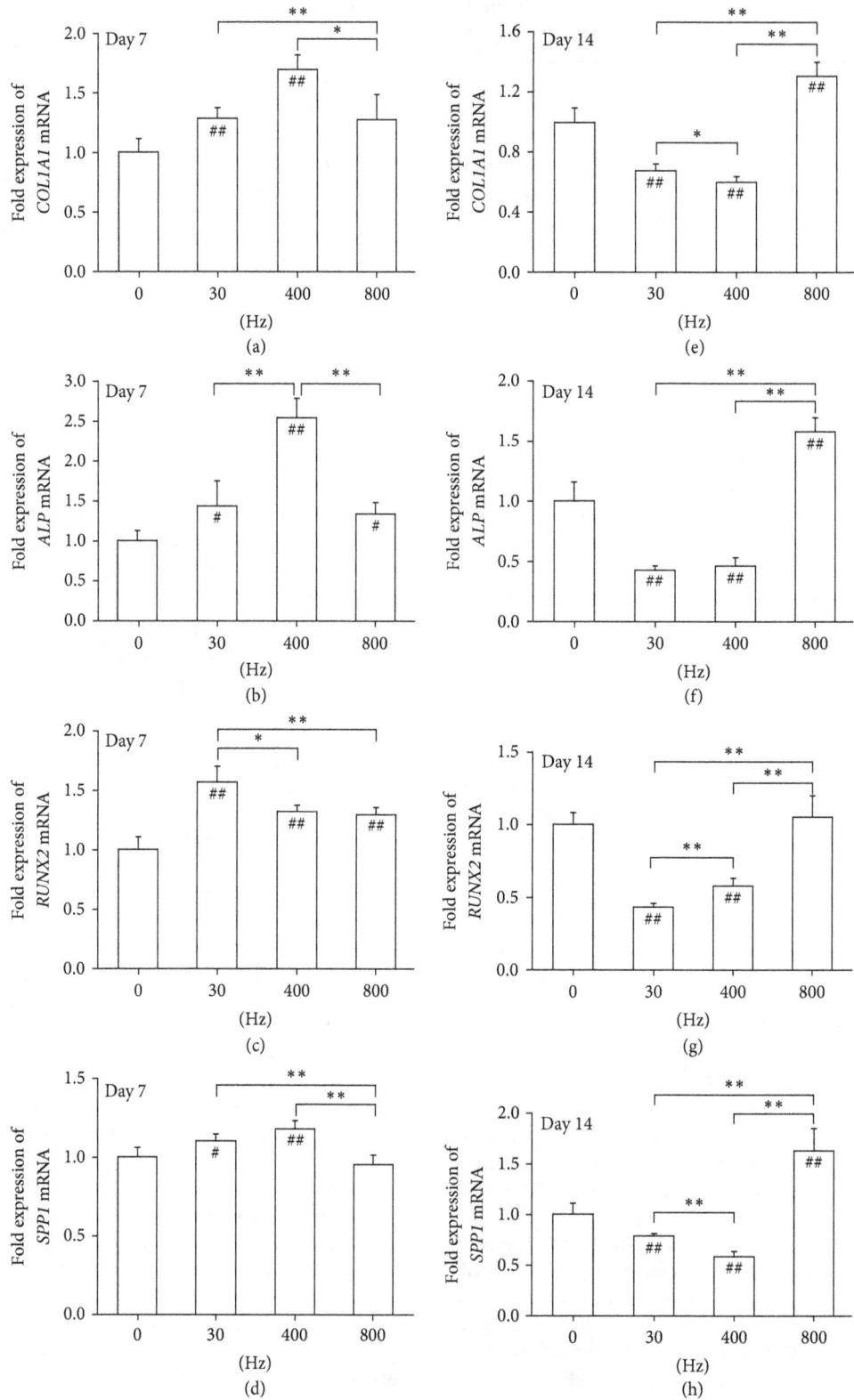

FIGURE 2: Acoustic-frequency vibratory stimulation (AFVS) modulates expression of mRNA encoding osteogenesis-specific markers in human bone marrow-derived mesenchymal stem cells (BM-MSCs) at the time points of day 7 ((a), (b), (c), and (d)) and day 14 ((e), (f), (g), and (h)). The mRNA levels of COL1A1 ((a), (e)), ALP ((b), (f)), RUNX2 ((c), (g)), and SPP1 ((d), (h)) were measured by real-time RT-PCR. Values are mean ± standard error of four independent experiments ($n = 4$). $^{*}P < 0.05$; $^{**}P < 0.01$ in the indicated groups from unpaired t-test. $^{#}P < 0.01$; $^{##}P < 0.01$ compared with the 0 Hz control group from unpaired t-test. From Chen et al. [10], CC BY 3.0.

suggest that additional factors, such as microarchitectural differences between scaffolds and TCP, may influence the mechanotransduction of vibration.

It is well known that cells in culture do not behave the same on TCP as they do in scaffolds or in their natural environments [44–48]. The same holds true for cells subjected to vibration [12, 13, 21, 43, 44]. Further, vibrated cells on TCP do not necessarily behave identically between similar investigations. In several LMHF studies, osteogenesis was increased significantly when MSCs on TCP were vibrated at accelerations of 0.3 g at 35 Hz or 45 Hz [49, 50]. Conversely, in other studies that vibrated MSCs on TCP with 0.3 g accelerations at 30 Hz or 60 Hz, osteogenesis was inhibited [13, 21]. Considering the variation in MSC response that TCP elicits, scaffolds may provide a more accurate and consistent in situ representation of the MSC response to vibration.

3.2. Two-Dimensional Scaffolds.
Scaffolds provide a more complex cellular interaction than a simple monolayer on TCP. A 2D scaffold has cells cultured on a flat surface while a 3D scaffold has cells embedded within or seeded on a multidimensional surface. Two-dimensional scaffolds for vibration studies are often membranes coated with osteogenic proteins or minerals [15, 16, 51]. Cell attachment and force transmission vary with different scaffolds or bound matrix proteins.

Edwards and Reilly seeded hMSCs in gelatin coated 12-well plates [14]. Plates were subjected to LMHF vibrations of 15, 30, 45, or 60 Hz at 0.02 g. Vibration occurred for 10 or 45 minutes. The authors also investigated the effect of osteogenic media (media containing dex) during their vibration studies. Alkaline phosphatase (ALP) activity was the only measurement of cell commitment. ALP activity was greatest following 45-minute vibrations of 60 Hz with osteogenic media. Forty-eight hours after stimulation, MSCs subjected only to vibration did not demonstrate a statistically relevant increase of ALP activity. ALP activity was greater in cells that had dex+ media compared to dex− media. For dex+ cells, 60 Hz stimulation had a statistically greater ALP activity compared to other frequencies.

Sen et al. investigated the inhibition of adipogenesis in mouse MSCs after vibration [15]. The authors subjected MSCs seeded on a collagen-coated silicon membrane to low intensity vibrations of <10 microstrain, at 90 Hz. The MSCs were stimulated for 20 minutes twice a day. Following vibration, cells were tested for the adipogenic markers adiponectin, PPARγ2, and aP2. The MSCs expressed decreased levels for all adipogenic markers and the development of lipid granules was inhibited. The authors also investigated the synergistic effect of vibration and adipogenic media. The MSCs subjected to vibration and adipogenic media had a statistically insignificant expression of adipogenic markers. The results indicate that adipogenesis was inhibited in cells subjected to vibration and that treatment with adipogenic media did not recover adipogenesis. Conversely, markers for osteogenic differentiation were promoted significantly after vibration, though this effect was only observed in MSCs treated with adipogenic media.

Tong et al. subjected hMSCs to 200 Hz acoustic vibration to replicate vocal cord vibrations [16]. Cells were seeded on PCL scaffolds coated with fibronectin and subjected to vibrations for 12 hrs/day over 7 days continuously or discontinuously. All vibrated cells expressed enhanced F-actin and a5b1 integrin expression. Levels of vocal fold extracellular matrix components were significantly elevated. Myogenic differentiation in MSCs were indicated by elevated levels of tenascin-C, collagen III, and procollagen I, while osteogenic markers were not expressed.

Edwards and Reilly, similar to Lau et al., found that a synergist effect of vibration and osteogenic media promoted osteogenesis [12, 14]. Sen et al. found adipogenesis to be inhibited after vibration, even with the addition of adipogenic media. Tong et al. found myogenic differentiation of hMSCs seeded on fibronectin-coated scaffolds [16]. Each of these studies used a different biological coating on a synthetic scaffold. It is important to consider that the different MSC responses to vibration observed by these studies may in part be explained by different MSC responses to the scaffolds. In short, the biological components of the scaffolds are potentially introducing a compounding biological variable to the investigations, contributing to the observed synergistic responses.

3.3. Three-Dimensional Scaffolds.
Few studies of MSC response to vibration in vitro have used three-dimensional substrates [10–13, 15, 16, 51]. Three-dimensional substrates translate mechanical force to cells via different mechanisms than 2D substrates [52]. Three-dimensional substrates may better model in situ cell attachment and the resultant effects of mechanical loading [53]. In 3D environments, there are increased cell-to-cell contact and cell-to-extracellular matrix interactions compared to 2D monolayers [53]. Due to these factors, cells within 3D scaffolds likely better model the in vivo response to vibration than cells on 2D substrates.

3.3.1. Natural Scaffolds.
Kim et al. vibrated hMSCs after inoculation on a collagen sponge [13]. Cells were exposed to 30 Hz vibration at 0.3 g. In their differentiation assay, the authors found that the osteogenic markers OPG, Col I, and VEGF expression were increased after MSCs were vibrated. These results differ from the previously described results of Chen et al., who found increased adipogenesis in TCP-monolayer hMSCs subjected to 30 Hz vibration at 0.3 g. The difference in the response of MSCs within a 3D scaffold and MSCs in a monolayer suggests that factors such as microarchitectural cues may mechanotransduce vibration.

Zhou et al. subjected rat MSCs seeded on 3D bone-derived scaffolds to LMHF vibration [17]. The hollow components of the scaffolds allowed cells to attach within the multidimensional matrix. Rat MSCS seeded on TCP were used as control groups. The authors vibrated cells with 40 Hz at 0.3 g for 6 hours. The vibrated MSCs demonstrated increased levels of osteogenic markers ALP, Coll I, and OC. ALP activity was significantly higher in cells vibrated within 3D scaffolds than cells vibrated on TCP (Figure 3). However, vibration resulted in lower proliferation after day 7. The increased response from vibrated MSCs within 3D scaffolds compared to MSCs on TCP should motivate further investigation of MSCs within 3D scaffolds.

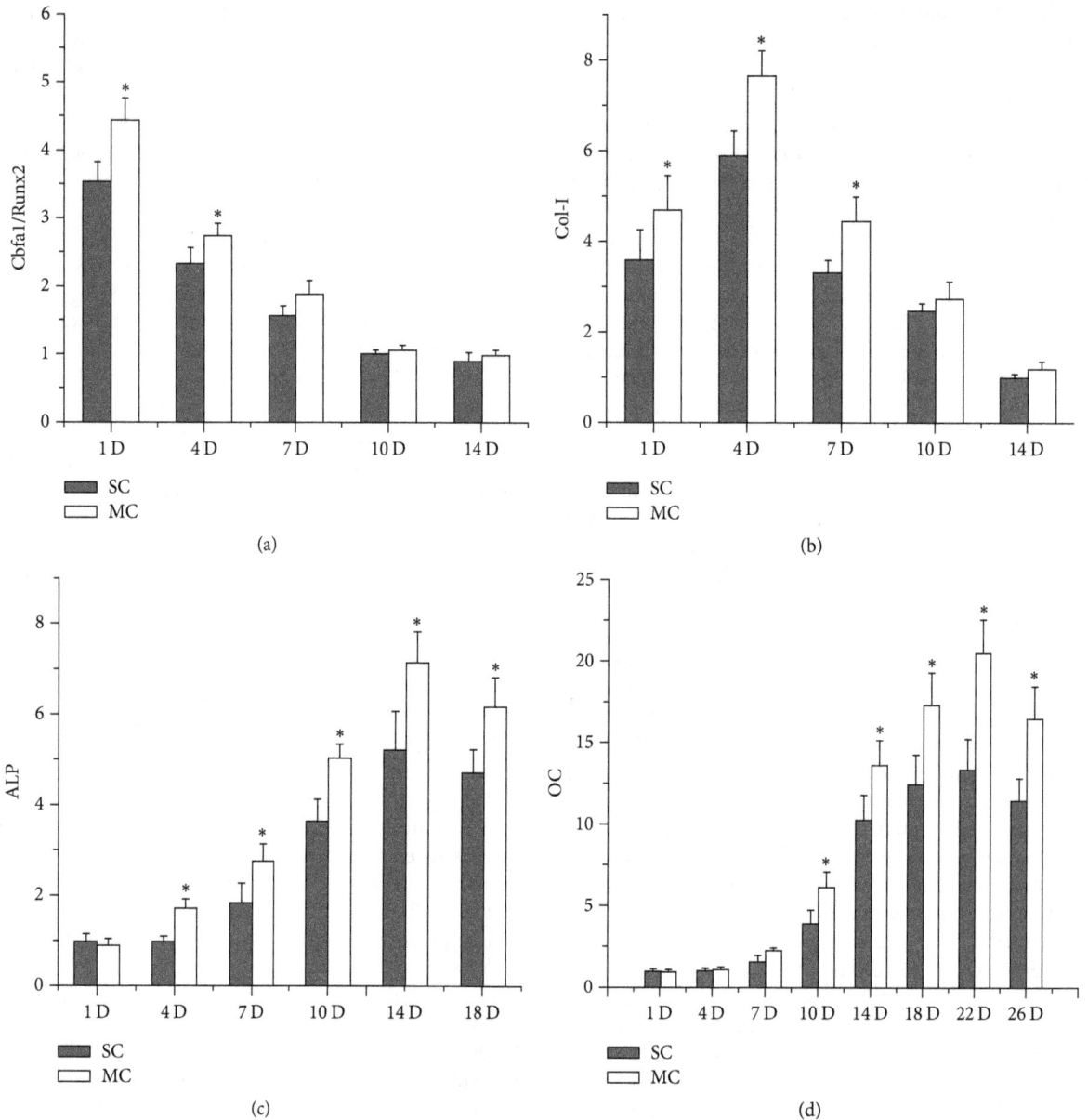

FIGURE 3: Effect of microvibration on osteogenic gene expressions in BMSC cellular scaffolds. Cbfa1/Runx2, Col I, ALP, and OC mRNA expressions were assayed on days 1, 4, 7, 10, 14, 18, 22, and 26. Data show that microvibration greatly upregulated these mRNA levels at different stages of osteogenesis. Each bar represents the mean ± standard deviation ($n = 3$); $^*P < 0.05$. SC, static culture. MC, microvibration culture. Col I, collagen I. ALP, alkaline phosphatase. OC, osteocalcin. From Zhou et al., [17] with kind permission from eCM journal (http://www.ecmjournal.org).

3.3.2. Synthetic Scaffolds. To the author's knowledge, our laboratory has conducted the only vibration studies on MSCs entrapped within a synthetic material [18]. Cells entrapped within PEGDA microspheres were subjected to vibrations of 100 Hz at 0.3 g, 3.0 g, or 6.0 g for 24 hours. Cells were subsequently tested for adipocyte, chondrocyte, and osteoblast differentiation. Osteogenic differentiation in MSCs was observed at 0.3 and 3.0 g accelerations, while 6.0 g accelerations were lethal to cells. Alkaline phosphatase activity was observed on day 4 in MSCs subjected to 0.3 and 3.0 g. Alizarin red staining was also significantly increased in 0.3 and 3.0 g MSCs compared to nonvibrated controls. Chondrogenesis and adipogenesis were not observed at any time point.

The vibration studies expose MSCs to a range of accelerations and frequencies. Accelerations from 0.02 g to 6.0 g were used in combination with frequencies ranging from 10 to 800 Hz (Table 1). With such variations in vibration parameters, it is unsurprising that there is a great variation in observed response. However, when probing further, the disparity in cell response is mostly observed in cells seeded on 2D scaffolds. MSCs on or within 3D scaffolds uniformly

TABLE 2: Effect of cyclic tensile strain on MSCs seeded on 2D scaffolds.

Scaffold		Cell	Strain (%)	Time (hrs)	Differentiation	Ref.
2D Natural	Collagen	hMSCs	5	24	Myogenic	[19]
		hMSCs	10	24	Myogenic*	[20]
		hMSCs	3, 10	8, 48	Tenogenic (10%)	[21]
		hMSCs	0.8, 5, 10, 15	48	Osteogenesis (<5%) Osteogenic inhibition (>5%)	[22]
2D Natural + synthetic	Elastin	hMSCs	10	24	Myogenic*	[20]
	Silicone	rMSCs	10	N/A	Tenogenic	[23]
	Elastin-silicone	rMSCs	5, 10, 15, 20	24	Myogenic (10%)	[24]
	Elastin-silicone	rMSCs	10	24, 48, 72	Myogenic	[24]

*Myogenic expression transiently increased; expression reduced to basal levels after cell alignment.

differentiated to bone. While 3D scaffolds were tested in a smaller range of vibration parameters (0.3–6 g; 30–100 Hz), these studies may point to a uniform response to MSCs when vibrated within a 3D environment.

4. The Effect of Cyclic Tensile Strain on MSCs

Cyclic uniaxial tensile strain can be applied to cells encapsulated in or seeded on a flexible scaffold. Rigid materials such as TCP cannot be used for tensile strain studies. Silicon scaffolds are regularly used as a synthetic, flexible scaffold. The effects of tensile strain on MSCs, inducing tenogenic, osteogenic, and myogenic responses, have been investigated and are reviewed below [3, 19–32].

4.1. Two-Dimensional Scaffolds

4.1.1. Natural Scaffolds. The following studies investigate the effects of cyclic uniaxial tensile strain with magnitudes between 0.8% and 15% at 1 Hz on MSCs seeded on collagen membranes and other scaffolds (Table 2) [19–22]. The strain magnitudes were selected by the authors to replicate the tensile strain experienced in situ by bone, muscle, and tendon.

Chen et al. seeded hMSCs on collagen type I coated scaffolds and then subjected them to 3% and 10% cyclic tensile strain at 1 Hz for 8 or 48 hours [21]. The authors were investigating osteogenic or tenogenic commitment of MSCs following such strain. For all strains, organized cell alignment was noted. Cells subjected to both strain rates became longer, slenderer in shape, and oriented perpendicular to the axis of strain. hMSCs subjected to 3% strain for 8 hours demonstrated an upregulation of osteoblastic markers with increased levels of ALP and Cbfa1. hMSCs strained at 10% for 48 hours demonstrated significant increases in type I collagen, type III collagen, and tenascin-C, indicating tenogenic differentiation. The authors suggested strain amplitude and duration of strain may influence tenogenic and osteogenic commitment of MSCs.

Park et al. aimed to replicate the strain conditions of vascular smooth muscle on hMSCs seeded on elastin or collagen-coated membranes [20]. MSCs were subjected to 10% uniaxial tensile strain at 1 Hz for 24 hours. Smooth muscle cell (SMC) and osteogenic markers were investigated.

Following strain, cells in both scaffolds increased collagen I expression; however, markers of osteogenic differentiation were not significant. After being subjected to strain, levels of smooth muscle markers α-actin, SM-22α, and β-actin were transiently increased in cells on both scaffolds (Figure 4). Expression of α-actin and SM-22α subsequently decreased shortly after cells aligned themselves perpendicular to the direction of strain. After 3 days, SM-22α decreased by 50% on collagen-coated scaffolds and 25% on elastin-coated membranes. Levels of Col I also decreased after alignment. The authors concluded that uniaxial strain may promote MSC differentiation into SMCs if the cell orientation is fixed. Interestingly, the decrease in gene expression after alignment described by Park et al. contradicts the stable gene expression described by Chen et al., who did not observe decreased gene expression from MSCs subjected to 10% strain [20, 21].

Khani et al. investigated the mechanical properties of hMSCs subjected to uniaxial strain with or without chondrogenic media (with TGF-β_1) [19]. hMSCs were seeded on poly(dimethyl siloxane) (PDMS) with a collagen coating to enable cell attachment. For 24 hours seeded hMSCs were subjected to uniaxial strain of 5% at 1 Hz, comparable to physiological levels within human arteries. Strained cells without TGF-β_1 had significantly increased Young's Moduli (E) and elevated levels of the smooth muscle markers ASMA, h1-Calponin, and SM22A. The strained hMSCs demonstrated increased myogenesis with or without TGF-β_1 in the media. After stimulation, these cells had also become aligned perpendicular to the axis of strain. The authors suggested that the realignment of these cells may reinforce the material, creating a stiffer composition, and may ultimately impact mechanotransduction.

Koike et al. subjected hMSCs seeded on collagen I coated membranes to 0.8%, 5%, 10%, and 15% cyclic strain at 1 Hz for 2 days [22]. Cell proliferation significantly increased at 5%, 10%, and 15% strain compared to unloaded controls. At 1-hour and 6-hour markers, Cbfa1/Runx2 increased at 0.8% and 5% strain but decreased at 15% strain. At 24 hours and 48 hours, cell proliferation and Col I increased at 5%, 10%, and 15% strain while Cbfa1/Runx2 expression, osteocalcin expression, and ALP activity were significantly decreased. ALP activity was increased at 0.8% strain. These results indicate that high magnitude mechanical strain will inhibit osteoblastic

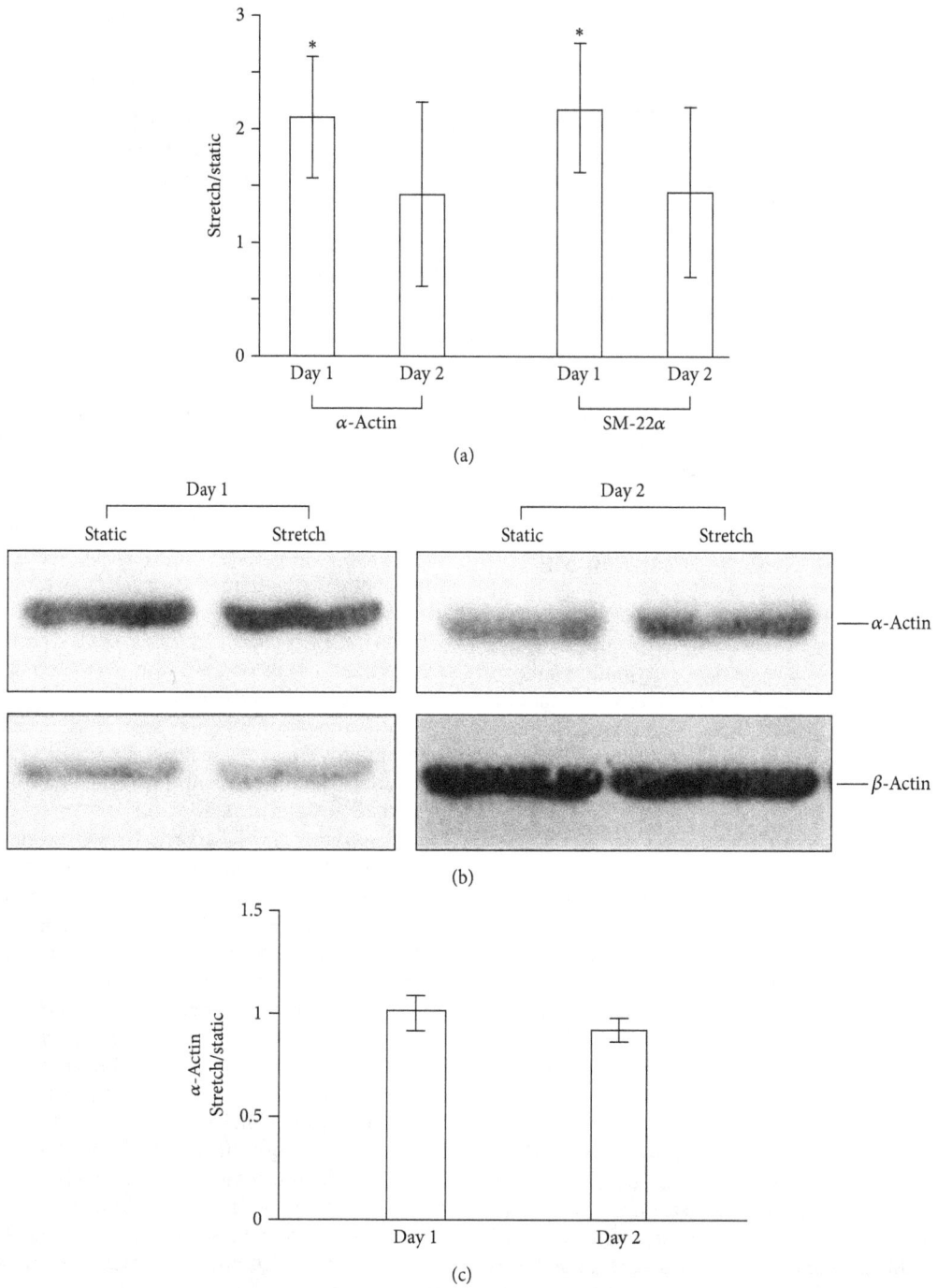

FIGURE 4: Effects of uniaxial strain on SM marker expression in MSCs. MSCs were cultured on collagen-coated elastic membranes for 1 day and subjected to 10% uniaxial strain at 1 Hz or kept as static controls for 1 and 2 days. (a) The RNA from each sample was reverse-transcribed into cDNA and the gene expression of SM a-actin, SM-22a, and GAPDH was analyzed by qPCR with their respective primers. The expression level of each gene was normalized with the level of GAPDH in the same sample. The ratio of the gene expression (stretch/static) is presented as mean ($\pm$) standard deviation from at least three experiments. *Significant difference ($P < 0.05$) from 1. (b) The protein expression of SM a-actin and actin was analyzed by immunoblotting with respective antibody. (c) Statistical analysis of protein expression. The protein expression was quantified, and the expression level of a-actin was normalized with the level of h-actin in the same samples. The ratio of the normalized protein expression (stretch/static) is presented as mean ($\pm$) standard deviation from at least three experiments. From Park et al., Biotechnology and Bioengineering [20].

differentiation, while strain at low magnitudes may enhance osteoblastic differentiation.

All studies described above used collagen scaffolds to investigate the response of hMSCs to uniaxial tensile strain [19–22]. On collagen scaffolds, MSCs subjected to low magnitude tensile strains (≤3%) underwent osteogenic differentiation [21, 22, 54]. At greater tensile strains (5%), myogenic differentiation was promoted [19, 22]. hMSCs subjected to high magnitude tensile strains (≥10%) showed an inhibition of osteogenic differentiation, transiently enhanced myogenic differentiation, and enhanced tenogenic differentiation [20–22]. The literature generally agrees that MSC osteodifferentiation occurs at lower strain magnitudes than MSC tenodifferentiation [19–22, 54].

4.1.2. Synthetic Scaffolds.

Park et al. noted a difference in MSC response to a synthetic scaffold compared to a collagen scaffold [20]. As described in Section 4.1.1, Park et al. subjected MSCs seeded on elastin-coated or collagen-coated scaffolds to 10% uniaxial tensile strain at 1 Hz for 24 hours. The hMSCs transiently upregulated smooth muscle markers. Smooth muscle gene expression was higher in cells seeded on the elastin scaffolds. The authors suggested that the MSCs sensed a difference in the mechanical loading of the two microenvironments.

Zhang et al. subjected rMSCs seeded on a silicone scaffold to 10% tensile strain at 1 Hz [23]. The authors investigated the effect of tensile strain on tenogenesis of hMSCs. The authors also investigated the effects of coculturing hMSCs with ligament fibroblasts. Only cells subjected to strain exhibited morphological changes. After tensile strain, rMSCs had a more elongated fibroblast-like cell type. The strain triggered an early upregulation of Col I and Col III. Tenascin-C expression was also upregulated in cells subjected to strain. Cells subjected to strain demonstrated greater levels of tenogenic gene expression than cells cocultured with fibroblasts but not subjected to strain.

Huang et al. subjected rMSCs seeded on an elastic-silicone membrane to 5%, 10%, 15%, and 20% tensile strain at 1 Hz for 24 hours [24]. The authors investigated the presence of cardiac-related gene expression using negative controls and positive controls. Cells subjected to cyclic strain expressed GATA-4, β-MHC, NKx2.5, and MEF2c. Gene expression was greatest in cells subjected to 10% strain. The researchers then subjected rMSCs to 10% strain at 1 Hz for 24, 48, and 72 hours. The expression of GATA-4, β-MHC, NKx2.5, and MEF2c was significantly increased for all durations of strain. The investigators suggested that cyclic mechanical strain of 10% at 1 Hz induces cardiomyogenic differentiation of MSCs.

Jagodzinski et al. applied tensile strain to hMSCs seeded on a silicone scaffold [55]. Cells were subjected to 2% or 8% strain, six hr/day for three days at 1 Hz. hMSCs were cultivated with (dex+) or without (dex−) dexamethasone. For both strain magnitudes, cells had significantly increased ALP secretion and collagen III upregulation. Cells subjected to 8% strain significantly upregulated Col I and Cbfa1. Cells that underwent strain had significantly greater gene expression, with or without dex, for all markers of gene expression. At both high and low magnitudes of cyclic strain,

hMSC osteogenic commitment was enhanced on silicon scaffolds, contrary to the observed trends of hMSCs on collagen scaffolds, which showed tenogenic commitment at high strain magnitudes and osteogenic commitment at low strain magnitudes.

After being subjected to 10% strain, MSCs seeded on synthetic scaffolds demonstrated enhanced myogenic differentiation, resulting in fibroblasts, smooth muscle cells, and cardiac cells [20, 23, 43]. For other strain magnitudes, MSCs seeded on synthetic scaffolds respond differently to tensile strain than MSCs seed on natural scaffolds [20, 23, 24, 55]. Cells on synthetic scaffolds had a greater expression of smooth muscle markers compared to cells on natural scaffolds [20]. Osteodifferentiation was induced at high magnitudes of strain in cells on silicone while osteogenesis was inhibited at high magnitudes of strain in cells on collagen [55]. Thus, demonstrating that the MSC response to strain differs on synthetic-based scaffolds compared to natural scaffolds.

While the preceding scaffolds were comprised of synthetic materials, only one was not modified with elastin. The MSCs on scaffolds comprised of or modified with elastin all exhibited myodifferentiation. The MSCs on truly synthetic scaffolds exhibited tenodifferentiation. Thus, for 2D scaffolds, scaffold type may be more important than the loading.

4.2. Three-Dimensional Scaffolds.

Cells entrapped within a scaffold or seeded throughout a structure are subjected to a different microenvironment than cells seeded on a planar scaffold. Subtle differences in loading or cell-cell communication may impact cell response to strain. The following studies investigate the effects of cyclic strain on cells entrapped within 3D scaffolds (Table 3).

4.2.1. Natural Scaffolds.

The first study to investigate hMSCs entrapped in 3D collagen matrix under cyclic strain was conducted by Sumanasinghe et al. [25]. The authors subjected hMSCs entrapped within a collagen matrix to 10% or 12% uniaxial cyclic tensile strain at 1 Hz for 4 hr/day. Strain was applied for 7 or 14 days. hMSCs remained highly viable for all strain conditions. hMSCs subjected to 10% strain demonstrated a significant increase in BMP-2 expression for both durations of strain. Cyclic strain of 12% induced a significant increase in BMP-2 expression only in cells subjected to 14 days of strain. The authors concluded that strain alone can induce osteogenic differentiation without the addition of osteogenic supplements.

Sumanasinghe et al. conduced a follow-up study to investigate the expression of proinflammatory MSC cytokines using identical strain conditions (10 or 12%; 1 Hz for 4 hr/day; 7 or 14 days) [26]. The authors also used osteogenic media to evaluate the combined effect of cyclic strain and osteogenic supplements. Initially, hMSCs undergoing strain had reduced viability. After day 6, hMSCs subjected to 10% strain had increased viability. Only strained cells receiving osteogenic media had increased levels of TNFα and IL-Iβ. The authors demonstrated that hMSCs entrapped within a collagen matrix maintain high viability after cyclic strain.

TABLE 3: Effect of cyclic tensile strain on MSCs within 3D scaffolds.

Scaffold		Cell	Strain (%)	Time (hrs)	Differentiation	Ref.
3D Natural	Collagen	hMSCs	10, 12	28, 56	Osteogenic	[25]
		hMSCs	10	56	Osteogenic	[26]
		hMSCs	10	168	Tenogenic	[27]
		rbMSCs	2.4	96	Inconclusive	[28]
3D Synthetic hydrogel	(RGD/RGE)	hMSCs	3	2, 24	Gene expression ↑ (RGD)	[29]
	PLA	hMSCs	2, 5	15	Inconclusive	[30]
	PEG	hMSCs	10	168	Tenogenic	[31]
	OPC	hMSCs	10	252	Tenogenic	[32]

Charoenpanich et al. entrapped hMSCs, specifically adipose derived stem cells, within a collagen I gel sheet [56]. The entrapped cells were subjected to 10% cyclic tensile at 1 Hz for 4 hr/day over 14 days. The authors performed a microarray analysis of 847 genes and found 184 transcripts affected by tensile strain. Network analysis suggested that strain may impact osteogenic differentiation by upregulation of proinflammatory cytokine regulator interleukin-1 receptor antagonist (IL1RN) and angiogenic inductors including fibroblast growth factor 2 (FGF-2) and vascular endothelial growth factor A (VEGF-A). Cells subjected to strain and osteogenic media resulted in significantly increased calcium deposits, suggesting a synergistic effect of the strain and media driving the cells towards osteogenic differentiation.

Qiu et al. applied cyclic strain to hMSCs seeded along collagen fibers to investigate fibroblastic differentiation [27]. The fibrous scaffold provided a nonplanar microenvironment. The hMSCs were subjected to 10% tensile strain at 1 Hz for 12 hrs/day over 14 days. Collagen I, collagen III, tenascin-C, and fibroblastic transcription factor scleraxis were all found to be significantly upregulated in cyclically strained hMSCs compared to unstrained control cells. Thus, 10% cyclic strain significantly promoted tenogenic differentiation of hMSCs.

Juncosa-Melvin et al. applied strain to rabbit MSCs seeded within collagen sponges [28]. MSCs were subjected to cyclic strain of 2.4% at 0.003 Hz for 8 hrs/day over 12 days. Tenogenic differentiation was not conclusively promoted by cyclic strain, but significant gene expression of collagen I and collagen III was induced by cyclic strain. Strained MSCs showed 3 or 4 times greater collagen I and collagen III production compared to unstrained controls. However, gene expression of fibronectin or decorin was not significantly increased in strained MSCs.

Few studies investigate both the effects of 3D scaffolds and cyclic tensile strain on hMSC differentiation without osteogenic supplements. In the few studies that do not use osteogenic media, most use scaffolds or coatings comprised of collagen, which is a major component in bone and hence provides a biological factor that induces its own cell response [57, 58].

4.2.2. Synthetic Scaffolds. Rathbone et al. investigated the response of hMSCs to cyclic tensile strain entrapped within 3D hydrogels with either the cell attachment tripeptide, arginylglycylaspartic acid (RGD), or a dummy tripeptide, arginyl-glycyl-glutamic acid (RGE) [29]. Cells were either entrapped within a hydrogel with cell attachment sites (RGD) or entrapped within a hydrogel without cell attachment sites (RGE). The authors' hydrogel was composed of Fmoc-FF:Fmoc-RGD/RGE. The hMSCs were subject to 3% strain 1 Hz for 1 hour or 24 hours and evaluated 2 hours or 24 hours after strain. Cells within hydrogels demonstrated high viability. The authors investigated CCNL2, WDR61, and BAHCC1 as potentially important mechanosensitive genes. After 1 hour of strain, hMSCs on monolayers significantly downregulated CCNL2, WDR61, and BAHCC1. After 24 hours of strain, hMSCs on monolayers significantly upregulated BAHCC1. BAHCC1 was not expressed by hMSCs in either of the 3D scaffolds. WDR61 was significantly upregulated by hMSCs in both 3D scaffolds after 1 hr of strain. CCNL2 was upregulated in hMSCs only in scaffolds with RGD. The cell response differed when in a monolayer or in a 3D scaffold and when within 3D scaffolds with or without RGD, thus indicating the impact of attachment sites on mechanotransduction in otherwise identical scaffolds.

Kreja et al. applied strain to hMSCs seeded throughout a novel textured PLA scaffold to investigate fibroblastic differentiation [30]. Cells were subjected to 2% or 5% strain at 1 Hz for 1 hr/day over 15 days. The authors analyzed the gene expression of ligament matrix markers: collagen I, collagen III, fibronectin, tenascin-C, decorin, MMP-1, MMP-2, and inhibitors TIMP-1 and TIMP-2. Cells subjected to strain did not demonstrate significant gene expression except in the downregulation of both MMP-1 and TIMP-2 in cells subjected to 5% strain. For both strain parameters, tenogenic differentiation was not promoted in hMSCs.

Yang et al. investigated the effects of strain on hMSCs entrapped within fast and slow degrading MMP-sensitive PEG hydrogels to investigate tenogenic differentiation [31]. Cells were subjected to 10% strain at 1 Hz for 12 hrs/day over 14 days. Cell realignment in response to the strain direction was not observed. hMSCs within the slow degrading hydrogel upregulated collagen III by 3.8-fold and upregulated tenascin-C by 2.5-fold while hMSCs within the fast degrading hydrogel upregulated collagen III by 2.1-fold and upregulated tenascin by 1.7-fold. The authors suggested that cyclic straining promoted tenogenic differentiation and that the presence of strain had a greater influence on cell differentiation than the difference in composition between the hydrogels.

Doroski et al. expanded the Yang et al. investigation where hMSCs entrapped with PEG-based hydrogels were cyclically strained [31, 32]. The researchers entrapped hMSCs within oligo(poly(ethylene glycol) fumarate) (OPF). Cells were then subjected to 10% strain at 1 Hz for 12 hrs/day over 21 days. By day 21, cyclic strain significantly upregulated the tenogenic markers collagen I, collagen III, and tenascin-C, while osteogenic, chondrogenic, and adipogenic markers were not increased. Thus, the cyclic strain promoted tenogenic differentiation.

It is interesting to note that collagen derived scaffolds mostly resulted in osteodifferentiation while synthetic scaffolds mostly resulted in tenodifferentiation. These results indicate that if seeking to differentiate MSCs using mechanical loading such as tensile strain, the scaffold material type will influence the MSC fate as much as the loading parameters will.

5. Summary

The response of MSCs subjected to cyclic strain and vibration appear to vary with loading parameter as much as it varies with culture conditions. Unsurprisingly, for similar loading conditions, TCP produced different results than 2D scaffolds, which in turn produced different results than 3D scaffolds. The varying effects of natural and synthetic scaffolds on MSC differentiation may be explained by the differing elastic moduli between these scaffolds or it may be the distinct microarchitecture of a natural scaffold compared to featureless synthetic scaffolds, or the dominating factor may be the native biochemical cues contained within natural scaffolds. These questions warrant further investigation if mechanical loading is to be pursued as an alternate method to induce MSC differentiation.

6. Future Perspectives

Future investigations of MSC differentiation using vibration and cyclic strain should explore varying types of natural scaffolds in concert with the loading parameters. Collagen is a primary component of bone and as such makes an ideal bone scaffold. Vibration within 3D collagen scaffolds ultimately led to osteogenic differentiation. Future studies utilizing scaffolds optimized for myogenic, adipogenic, or tenogenic differentiation should be explored to tease out whether vibration and/or cyclic strain in 3D scaffolds consistently leads to osteogenic differentiation or whether vibration and/or cyclic strain in 3D scaffolds leads to tissue optimized for that 3D scaffold.

Conflicts of Interest

The authors declare that they have no conflicts of interest.

References

[1] H. Lv, L. Li, M. Sun et al., "Mechanism of regulation of stem cell differentiation by matrix stiffness," *Stem Cell Research & Therapy*, vol. 6, no. 1, article no. 103, 2015.

[2] C. Vater, P. Kasten, and M. Stiehler, "Culture media for the differentiation of mesenchymal stromal cells," *Acta Biomaterialia*, vol. 7, no. 2, pp. 463–477, 2011.

[3] R. M. Delaine-Smith and G. C. Reilly, "Mesenchymal stem cell responses to mechanical stimuli," *Muscle, Ligaments and Tendons Journal*, vol. 2, no. 3, pp. 169–180, 2012.

[4] R. Tuan, G. Boland, and R. Tuli, "Adult mesenchymal stem cells and cell-based tissue engineering," *Arthritis Research & Therapy*, vol. 5, no. 1, 2003.

[5] C. Nombela-Arrieta, J. Ritz, and L. E. Silberstein, "The elusive nature and function of mesenchymal stem cells," *Nature Reviews Molecular Cell Biology*, vol. 12, no. 2, pp. 126–131, 2011.

[6] Y.-R. V. Shih, K.-F. Tseng, H.-Y. Lai, C.-H. Lin, and O. K. Lee, "Matrix stiffness regulation of integrin-mediated mechanotransduction during osteogenic differentiation of human mesenchymal stem cells," *Journal of Bone and Mineral Research*, vol. 26, no. 4, pp. 730–738, 2011.

[7] P. M. Tsimbouri, P. G. Childs, G. D. Pemberton et al., "Stimulation of 3D osteogenesis by mesenchymal stem cells using a nanovibrational bioreactor," *Nature Biomedical Engineering*, vol. 1, no. 9, pp. 758–770, 2017.

[8] L. Tan, B. Zhao, F. Ge, D. Sun, and T. Yu, "Shockwaves inhibit chondrogenic differentiation of human mesenchymal stem cells in association with adenosine and A2B receptors," *Scientific Reports*, vol. 7, no. 1, article 14377, 2017.

[9] U. A. Gurkan and O. Akkus, "The mechanical environment of bone marrow: A review," *Annals of Biomedical Engineering*, vol. 36, no. 12, pp. 1978–1991, 2008.

[10] X. Chen, F. He, D.-Y. Zhong, and Z.-P. Luo, "Acoustic-frequency vibratory stimulation regulates the balance between osteogenesis and adipogenesis of human bone marrow-derived mesenchymal stem cells," *BioMed Research International*, vol. 2015, Article ID 540731, 10 pages, 2015.

[11] L. Demiray and E. Özçivici, "Bone marrow stem cells adapt to low-magnitude vibrations by altering their cytoskeleton during quiescence and osteogenesis," *Turkish Journal of Biology*, vol. 39, no. 1, pp. 88–97, 2015.

[12] E. Lau, W. D. Lee, J. Li et al., "Effect of low-magnitude, high-frequency vibration on osteogenic differentiation of rat mesenchymal stromal cells," *Journal of Orthopaedic Research*, vol. 29, no. 7, pp. 1075–1080, 2011.

[13] I. S. Kim, Y. M. Song, B. Lee, and S. J. Hwang, "Human mesenchymal stromal cells are mechanosensitive to vibration stimuli," *Journal of Dental Research*, vol. 91, no. 12, pp. 1135–1140, 2012.

[14] J. Edwards and G. C. Reilly, "Low magnitude, high frequency vibration modulates mesenchymal progenitor differentiation," *Journal of Orthopaedic Research*, vol. 36, article 2186, 2011.

[15] B. Sen, Z.-H. Xie, N. Case, M. Styner, C. T. Rubin, and J. Rubin, "Mechanical signal influence on mesenchymal stem cell fate is enhanced by incorporation of refractory periods into the loading regimen," *Journal of Biomechanics*, vol. 44, no. 4, pp. 593–599, 2011.

[16] Z. Tong, R. L. Duncan, and X. Jia, "Modulating the behaviors of mesenchymal stem cells via the combination of high-frequency vibratory stimulations and fibrous scaffolds," *Tissue Engineering Part: A*, vol. 19, no. 15-16, pp. 1862–1878, 2013.

[17] Y. Zhou, X. Guan, Z. Zhu et al., "Osteogenic differentiation of bone marrow-derived mesenchymal stromal cells on bone-derived scaffolds: effect of microvibration and role of ERK1/2 activation," *European Cells and Materials*, vol. 22, pp. 12–25, 2011.

[18] S. Mehta, "Hydrogel encapsulation of cells mimics the whole body response to LMHF vibrations," Available from Dissertations & Theses @ Rutgers University; ProQuest Dissertations & Theses Global, 2015.

[19] M.-M. Khani, M. Tafazzoli-Shadpour, Z. Goli-Malekabadi, and N. Haghighipour, "Mechanical characterization of human mesenchymal stem cells subjected to cyclic uniaxial strain and TGF-β1," Journal of the Mechanical Behavior of Biomedical Materials, vol. 43, pp. 18–25, 2015.

[20] J. S. Park, J. S. F. Chu, C. Cheng, F. Chen, D. Chen, and S. Li, "Differential effects of equiaxial and uniaxial strain on mesenchymal stem cells," Biotechnology and Bioengineering, vol. 88, no. 3, pp. 359–368, 2004.

[21] Y. J. Chen, C. H. Huang, I. C. Lee, Y. T. Lee, M. H. Chen, and T. H. Young, "Effects of cyclic mechanical stretching on the mRNA expression of tendon/ ligament-related and osteoblast-specific genes in human mesenchymal stem cells," Connective Tissue Research, vol. 49, no. 1, pp. 7–14, 2008.

[22] M. Koike, H. Shimokawa, Z. Kanno, K. Ohya, and K. Soma, "Effects of mechanical strain on proliferation and differentiation of bone marrow stromal cell line ST2," Journal of Bone and Mineral Metabolism, vol. 23, no. 3, pp. 219–225, 2005.

[23] L. Zhang, X. Wang, H. Chen, and N. Tran, "Cyclic stretching and co-culture with fibroblasts promote the differentiation of rat mesenchymal stem cells to ligament fibroblasts," Journal of Biomechanics, vol. 39, pp. S579–S580, 2006.

[24] Y. Huang, L. Zheng, X. Gong et al., "Effect of cyclic strain on cardiomyogenic differentiation of rat bone marrow derived mesenchymal stem cells," PLoS ONE, vol. 7, no. 4, Article ID e34960, 2012.

[25] R. D. Sumanasinghe, S. H. Bernacki, and E. G. Loboa, "Osteogenic differentiation of human mesenchymal stem cells in collagen matrices: Effect of uniaxial cyclic tensile strain on bone morphogenetic protein (BMP-2) mRNA expression," Tissue Engineering Part A, vol. 12, no. 12, pp. 3459–3465, 2006.

[26] R. D. Sumanasinghe, T. W. Pfeiler, N. A. Monteiro-Riviere, and E. G. Loboa, "Expression of proinflammatory cytokines by human Mesenchymal stem cells in response to cyclic tensile strain," Journal of Cellular Physiology, vol. 219, no. 1, pp. 77–83, 2009.

[27] Y. Qiu, J. Lei, T. J. Koob, and J. S. Temenoff, "Cyclic tension promotes fibroblastic differentiation of human MSCs cultured on collagen-fibre scaffolds," Journal of Tissue Engineering and Regenerative Medicine, vol. 10, no. 12, pp. 989–999, 2016.

[28] N. Juncosa-Melvin, K. S. Matlin, R. W. Holdcraft, V. S. Nirmalanandhan, and D. L. Butler, "Mechanical stimulation increases collagen type I and collagen type III gene expression of stem cell-collagen sponge constructs for patellar tendon repair," Tissue Engineering Part A, vol. 13, no. 6, pp. 1219–1226, 2007.

[29] S. R. Rathbone, J. R. Glossop, J. E. Gough, and S. H. Cartmell, "Cyclic tensile strain upon human mesenchymal stem cells in 2D and 3D culture differentially influences CCNL2, WDR61 and BAHCC1 gene expression levels," Journal of the Mechanical Behavior of Biomedical Materials, vol. 11, pp. 82–91, 2012.

[30] L. Kreja, A. Liedert, H. Schlenker et al., "Effects of mechanical strain on human mesenchymal stem cells and ligament fibroblasts in a textured poly(L-lactide) scaffold for ligament tissue engineering," Journal of Materials Science: Materials in Medicine, vol. 23, no. 10, pp. 2575–2582, 2012.

[31] P. J. Yang, M. E. Levenston, and J. S. Temenoff, "Modulation of mesenchymal stem cell shape in enzyme-sensitive hydrogels is decoupled from upregulation of fibroblast markers under cyclic tension," Tissue Engineering Part A, vol. 18, no. 21-22, pp. 2365–2375, 2012.

[32] D. M. Doroski, M. E. Levenston, and J. S. Temenoff, "Cyclic tensile culture promotes fibroblastic differentiation of marrow stromal cells encapsulated in poly(ethylene glycol)-based hydrogels," Tissue Engineering Part A, vol. 16, no. 11, pp. 3457–3466, 2010.

[33] C. Snow-Harter, M. L. Bouxsein, B. T. Lewis, D. R. Carter, and R. Marcus, "Effects of resistance and endurance exercise on bone mineral status of young women: A randomized exercise intervention trial," Journal of Bone and Mineral Research, vol. 7, no. 7, pp. 761–769, 1992.

[34] D. R. Taaffe, T. L. Robinson, C. M. Snow, and R. Marcus, "High-impact exercise promotes bone gain in well-trained female athletes," Journal of Bone and Mineral Research, vol. 12, no. 2, pp. 255–260, 1997.

[35] C. T. Rubin, R. Recker, D. Cullen, J. Ryaby, J. McCabe, and K. McLeod, "Prevention of postmenopausal bone loss by a low-magnitude, high-frequency mechanical stimuli: a clinical trial assessing compliance, efficacy, and safety," Journal of Bone and Mineral Research, vol. 19, no. 3, pp. 343–351, 2004.

[36] H. Merriman and K. Jackson, "The effects of whole-body vibration training in aging adults: A systematic review," Journal of Geriatric Physical Therapy, vol. 32, no. 3, pp. 134–145, 2009.

[37] H. Seidel and R. Heide, "Long-term effects of whole-body vibration: a critical survey of the literature," International Archives of Occupational and Environmental Health, vol. 58, no. 1, pp. 1–26, 1986.

[38] M. Cardinale and J. Rittweger, "Vibration exercise makes your muscles and bones stronger: Fact or fiction?" British Menopause Society, vol. 12, no. 1, pp. 12–18, 2006.

[39] J. Wang, K. Leung, S. Chow, and W. Cheung, "The effect of whole body vibration on fracture healing – a systematic review," European Cells and Materials, vol. 34, pp. 108–127, 2017.

[40] F. Tian, Y. Wang, and D. D. Bikle, "IGF-1 signaling mediated cell-specific skeletal mechano-transduction," Journal of Orthopaedic Research, 2017.

[41] P. G. Childs, C. A. Boyle, G. D. Pemberton et al., "Use of nanoscale mechanical stimulation for control and manipulation of cell behaviour," Acta Biomaterialia, vol. 34, pp. 159–168, 2016.

[42] S. Judex, X. Lei, D. Han, and C. Rubin, "Low-magnitude mechanical signals that stimulate bone formation in the ovariectomized rat are dependent on the applied frequency but not on the strain magnitude," Journal of Biomechanics, vol. 40, no. 6, pp. 1333–1339, 2007.

[43] S. J. Hwang, Y. M. Song, T. H. Cho et al., "The implications of the response of human mesenchymal stromal cells in three-dimensional culture to electrical stimulation for tissue regeneration," Tissue Engineering Part: A, vol. 18, no. 3-4, pp. 432–445, 2012.

[44] M. Kabiri, B. Kul, W. B. Lott et al., "3D mesenchymal stem/stromal cell osteogenesis and autocrine signalling," Biochemical and Biophysical Research Communications, vol. 419, no. 2, pp. 142–147, 2012.

[45] E. Cukierman, R. Pankov, and K. M. Yamada, "Cell interactions with three-dimensional matrices," Current Opinion in Cell Biology, vol. 14, no. 5, pp. 633–639, 2002.

[46] J. A. Pedersen and M. A. Swartz, "Mechanobiology in the third dimension," Annals of Biomedical Engineering, vol. 33, no. 11, pp. 1469–1490, 2005.

[47] A. Bartholomew, C. Sturgeon, M. Siatskas et al., "Mesenchymal stem cells suppress lymphocyte proliferation in vitro and prolong skin graft survival in vivo," *Experimental Hematology*, vol. 30, no. 1, pp. 42–48, 2002.

[48] J. M. Gimble, F. Guilak, M. E. Nuttall, S. Sathishkumar, M. Vidal, and B. A. Bunnell, "*In vitro* differentiation potential of mesenchymal stem cells," *Transfusion Medicine and Hemotherapy*, vol. 35, no. 3, pp. 228–238, 2008.

[49] M. Vanleene and S. J. Shefelbine, "Therapeutic impact of low amplitude high frequency whole body vibrations on the osteogenesis imperfecta mouse bone," *Bone*, vol. 53, no. 2, pp. 507–514, 2013.

[50] H.-F. Shi, W.-H. Cheung, L. Qin, A. H.-C. Leung, and K.-S. Leung, "Low-magnitude high-frequency vibration treatment augments fracture healing in ovariectomy-induced osteoporotic bone," *Bone*, vol. 46, no. 5, pp. 1299–1305, 2010.

[51] J. H. Edwards, "Vibration stimuli and the differentiation of musculoskeletal progenitor cells: Review of results," *World Journal of Stem Cells*, vol. 7, no. 3, pp. 568–582, 2015.

[52] J. Starke, K. Maaser, B. Wehrle-Haller, and P. Friedl, "Mechanotransduction of mesenchymal melanoma cell invasion into 3D collagen lattices: Filopod-mediated extension-relaxation cycles and force anisotropy," *Experimental Cell Research*, vol. 319, no. 16, pp. 2424–2433, 2013.

[53] F. Pampaloni, E. G. Reynaud, and E. H. K. Stelzer, "The third dimension bridges the gap between cell culture and live tissue," *Nature Reviews Molecular Cell Biology*, vol. 8, no. 10, pp. 839–845, 2007.

[54] J. Tan, X. Xu, Z. Tong et al., "Decreased osteogenesis of adult mesenchymal stem cells by reactive oxygen species under cyclic stretch: a possible mechanism of age related osteoporosis," *Bone Research*, vol. 3, article 15003, 2015.

[55] M. Jagodzinski, M. Drescher, J. Zeichen et al., "Effects of cyclic longitudinal mechanical strain and dexamethasone on osteogenic differentiation of human bone marrow stromal cells," *European Cells and Materials*, vol. 7, pp. 35–41, 2004.

[56] A. Charoenpanich, M. E. Wall, C. J. Tucker, D. M. K. Andrews, D. S. Lalush, and E. G. Loboa, "Microarray analysis of human adipose-derived stem cells in three-dimensional collagen culture: osteogenesis inhibits bone morphogenic protein and wnt signaling pathways, and cyclic tensile strain causes upregulation of proinflammatory cytokine regulators and angiogenic factors," *Tissue Engineering Part A*, vol. 17, no. 21-22, pp. 2615–2627, 2011.

[57] E. Donzelli, A. Salvadè, P. Mimo et al., "Mesenchymal stem cells cultured on a collagen scaffold: In vitro osteogenic differentiation," *Archives of Oral Biolog*, vol. 52, no. 1, pp. 64–73, 2007.

[58] G. Papavasiliou, S. Sokic, and M. Turturro, "Synthetic PEG hydrogels as extracellular matrix mimics for tissue engineering applications," in *Biotechnology Molecular Studies and Novel Applications for Improved Quality of Human Life*, 2012.

10

Injectable Hydrogel versus Plastically Compressed Collagen Scaffold for Central Nervous System Applications

Magdalini Tsintou,[1] Kyriakos Dalamagkas,[1] and Alexander Seifalian ![ORCID][2]

[1]Centre for Nanotechnology & Regenerative Medicine, Division of Surgery and Interventional Science,
  University College of London, London, UK
[2]Nanotechnology & Regenerative Medicine Commercialisation Centre Ltd., The London BioScience Innovation Centre, London, UK

Correspondence should be addressed to Alexander Seifalian; a.seifalian@gmail.com

Academic Editor: Weifeng Zhao

Central Nervous System (CNS) repair has been a challenge, due to limited CNS tissue regenerative capacity. The emerging tools that neural engineering has to offer have opened new pathways towards the discovery of novel therapeutic approaches for CNS disorders. Collagen has been a preferable material for neural tissue engineering due to its similarity to the extracellular matrix, its biocompatibility, and antigenicity. The aim was to compare properties of a plastically compressed collagen hydrogel with the ones of a promising collagen-genipin injectable hydrogel and a collagen-only hydrogel for clinical CNS therapy applications. The focus was demonstrating the effects of genipin cross-linking versus plastic compression methodology on a collagen hydrogel and the impact of each method on clinical translatability. The results showed that injectable collagen-genipin hydrogel is better clinical translation material. Full collagen compression seemed to form extremely stiff hydrogels (up to about 2300 kPa) so, according to our findings, a compression level of up to 75% should be considered for CNS applications, being in line with CNS stiffness. Taking that into consideration, partially compressed collagen 3D hydrogel systems may be a good tunable way to mimic the natural hierarchical model of the human body, potentially facilitating neural repair application.

1. Introduction

The limited regenerative capacity of the Central Nervous System (CNS) is what makes the neurological conditions devastating, offering poor therapeutic options to the patients. It is not only the mechanical gap that disturbs the neuronal function, but also the triggered cascade of events that leads to secondary neuronal degeneration and death. Therefore, there is a real, pressing clinical need for the development of therapeutic strategies for the currently untreatable disorders of the CNS. The advances in neural tissue engineering have provided several tools that may help in addressing those problems in the future.

1.1. Injectable Hydrogel Systems. Several biodegradable preformed polymeric implants have been used as drug delivery systems for sustained release, although they require invasive

surgical techniques for implantation [1, 2]. Injectable hydrogels, with in situ gelling properties, provide the advantage of injection through a thin needle in a less invasive way than implantation, and, if a biodegradable polymer is used, the need for surgical removal is also eliminated.

Hydrogels have been used in 3D model technology for several years and in a wide range of tissues (e.g., bones, cartilage, and nerves) [3–7]. Hydrogels closely mimic the tissue environment because of their high water consistency and materials used, while, in parallel, their tunability makes them a kind of very flexible, highly controlled microenvironment [3, 8]. Because of the aforementioned advantages, several products are released in the market such as Matrigel [9] or Extracel [10]. All natural materials have the advantage of cell binding sites and adhesion molecules, creating a microenvironment that closely mimics the extracellular matrix (ECM) and this is why 3D models based on natural materials have attracted attention [11–13]. In this study, we chose to use collagen, which has been a material of choice for several tissue regenerative applications due to its properties.

1.2. Collagen Gels. Collagen-based matrices are widely used in tissue regenerative applications due to the collagen's ubiquitous presence in the human body (i.e., skin, bone, cartilage, and tendons), antigenic behaviour, and biodegradability [14–16]. Thus, it is critical to be able to utilise the mechanical properties of collagen hydrogels by cross-linking mechanisms.

However, the effectiveness of collagen-based tissue engineered materials has been severely limited by their lack of mechanical strength. A variety of methods exists to cross-link collagen gels. Physical treatments such as ultraviolet (UV) and γ-irradiation and dehydrothermal treatments are not practical, because of their limited use in cellular tissues. Chemical treatments with aldehydes are used to preserve and stiffen the tissues, but these treatments are highly cytotoxic. In vivo, tissues are naturally cross-linked by enzymes. However, use of these enzymes for bulk changes in mechanical properties in 3D models is not cost-effective. Chemical aldehydes are used as a fixative to preserve tissues but are highly toxic [17, 18].

1.3. Genipin. Genipin has been investigated to modulate mechanical stiffness of collagen and gelatin. Genipin is the active compound found in *Gardenia jasminoides* fruit extract and it cross-links collagen through cross-linking of amine groups on lysine and arginine residues, resulting in a gel strength comparable to glutaraldehyde, but it is 10,000-fold less cytotoxic [19]. In addition to an increase in mechanical strength of collagen, genipin cross-linking is associated with a colour change in which opaque collagen turns blue.

Genipin may cross-link collagen in a variety of different mechanisms. Genipin molecules may react with amino groups within a tropocollagen molecule or between adjacent tropocollagen molecules to form intrahelical and interhelical cross-links in the genipin-fixed tissue [20]. In addition, inter-microfibrillar cross-links may be formed between collagen microfibrils via polymerization of genipin molecules before cross-linking (oligomeric cross-link).

Furthermore, the degradation rate of genipin cross-linked gelatin has been found to be significantly slower than the one of glutaraldehyde-cross-linked counterparts [21]. The mechanical and rheological behaviour of genipin cross-linked gelatin has been investigated, revealing that, with an increase in genipin concentration and temperature, the gelatin network shifts from being dominated by hydrogen bonds (physical cross-links) to covalent cross-linking (chemical cross-links) [22]. Although genipin is an attractive cross-linker for collagen, its cytotoxicity at high concentrations (5 mM) limits its usage to small concentrations [23].

1.4. Plastic Compression. The method of plastic compression of collagen has been originally reported by Brown et al. in 2005 [24]. The method is based on the uniaxial removal of unbound water from hyper-hydrated collagen gels, reconstituted from acidic solution. As a result, collagen sheets are produced which, dependent on the application, can contain a known number of viable embedded cells.

The word "plastic" refers to the irreversible nature of the process, that is, the thickness of collagen sheets does not change (i.e., reswell) significantly in fluid once the load is removed. The main advantages of this method are simplicity, speed, and reproducibility, calculable, predictable physical, and concentration parameters, and compatibility with a viability of a resident cell population. Thus, in contrast to other techniques, the improved mechanical properties, achieved using this method, are controlled by the researcher rather than cells, but, above all, without loss of cell viability.

The advantage of multilayered compressed collagen hydrogels is that there are no progressive restrictions of the fluid leaving surface (FLS). Each new gel layer forms a new FLS when compressed. Theoretically, the thickness of each single layer is the main restriction, not their total number, as each layer is compressed individually. Also, it should be noted that this model gives an opportunity to fabricate multiple constructs simultaneously (6 to 96, depending on the well-plate format used). Additionally, by using this method, it is possible to fabricate complex multilayered tissues with different cell types or densities in each layer. It may also be possible to control cell infiltration between the layers, as it is known that increased stiffness of the matrix enhances motility of some cell types [25].

In this study, we have fabricated a promising collagen-genipin injectable hydrogel that may be friendly for CNS applications and we are comparing that to a pure collagen hydrogel and to an inherently stronger compressed collagen hydrogel in different levels of compression. Our goal is to determine the effects of genipin cross-linking compared to the effects of the plastic compression technique on the inherent properties of collagen hydrogels, as well as to check the biocompatibility of the described systems with the CNS. This will help in accomplishing the optimal in vivo functional results in later studies in order to maximise clinical potentials.

2. Materials and Methods

2.1. Injectable Collagen and Collagen-Genipin Hydrogel Fabrication. All the solutions should be kept on ice 30 minutes before the initiation of the experiment to avoid premature gelation of the collagen hydrogel solution.

The initial collagen solution was made by mixing 10% 10x Minimum Essential Medium (MEM) and 70% of rat tail collagen 2 mg/ml (Type I) (FirstLink, UK) solution in a wide base flask. This solution was then neutralised with 1 M NaOH until the colour got stabilised to a bright fuchsia (pink) colour (changed from yellow). Gelation can incur prematurely at this stage so the remaining steps of the protocol should be done very fast. The 10% cross-linking genipin diluted in phosphate-buffered saline (PBS) solutions was added (0.5 mM concentration according to the literature review) to our samples, but PBS alone was added in the control collagen-only gels (10%). PBS was added again (10%) after that step. After swirling and mixing the solution, the solutions were either kept on ice for rheometry or put in the incubator at 37°C in a 12-well plate. 3 ml was the quantity chosen for the collagen hydrogels in the 12-well plates, according to the original paper directions. All materials were purchased from Sigma-Aldrich unless otherwise stated.

All the values reported in the following sections of the paper are the averages of at least three samples for each hydrogel type and/or compression level.

2.2. Rheology.

The mechanical properties of the hydrogels were tested through rheometry in the oscillatory mode. To get the gelation point of the hydrogel the temperature of the rheometer's plate was set to 4°C to avoid premature gelation that could alter the results.

Shear Pa was set to 0.1 and frequency to 1 rad/sec, while the temperature was set up to give some measurements for the stabilized solution at 4°C (10 seconds) then rise fast to 37°C obtaining measurements for 10 seconds and then get stabilized at 37°C (body temperature) until the storage modulus becomes equal to the loss modulus, signifying the gelation point. Mineral oil or silicon oil was applied around the hydrogel solution on the rheometry plate to avoid evaporation and questionable results. When storage and loss moduli plateaued, the test got aborted since gelation had already taken place. Using the same gelled hydrogel, the preshear oscillation with frequency sweep (0.1 rad/sec–50 rad/sec) was tested at 37°C for Shear Stress 1 Pa and frequency 1 rad/sec to get some insight on the hydrogels mechanical properties. In an attempt to simulate the room temperature versus body temperature, we also tested the same properties raising the temperature to 25°C (room temperature) instead of 37°C. Gel point was thought to be the time at which the shear storage modulus G' = the shear loss modulus G''.

Young's modulus was determined by calculating the slope of the steepest region of the stress-strain curve.

The expression in mathematical terms to calculate Young's modulus is the following:

$$E = \frac{\text{Stress}}{\text{Strain}} = \frac{F}{A} * \frac{L}{\Delta L}, \qquad (1)$$

where F is the force applied on the sample, A is the unstressed cross-sectional area through which the force is applied, L is the unstressed length, and ΔL is the change in length.

2.3. Degradation Assays.

After the hydrogels were left in the incubator to "mature" at 37°C for either 30 minutes or 24 hours the initial weight of the gels was measured. Artificial cerebrospinal fluid (aCSF)-0.1% collagenase mixture was made and was added on the top of those gels. The solution of aCSF-0.1% collagenase was replaced every 30 minutes for the next 4 hours. For control gels, only PBS was added on the top of the gels instead of the aCSF-0.1% collagenase mixture. At the end, the collagenase solution or the PBS for the controls was removed and the wells were washed with PBS solution and were put on a shaker table for 20 minutes. The remaining partially digested gels were collected and weighed to obtain the final wet weight. The samples were freeze-dried to obtain the dry weight as well. The calculations were based on the following:

$$\text{Gel Remaining}\,(\%) = \frac{(\text{Final wet weight})}{(\text{Initial wet weight})} * 100. \qquad (2)$$

2.4. Swelling Ratio.

The hydrogels were incubated for 24 hours in a 12-well plate. Swelling with PBS for further 24 hours followed. Disks were scooped with tweezers and weighed immediately afterwards to obtain the swollen weight of the gels. Then, after lyophilization, the dry weight was measured and the swelling ratio [26] was calculated according to the following equation:

$$\text{Swelling ratio} = \frac{\text{Swollen weight}}{\text{Dry weight}}. \qquad (3)$$

2.5. Compressed Collagen Method.

To compress the collagen hydrogels small chromatography rolls of the same diameter as the diameter of the well plate were applied on the top of each gelled hydrogel and left to be compressed.

To fabricate a 3D model, after the compression, a new collagen gel solution was added on the top of the compressed gel and it was then compressed again along with the previously compressed one. We could have added as many layers as we pleased but, for our purpose, two layers were considered to be enough, given that this study is a preliminary study focusing on CNS applications that do not require denser constructs [27]. An alternative method would have been to have one gel on the top of the other in the well and compress them all together at the end, but this would be limited by the height of the well.

2.6. Measurement of Fluid Loss and Time of Compression.

Absorbent paper rolls were weighed on the electronic balance to two decimal points every minute for the first 5 minutes after compression and then 5 minutes until no measurable change of weight was noted. For example, if after 10 minutes of compression the weight of the paper roll did not change compared to the previous reading, which was measured after 5 minutes, time point of 5 minutes was taken as the time of full compression.

Weight gain in the absorbent paper rolls was recorded as fluid loss from the collagen gel during compression. The weight of the water removed from the gels was calculated according to the equation below:

$$\Delta W = \text{Weight } t_n - \text{Weight } t_0, \qquad (4)$$

where Weight t_n is the weight at time point n and Weight t_0 is initial weight of paper roll (4.4 ± 0.3 g). The rate of fluid loss from the gels of different heights was calculated as

$$Q = \frac{\text{Weight } t_{n+1} - \text{Weight } t_n}{\text{Time}_{n+1} - \text{Time}_n}, \qquad (5)$$

where Q is the rate of fluid loss, weight t_{n+1} is the weight of the paper roll at the time point $n + 1$, weight t_n is the weight of the paper roll at the time point n, time$_{n+1}$ is the time point $n + 1$, and time$_n$ is the time point n.

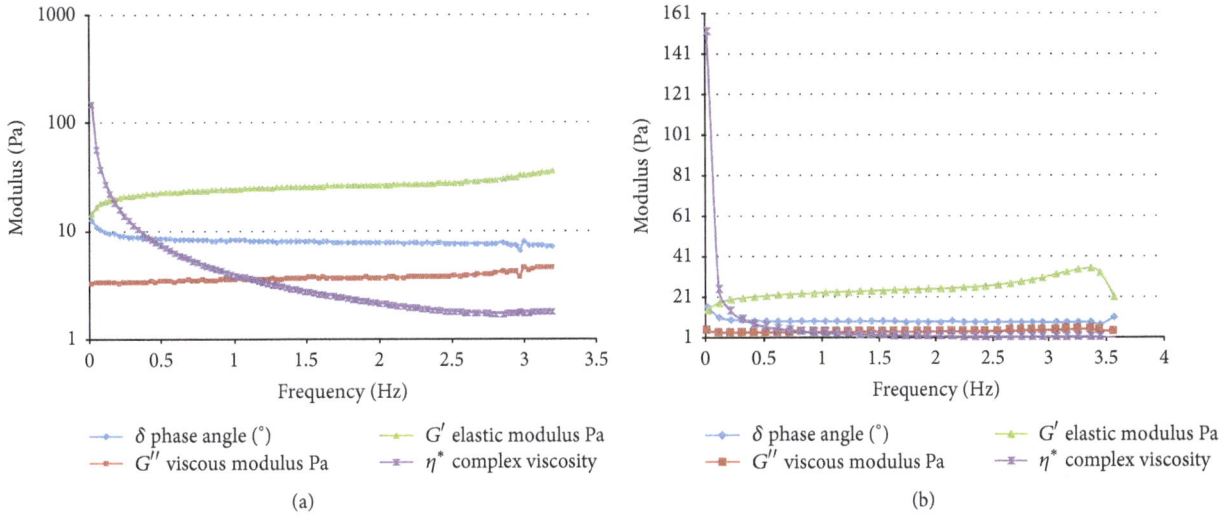

FIGURE 1: Rheometry frequency sweep for gelled collagen-genipin hydrogels in oscillatory mode (a) and gelled collagen hydrogels in oscillatory mode (b). It is illustrated that δ drops, so those hydrogels are viscoelastic materials. $G' > G''$ so this regards a well-structured (gelled) system. We can also see that G' and G'' are almost independent of frequency so, in general, sedimentation is unlikely to occur and particles are strongly associated. G' of collagen-genipin hydrogels $> G'$ of collagen hydrogels and G'' of collagen-genipin hydrogels $< G''$ of collagen hydrogels.

2.7. Hydraulic Resistance of the Fluid Leaving Surface (FLS). The hydraulic resistance of the FLS (RFLS) has been thought to be increased during the plastic compression process. R_{FLS} was calculated according to the following equation:

$$R_{FLS} = \frac{A \times P}{\mu \times Q}, \qquad (6)$$

where R_{FLS} is the hydraulic resistance of fluid leaving surface (FLS), Q is the rate of flow (in ml/min), A is surface area (in cm^2), P is pressure over the surface (in N), and μ is the dynamic viscosity of water (1.002×10^{-3} Ns/m^2).

2.8. Correlation of Compression Level to Rheology. The stiffness of the compressed collagen hydrogel was estimated according to Young's modulus measurements of the rheometer for the different % levels of plastic compression, namely, for 50, 75, and 99% collagen compression, as described in Section 2.2. for "Rheology."

3. Results

3.1. Mechanical and Gelation Studies for Injectable Collagen Hydrogels

3.1.1. Rheometry. A typical example of the rheological data at 37°C is shown in Figure 1. The gel point for the genipin cross-linked hydrogel at 37°C was on average approximately 38 seconds, whereas for the non-cross-linked collagen hydrogel it was 42 seconds. At 25°C, genipin cross-linked hydrogels gelled in 86 seconds and non-cross-linked hydrogels in 81 seconds. The fact that the genipin cross-linked hydrogels required less time to gel at 37°C compared to 25°C might be due to the unique properties of genipin and the effect of

temperature in the gelation time and compressive strength, as described before [28], even though those results need to be validated with future studies. G' for the collagen gels cross-linked with 0.5 mM genipin was higher than G' for the collagen hydrogel and G'' for cross-linked hydrogels was lower than G'' for the collagen hydrogels; the results were not statistically significant though for either condition ($p = 0.17$ and $p = 0.09$, resp., for 37°C and $p = 0.21$ and $p = 0.12$). Young's moduli were ranging from 14.98 kPa up to 22.46 kPa for the first minutes after gelation both for collagen and collagen-genipin hydrogels, which is within the "CNS-friendly" range of Young modulus.

Nevertheless, it seemed like the cross-linked hydrogels kept increasing their modulus over time. After the gels were left to mature for 24 hours, rheology measurements suggested that Young's modulus significantly increased, reaching up to 110 kPa ± 21 ($p < 0.05$) for collagen-genipin hydrogels and up to 65 kPa ± 11 for collagen hydrogels ($p < 0.05$). Thus, approximately a 6-fold increase was observed in the collagen-genipin hydrogels modulus. A 48- or 72-hour testing of the mechanical properties through rheology might be useful in the future, in order to establish how long it takes for G' to be saturated in collagen-genipin hydrogels, indicating the end of cross-linking. This would give us the final Young's modulus for the fully cross-linked collagen-genipin hydrogel to allow a more accurate understanding of the mechanical interaction of the hydrogel system with the CNS.

3.1.2. Hydrogel Degradation Studies. The hydrogels were left to "mature" in the incubator at 37°C for 24 hours before the initiation of the degradation assay, to allow for the mechanical properties of the hydrogels to stabilize after the cross-linking with genipin. The degradation assay lasted for 4 hours and every 30 minutes measurements were conducted to check the

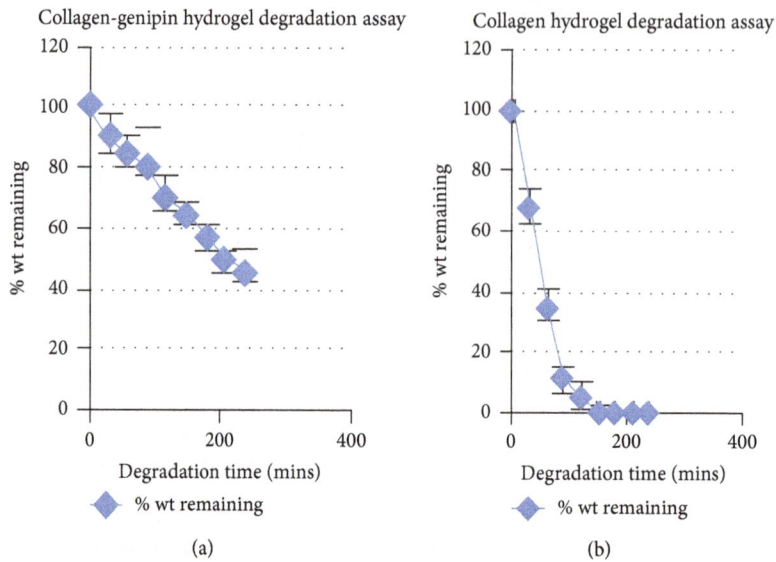

FIGURE 2: On (a), the results of the collagen-genipin hydrogel results are depicted showing a gradual decline in the % weight remaining of the hydrogel over the time due to the exposure to the 0.1% collagenase solution. On (b), the results of the same assay for a collagen hydrogel are depicted, indicating a much more rapid decline in the % weight remaining. It is noted that all gels were left in the incubator to "mature" at 37°C for 24 hours before the initiation of the assay.

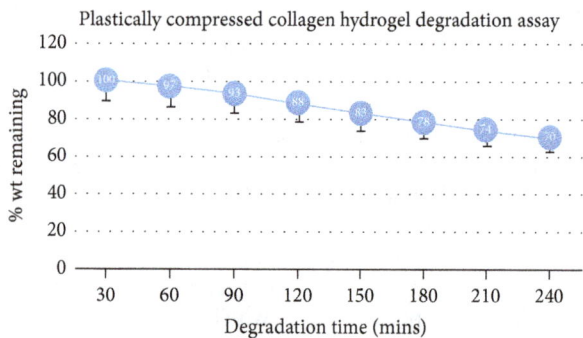

FIGURE 3: This graph depicts the degradation assay results for the plastically compressed collagen hydrogels. There is a marked resistance to the 0.1% collagenase solution compared to the other hydrogels tested.

resistance of the gels to degradation induced by the collagenase solution. PBS was used in place of the collagenase 0.1% solution for our control hydrogels. There was no difference in the weights measured during this time period for the control groups. Figure 2 illustrates the % weight remaining after 4 hours of exposure to 0.1% collagenase for both collagen and collagen-genipin hydrogels. The results suggest that the cross-linking with genipin significantly increases the resistance of the hydrogels to the collagenase-induced degradation ($p <$ 0.05).

Similarly, the degradation rate for our plastically compressed collagen hydrogels was tested. The results for the fully compressed hydrogels are shown in Figure 3. It is evident that plastic compression of the collage hydrogel significantly increased the resistance to the collagenase-induced degradation.

Swelling Ratio. The swelling ratios of the hydrogels were calculated according to (3) previously described, after 24-hour incubation in PBS solution. The swelling ratio of the collagen gels was 82.1 ± 2.3 and the swelling ratio of collagen-genipin gels was 103 ± 4.5. On the other hand, the swelling ratio of the fully compressed collagen hydrogels was between the two aforementioned values, but not significantly higher than the one of the uncompressed collagen hydrogels. In particular, the swelling ratio for the fully compressed collagen hydrogels was 89.4 ± 3.1.

3.2. Mechanical and Gelation Studies for the Compressed Collagen Hydrogels. Gels were fabricated in accordance with the aforementioned protocols and were left in the incubator at 37°C for 24 hours to "mature." Due to the observed rheology measurements that we mentioned above, suggesting that during the first 24 hours the stiffness of collagen-genipin hydrogels significantly increases, approaching Young's modulus values of around 110 kPa, collagen-genipin hydrogels were not considered appropriate for testing through plastic compression, after taking into account our results from the compression of the collagen-only hydrogels. Besides, the primary purpose of this experiment was to compare the impact of the cross-linking of collagen with genipin to the mechanical effect of the plastic compression of collagen, as means of strengthening the inherently weak collagen hydrogels for CNS applications.

3.2.1. Measurement of Fluid Loss and Time of Compression. It is known that >90% of fluid will be extracted from the hydrogels with the method of plastic compression. Using absorbent paper rolls as described before, we performed sequential weight measurements to quantify the fluid loss

TABLE 1: Correlation of the hydrogels' stiffness with the % compression rate.

% fluid loss	Young's modulus (kPa)
0	20.86 ± 3.20
50	40.14 ± 17.10
75	230.51 ± 41.00
99	2238 ± 776

over time and find the compression rates of our hydrogels until the gels get fully compressed.

According to (4) mentioned above, the weight of the water that was removed from the hydrogels was calculated. The initial weight of the paper rolls was found to be around (4.32 ± 0.24 g). For the 3 ml collagen-only plastically compressed hydrogels, the initial gel height in the moulds with constant surface areas of 379.9 mm^2 was 7.9 mm. Approximately 10 minutes were needed in order to reach full compression of those hydrogels. 92.5 ± 1.4 was the calculated percentage of total fluid loss (±SD), while Young's modulus was found to reach extremely high values after full compression (2238 ± 776 kPa).

The Young modulus was also tested for intermediate levels of fluid losses from the hydrogels in order to get a better understanding on how to tune the compression level in accordance with the desired biomedical application. Table 1 summarises the findings.

It is noted that the difference of % fluid loss was statistically significant for the time periods 0-1 minutes and 3-4 minutes ($p < 0.05$).

3.2.2. Rate of Fluid Loss. The rate of fluid loss was calculated according to (5). The rate of fluid loss for the period 0-1 minutes was obviously significantly higher in comparison to all the other time points examined (1.24 ± 0.27 ml/min, $p < 0.05$) (Figure 4). For the 1-2-minute time period, the rate of fluid loss dropped significantly to 0.37 ± 0.18 ml/min ($p < 0.05$). Next, the rate of the fluid loss dropped further for the time period 2-3 minutes (0.31 ± 0.09 ml/min, $p > 0.05$). There was a statistically significant reduction of the rate of fluid loss for the remaining time periods (3-4 minutes and 4-5 minutes) in comparison to the period 2-3 minutes. In specific, the rate of fluid loss was 0.21 ± 0.019 for the time period 3-4 minutes and 0.19 ± 0.018 ml/min for the period 4-5 minutes.

3.2.3. Measurement of the Hydraulic Resistance of FLS (R_{FLS}). It has already been accepted that the FLS of the compressed hydrogel does not have the same properties as the surface opposite to FLS. Collagen fibrils are being accumulated at the FLS site since they keep being moved during the plastic compression process towards the fluid exit point.

Utilising equation (6) mentioned above, it was found that R_{FLS} increased exponentially with time during compression. After 1 minute of plastic compression RFLS was $7.9 \times 10^4 \pm 1.1 \times 10^4$ cm^{-1}, while during the 2nd minute the RFLS significantly increased to $10.04 \times 10^4 \pm 0.6 \times 10^4$ cm^{-1} ($p < 0.05$). The 3rd minute of compression led to another significant increase

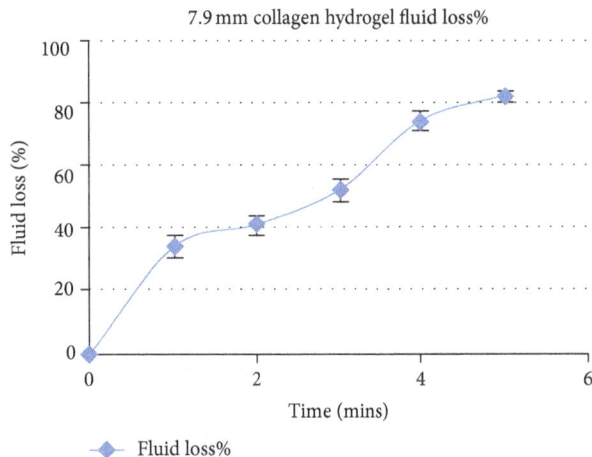

FIGURE 4: Illustration of the % fluid loss over time for collagen-only hydrogels (7.9 mm, 3 ml) for five minutes. Results are depicted as average ± SD.

of R_{FLS} ($13.29 \times 10^4 \pm 0.9 \times 10^4$ cm^{-1}). The last two minutes resulted in R_{FLS} of $17.4 \times 10^4 \pm 1.1 \times 10^4$ cm^{-1} (4th minute) and $18.87 \times 10^4 \pm 0.53 \times 10^4$ cm^{-1} (5th minute).

3.3. Mechanical and Gelation Studies for Compressed Collagen 3D Hydrogel Models

3.3.1. Effect of Multilayering on Dynamics of Fluid Loss from Collagen Gels during Plastic Compression. With the same process as described above for the single collagen hydrogels, the % of fluid loss was calculated over a 5-minute period of time until no change was observed in the weight of the paper roll, indicating full compression. The methodology for developing multilayered plastically compressed collagen 3D hydrogel models has already been described before in Section 2.5. Figure 5 summarises our findings.

4. Discussion

4.1. Collagen and Collagen-Genipin Hydrogels. After neutralising acid solubilized collagen solution with NaOH, a modelling material is developed, which is about to gel when the temperature rises above 4°C, to form a fibrillar gel network [29]. When the collagen mixture is exposed to body temperature (37°C), the reaction kinetics have been found to occur after some seconds up to minutes [30]. Collagen has about 90 amine groups in every collagen molecule, which act as cross-linking sites for genipin and are dispersed throughout the triple helical and telopeptide regions of collagen [31].

The initial reaction of the primary amine groups with genipin triggers the formation of a second activated form of genipin, which, in turn, induces the further polymerization of genipin molecules [32].

The high resistance of the collagen-genipin hydrogels to collagenase degradation is justified due to the wide variety of intrahelical, interhelical, and intermicrofibrillar covalent cross-links throughout the collagen hydrogel [33].

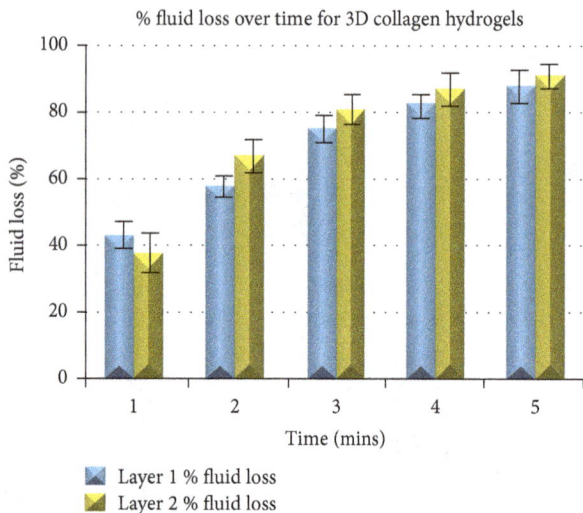

FIGURE 5: % fluid loss of both collagen layers over time for the 3D compressed collagen hydrogel model. It is noticeable that % fluid loss is always higher for the 2nd layer after the 1st minute of compression.

It has been suggested that the impact of the genipin cross-linking on the hydrogel systems is concentration-sensitive. Based on our data the use of this genipin concentration in the described collagen hydrogel might lead to improved biocompatibility and cell viability, given the higher swelling ratio reported when compared to the collagen gels after 24 hours. This observation may be attributed to the increased mechanical strength of the gel, which allowed it to stand under its own weight and entrap water as opposed to collagen hydrogels, which sagged under their own weight. Further studies would be helpful to verify that hypothesis.

The gel point where $G' = G''$ was found to be slightly reduced for collagen gels that were cross-linked with genipin at 37°C. Thus, genipin seemed to accelerate the collagen solution gelation, even though this was not a statistically significant result.

A significant change in the shear storage modulus (G') though was noted, which results in approximately 50% stiffer gels about half an hour after the cross-linking with genipin, in comparison with the collagen-only hydrogels which maintain similar moduli.

The mechanical properties of the hydrogel, including Young's modulus of the gel, are highly important for the biomedical application since it has been known that the mechanical strength of the material can affect cell growth/differentiation. Therefore, choosing the right material with the appropriate mechanical properties in accordance with the suggested application is the first step for a successful approach. The optimal material shall have an elastic modulus matching the one of the native growth environment, meaning that for our purpose we need to use a system with mechanical properties matching the ones of the spinal cord/brain tissue [34, 35].

Nevertheless, the elastic modulus of the CNS tissue is not clear, given that there is some disagreement among different studies. In general, the typical values which are mentioned lie within the range of ~3 kPa–300 kPa for spinal cord tissue [36] and ~500 Pa for brain tissue [35]. Given the soft nature of the spinal cord and of the CNS tissue in general, hydrogels are considered ideal candidates for therapeutic approaches in CNS due to the highly swollen and weak nature that matches the environment.

The moduli of all of our gels lie within the aforementioned range, showing promise for the hydrogels application in the CNS. It is noted though that G' was found to gradually increase during the first 24 hours. Even though it was beyond the scope of this study to investigate that further, future work might need to reach up to the point that G' gets saturated. It is definitely an interesting preliminary finding though, suggesting that collagen-genipin hydrogels need some time to "mature," since cross-linking is still taking place, altering the mechanical properties of the hydrogels several hours after the gels' formation. On the contrary, G' of collagen-only hydrogels has been found to get saturated after less than an hour. Despite that they still remain within the modulus of the CNS tissue.

Degradation experiments over a 4-hour period of time were conducted on gels matured for 24 hours to establish a kinetic profile for gel degradation. Significant increases in gel strength were obtained with 0.5 mM genipin when compared to collagen-only hydrogels. The degradation resistance increased with genipin concentration due to the increased rate of reaction at higher concentrations as analysed above, even though the resistance level was lower than the one depicted in our plastically compressed collagen hydrogels.

Overall, it is concluded that genipin shows a significant effect on the degradation properties of the hydrogel, in comparison to the weak impact on the collagen gel's mechanical properties. Taking into account that the stiffness of the gel is mainly dependent on the short range cross-links, which help in opposing the collagen fibres' motion, it is hypothesized that the existence of a wide range of short and long cross-links throughout the collagen gel induces a more significant impact on the enzyme degradation resistance of the hydrogel in comparison to the impact on the stiffness. Future work might benefit from prolonging the degradation experiment in order to validate the results of this preliminary study.

In terms of the potential cytotoxicity of genipin, this seems to be a concentration-dependent issue. Even though there are studies suggesting that cell viability can be affected when genipin is the cross-linker, potentially due to the cross-linking of the non-cross-linked genipin to the amine acids of the medium [37], tuning the genipin concentration or proceeding with frequent change of the medium, collagen-genipin hydrogels can overcome that limitation. The collagen-genipin hydrogel seems to be promising for future neural engineering applications and might enable minimally invasive therapeutic techniques in the future, but in depth in vitro and in vivo studies need to be conducted first to verify these preliminary findings.

4.2. Compressed Collagen Hydrogels. Collagen compressed hydrogels were developed according to the novel technique of

Professor Brown [24]. The concept of the plastic compression process regards the uniaxial rapid expulsion of more than 98% interstitial fluid from the collagen hydrogels under load. Fluid loss and collagen compressed construct thickness were in agreement with the results of the original paper.

It is of great importance that, among all the tested hydrogels, the plastically compressed collagen hydrogels were the ones that demonstrated even more raised levels of resistance, significantly affecting the degradation of the plastically compressed collagen hydrogels.

The full compression of the collagen hydrogels has been previously proven not to significantly affect cell viability [38] but, according to our findings, Young's modulus has been found to increase drastically up to about 2300 kPa which goes far beyond Young's modulus value of the CNS. According to our findings, a compression level of up to 75% would be in line with the stiffness of the CNS (see Table 1); this would facilitate the application of such a hydrogel for CNS tissue engineering applications.

What is of great importance though and should be taken into account is that after the reexposure of the compressed hydrogel to fluids (e.g., PBS), it will reswell, without reaching its initial hydration status. This would potentially decrease Young's modulus but, even though this was not tested in this experiment, it is hypothesised that the compressed collagen stiffness was so high that any change would not be adequate to approach the low stiffness values of the CNS tissue. Besides, the swelling ratio was still found to be comparable to the one of the uncompressed collagen-only hydrogels, supporting our latter hypothesis. In depth analysis though might be worth being conducted in future studies.

4.3. Compressed-Collagen 3D Hydrogels Systems. The human organism is a hierarchical system where each tissue is a result of the assembly of many separate layers with specialised residing cells. The ultimate goal of biomedical engineering is the mimicking of such a hierarchical model, trying to assimilate the natural environment of the body to optimize the therapeutic effects. Several techniques have been developed (both cell- and biomaterial-based), but all of them have limitations and complexities that discourage their wide usage to fabricate a functional 3D model.

The method of plastic compression that is used here is easy to use, while it can be tuned to match the stiffness and layers of the natural body tissues. The interlayer connection facilitates the biomedical engineering applications, using natural polymers and vivid cells in situ.

The injectable hydrogels, on the other hand, are also 3D models, which are even friendlier for the CNS tissue; they are easy to make and can be highly tunable in order to match the needed properties to optimize our results. In addition, they do not require an open surgery, so neurosurgeons favour their use due to the less invasive nature, which could potentially lack unnecessary complications. Further tests though need to be conducted in order to establish the better therapeutic approach for CNS-related conditions.

5. Conclusion

To conclude, the method of collagen plastic compression is a very promising and easy to use technique with tunable properties, but it seems that, in order for such hydrogels systems to be used for CNS repair, full plastic compression of the collagen hydrogels should probably be avoided due to the tremendous increase in the hydrogel's stiffness after full compression. The stiffness that was found to be in line with the stiffness of the natural CNS environment corresponds to % level of compression up to 75%. If that is taken into consideration, the 3D compressed models could be a great alternative approach for neural engineering strategies in order to accomplish a model that resembles the natural hierarchical model of the human tissues, reaching a better regenerative potential.

Collagen-genipin injectable hydrogels on the other hand have been found to be very easy to use, while, in parallel, they are favoured by the neurosurgeons due to the less interventional therapeutic approach that can be implemented. The gels exhibit mechanical properties similar to the CNS tissue and they have previously been found to adequately support cells growth, facilitating neural regenerative processes. The genipin concentration though should probably be optimized and longer-term studies should test the degradation rate of the optimized hydrogel over time for optimal structural support that will allow adequate regeneration.

Overall, the tested hydrogel systems hold promise for CNS applications, but it is still very soon to conclude on which system is the best for clinical applications, since both hydrogel systems need to be further optimized and tuned. Longer in vitro, as well as in vivo, studies need to be conducted to check the efficiency of those systems more accurately. It seems that the clinical potential of neural engineering strategies is endless and is about to improve in the near future.

Conflicts of Interest

The authors have no conflicts of interest to declare.

Authors' Contributions

Magdalini Tsintou, Kyriakos Dalamagkas, and Alexander Seifalian contributed equally to this work.

Acknowledgments

This work is in memory of Professor Robert Brown, who sadly passed away on February 4, 2016. His research innovation work on compression of biomimetic fabrication process for collagen tissues established him as an international leadingscientist for biomimetic tissue engineering. The authors would also like to thank Dr. Bala Subramaniyam Ramesh for the helpful pieces of advice and technical/scientific support provided. The project was cofunded by the Project "IKY Scholarships" from resources of the OP "Education and Lifelong Learning," the European Social Fund (ESF) NSRF

2007–2013, Leventis Foundation, and the University College of London.

References

[1] B. D. Ulery, L. S. Nair, and C. T. Laurencin, "Biomedical applications of biodegradable polymers," *Journal of Polymer Science Part B: Polymer Physics*, vol. 49, no. 12, pp. 832–864, 2011.

[2] H. Brem and R. Langer, "Polymer-based drug delivery to the brain," *Social Science and Medicine*, vol. 3, pp. 52–61, 1996.

[3] Y. Park, M. Sugimoto, A. Watrin, M. Chiquet, and E. B. Hunziker, "BMP-2 induces the expression of chondrocyte-specific genes in bovine synovium-derived progenitor cells cultured in three-dimensional alginate hydrogel," *Osteoarthritis and Cartilage*, vol. 13, no. 6, pp. 527–536, 2005.

[4] J. P. Fisher, S. Jo, A. G. Mikos, and A. H. Reddi, "Thermoreversible hydrogel scaffolds for articular cartilage engineering," *Journal of Biomedical Materials Research Part A*, vol. 71, no. 2, pp. 268–274, 2004.

[5] S. E. Sakiyama, J. C. Schense, and J. A. Hubbell, "Incorporation of heparin-binding peptides into fibrin gels enhances neurite extension: an example of designer matrices in tissue engineering," *Journal of Federation of American Societies for Experimental Biology*, vol. 13, no. 15, pp. 2214–2224, 1999.

[6] X. Yu, G. P. Dillon, and R. V. Bellamkonda, "A laminin and nerve growth factor-laden three-dimensional scaffold for enhanced neurite extension," *Tissue Engineering Part A*, vol. 5, no. 4, pp. 291–304, 1999.

[7] S. E. Blackshaw, S. Arkison, C. Cameron, and J. A. Davies, "Promotion of regeneration and axon growth following injury in an invertebrate nervous system by the use of three-dimensional collagen gels," *Proceedings of the Royal Society B Biological Science*, vol. 264, no. 1382, pp. 657–661, 1997.

[8] G. P. Dillon, X. Yu, A. Sridharan, J. P. Ranieri, and R. V. Bellamkonda, "The influence of physical structure and charge on neurite extension in a 3D hydrogel scaffold," *Journal of Biomaterials Science, Polymer Edition*, vol. 9, no. 10, pp. 1049–1069, 1998.

[9] Matrigel Matrix, Extracellular Matrix, Corning https://www.corning.com/worldwide/en/products/life-sciences/products/surfaces/matrigel-matrix.html.

[10] ExtracelTM Product Data Sheet, 2007, https://www.advanced-biomatrix.com/extracel%E2%84%A2-product-data-sheet/.

[11] W. Koh, A. N. Stratman, A. Sacharidou, and G. E. Davis, "In vitro three dimensional collagen matrix models of endothelial lumen formation during vasculogenesis and angiogenesis," *Methods in Enzymology*, vol. 443, no. 8, pp. 83–101, 2008.

[12] H. Zimmermann, S. G. Shirley, and U. Zimmermann, "Alginate-based encapsulation of cells: past, present, and future," *Current Diabetes Reports*, vol. 7, no. 4, pp. 314–320, 2007.

[13] M. Dvir-Ginzberg, T. Elkayam, and S. Cohen, "Induced differentiation and maturation of newborn liver cells into functional hepatic tissue in macroporous alginate scaffolds," *The FASEB Journal*, vol. 22, no. 5, pp. 1440–1449, 2008.

[14] M. Antman-Passig and O. Shefi, "Remote magnetic orientation of 3D collagen hydrogels for directed neuronal regeneration," *Nano Letters*, vol. 16, no. 4, pp. 2567–2573, 2016.

[15] A. P. G. Castro, P. Laity, M. Shariatzadeh, C. Wittkowske, C. Holland, and D. Lacroix, "Combined numerical and experimental biomechanical characterization of soft collagen hydrogel substrate," *Journal of Materials Science: Materials in Medicine*, vol. 27, no. 4, article no. 79, pp. 1–9, 2016.

[16] E. A. A. Neel, U. Cheema, J. C. Knowles, R. A. Brown, and S. N. Nazhat, "Use of multiple unconfined compression for control of collagen gel scaffold density and mechanical properties," *Soft Matter*, vol. 2, no. 11, pp. 986–992, 2006.

[17] J. Skopinska-Wisniewska, J. Kuderko, A. Bajek, M. Maj, A. Sionkowska, and M. Ziegler-Borowska, "Collagen/elastin hydrogels cross-linked by squaric acid," *Materials Science and Engineering C: Materials for Biological Applications*, vol. 60, pp. 100–108, 2016.

[18] R. Parenteau-Bareil, R. Gauvin, and F. Berthod, "Collagen-based biomaterials for tissue engineering applications," *Materials*, vol. 3, no. 3, pp. 1863–1887, 2010.

[19] J. Jin, M. Song, and D. J. Hourston, "Novel chitosan-based films cross-linked by genipin with improved physical properties," *Biomacromolecules*, vol. 5, no. 1, pp. 162–168, 2004.

[20] H. Sung, W. Chang, C. Ma, and M. Lee, "Crosslinking of biological tissues using genipin and/or carbodiimide," *Journal of Biomedical Materials Research Part B: Applied Biomaterials*, vol. 64A, no. 3, pp. 427–438, 2003.

[21] H. Liang, W. Chang, K. Lin, and H. Sung, "Genipin-crosslinked gelatin microspheres as a drug carrier for intramuscular administration: in vitro andin vivo studies," *Journal of Biomedical Materials Research Part B: Applied Biomaterials*, vol. 65A, no. 2, pp. 271–282, 2003.

[22] M. T. Nickerson, J. Patel, D. V. Heyd, D. Rousseau, and A. T. Paulson, "Kinetic and mechanistic considerations in the gelation of genipin-crosslinked gelatin," *International Journal of Biological Macromolecules*, vol. 39, no. 4-5, pp. 298–302, 2006.

[23] H. G. Sundararaghavan, G. A. Monteiro, B. L. Firestein, and D. I. Shreiber, "Neurite growth in 3D collagen gels with gradients of mechanical properties," *Biotechnology and Bioengineering*, vol. 102, no. 2, pp. 632–643, 2009.

[24] R. A. Brown, M. Wiseman, C.-B. Chuo, U. Cheema, and S. N. Nazhat, "Ultrarapid engineering of biomimetic materials and tissues: Fabrication of nano- and microstructures by plastic compression," *Advanced Functional Materials*, vol. 15, no. 11, pp. 1762–1770, 2005.

[25] E. Hadjipanayi, V. Mudera, and R. A. Brown, "Guiding cell migration in 3D: a collagen matrix with graded directional stiffness," *Cell Motility and the Cytoskeleton*, vol. 66, no. 3, pp. 121–128, 2009.

[26] H. Park, X. Guo, J. S. Temenoff et al., "Effect of swelling ratio of injectable hydrogel composites on chondrogenic differentiation of encapsulated rabbit marrow mesenchymal stem cells in vitro," *Biomacromolecules*, vol. 10, no. 3, pp. 541–546, 2009.

[27] U. Cheema and R. A. Brown, "Rapid fabrication of living tissue models by collagen plastic compression: understanding three-dimensional cell matrix repair," *Advances in Wound Care*, vol. 2, no. 4, pp. 176–184, 2013.

[28] L. Bi, Z. Cao, Y. Hu et al., "Effects of different cross-linking conditions on the properties of genipin-cross-linked chitosan/collagen scaffolds for cartilage tissue engineering," *Journal of Materials Science: Materials in Medicine*, vol. 22, no. 1, pp. 51–62, 2011.

[29] G. Forgacs, S. A. Newman, B. Hinner, C. W. Maier, and E. Sackmann, "Assembly of collagen matrices as a phase transition revealed by structural and rheologic studies," *Biophysical Journal*, vol. 84, no. 2, pp. 1272–1280, 2003.

[30] M. P. Palmer, E. L. Abreu, A. Mastrangelo, and M. M. Murray, "Injection temperature significantly affects in vitro and in vivo performance of collagen-platelet scaffolds," *Journal of Orthopaedic Research*, vol. 27, no. 7, pp. 964–971, 2009.

[31] J. S. Pieper, T. Hafmans, J. H. Veerkamp, and T. H. van Kuppevelt, "Development of tailor-made collagen-glycosaminoglycan matrices: EDC/NHS crosslinking, and ultrastructural aspects," *Biomaterials*, vol. 21, no. 6, pp. 581–593, 2000.

[32] H. G. Sundararaghavan, G. A. Monteiro, N. A. Lapin, Y. J. Chabal, J. R. Miksan, and D. I. Shreiber, "Genipin-induced changes in collagen gels: correlation of mechanical properties to fluorescence," *Journal of Biomedical Materials Research Part A*, vol. 87, no. 2, pp. 308–320, 2008.

[33] L. Ma, C. Gao, Z. Mao, J. Zhou, and J. Shen, "Enhanced biological stability of collagen porous scaffolds by using amino acids as novel cross-linking bridges," *Biomaterials*, vol. 25, no. 15, pp. 2997–3004, 2004.

[34] J. B. Leach, X. Q. Brown, J. G. Jacot, P. A. Dimilla, and J. Y. Wong, "Neurite outgrowth and branching of PC12 cells on very soft substrates sharply decreases below a threshold of substrate rigidity," *Journal of Neural Engineering*, vol. 4, no. 2, article 003, pp. 26–34, 2007.

[35] K. Saha, A. J. Keung, E. F. Irwin et al., "Substrate modulus directs neural stem cell behavior," *Biophysical Journal*, vol. 95, no. 9, pp. 4426–4438, 2008.

[36] R. J. Oakland, R. M. Hall, R. K. Wilcox, and D. C. Barton, "The biomechanical response of spinal cord tissue to uniaxial loading," *Proceedings of the Institution of Mechanical Engineers, Part H: Journal of Engineering in Medicine*, vol. 220, no. 4, pp. 489–492, 2006.

[37] G. Fessel, J. Cadby, S. Wunderli, R. van Weeren, and J. G. Snedeker, "Dose and time dependent effects of genipin crosslinking on cell viability and tissue mechanics—toward clinical application for tendon repair," *Acta Biomaterialia*, vol. 10, no. 5, pp. 1897–1906, 2014.

[38] K. Hu, H. Shi, J. Zhu et al., "Compressed collagen gel as the scaffold for skin engineering," *Biomedical Microdevices*, vol. 12, no. 4, pp. 627–635, 2010.

11

Magnesium Oxide Nanoparticles Reinforced Electrospun Alginate-Based Nanofibrous Scaffolds with Improved Physical Properties

**R. T. De Silva,[1] M. M. M. G. P. G. Mantilaka,[1] K. L. Goh,[2]
S. P. Ratnayake,[1] G. A. J. Amaratunga,[1,3] and K. M. Nalin de Silva[1,4]**

[1]Nanotechnology and Science Park, Sri Lanka Institute of Nanotechnology (SLINTEC), Pitipana, Homagama, Sri Lanka
[2]School of Mechanical and Systems Engineering, Newcastle University, Newcastle Upon Tyne, UK
[3]Electrical Engineering Division, Department of Engineering, University of Cambridge, 9 J. J. Thomson Avenue, Cambridge CB3 0FA, UK
[4]Department of Chemistry, University of Colombo, Colombo 3, Sri Lanka

Correspondence should be addressed to R. T. De Silva; rangikadsilva@gmail.com

Academic Editor: Traian V. Chirila

Mechanically robust alginate-based nanofibrous scaffolds were successfully fabricated by electrospinning method to mimic the natural extracellular matrix structure which benefits development and regeneration of tissues. Alginate-based nanofibres were electrospun from an alginate/poly(vinyl alcohol) (PVA) polyelectrolyte complex. SEM images revealed the spinnability of the complex composite nanofibrous scaffolds, showing randomly oriented, ultrafine, and virtually defects-free alginate-based/MgO nanofibrous scaffolds. Here, it is shown that an alginate/PVA complex scaffold, blended with near-spherical MgO nanoparticles (ø 45 nm) at a predetermined concentration (10% (w/w)), is electrospinnable to produce a complex composite nanofibrous scaffold with enhanced mechanical stability. For the comparison purpose, chemically cross-linked electrospun alginate-based scaffolds were also fabricated. Tensile test to rupture revealed the significant differences in the tensile strength and elastic modulus among the alginate scaffolds, alginate/MgO scaffolds, and cross-linked alginate scaffolds ($P < 0.05$). In contrast to cross-linked alginate scaffolds, alginate/MgO scaffolds yielded the highest tensile strength and elastic modulus while preserving the interfibre porosity of the scaffolds. According to the thermogravimetric analysis, MgO reinforced alginate nanofibrous scaffolds exhibited improved thermal stability. These novel alginate-based/MgO scaffolds are economical and versatile and may be further optimised for use as extracellular matrix substitutes for repair and regeneration of tissues.

1. Introduction

Polymeric nanofibres have gained enormous attention in the recent past due to those of particular interest in tissue engineering applications [1]. Typically, the artificial scaffolds which are being used in tissue engineering applications should mimic the spatial-porous-structured morphology of extracellular matrices (ECM) which can be found in native tissues and organs of human body to facilitate the cell growth and proliferation. Artificial ECM scaffolds which are used for tissue engineering applications can be fabricated using different techniques such as freeze-drying [2, 3], template-based solution casting [4], 3D printing [5], wet-spinning [6], and electrospinning [1, 3]. Among these techniques, electrospinning is one of the most feasible methods to produce scaffolds due to its versatility and robustness.

Electrospinning enables the production of three-dimensional porous-structured fibrous mat which mimics the natural structure of ECM and helps to promote cell adhesion and permit sufficient gases to exchange [1]. Recently, biopolymer based electrospun scaffolds have been extensively studied for tissue engineering applications. To date, a number of biopolymers, notably chitosan [7], alginate [8], poly(lactic acid) (PLA) [9], and poly(ethylene oxide) (PEO) [10], have been used in fabricating nanofibrous scaffolds by electrospinning. In particular, alginate takes a predominant place due to its biocompatibility, biodegradability, and relatively low cost for mass production [11]. These unique properties have enabled alginate to be used in many biomedical applications such as drug delivery and skin/bone scaffolds [11, 12]. Alginate is a linear polysaccharide copolymer composed of two sterically different repeating units of β-d-mannuronate (M unit) and α -L-glucuronate (G unit) in various M/G ratios. Although a few aspects of alginate-based electrospun scaffolds related to ECM tissues such as cell adhesion [13], alterations in scaffold fibre dimensions [14] and scaffold degradation [15] have been extensively studied, and only a limited number of studies have been carried out on reinforced alginate nanofibrous mats to ensure the required mechanical strength and structural properties.

Typically the mechanical properties of biopolymer scaffolds are enhanced by either cross-linking or incorporating micro-/nanofillers. Although cross-linking biopolymer scaffolds is a promising method, it reduces the in vivo degradation rate of the biopolymeric scaffold and changes the host tissue responses [16]. On the other hand, incorporating micro-/nanofillers into polymeric fibres enables the production of multifunctional scaffolds with enhanced mechanical properties and other vital characteristics such as antimicrobial and anti-inflammatory characteristics of ECM scaffolds. To date, different types of micro-/nanofillers such as hydroxyapatite (HA) [12, 17], chitin whiskers [18], ZnO [19], and Ag nanoparticles [20] have been used to reinforce electrospun alginate nanofibrous scaffolds. With the aforementioned requirements in reinforcing alginate nanofibrous scaffolds, it is essential to widen the research scope by evaluating different types of nanofillers to enhance the mechanical strength of alginate scaffolds. In this study MgO nanoparticles have been utilized for the first time to reinforce electrospun alginate fibrous scaffolds and their performances were evaluated.

MgO nanoparticles have gained much interest in recent years due to their attractive properties including large surface area-to-volume ratio, thermal and electrical insulation, strong adsorption ability of dye wastes and toxic gases, antimicrobial activity, nontoxicity, and biocompatibility [21–25]. With these outstanding features, MgO nanoparticles have been vastly used in applications such as a catalyst, ceramic material, thermal and electrical insulator, bactericide, material to treat toxic liquid and gaseous wastes, multifunctional composites, and a refractory material [21–25]. MgO nanoparticles are conveniently synthesised with economical routes using low-cost raw materials including magnesium salts, brines, and naturally occurring minerals such as dolomite and magnesite [23]. MgO nanoparticles are mainly synthesised by calcination of nanometre scale magnesium carbonates, magnesium hydroxide, and their composites. In the calcination method, precursor nanoparticles are basically kept in the nanometre scale using polymers and surfactants [22, 23, 25, 26]. Recently MgO nanoparticles have been used to reinforce a number of biopolymers. Zhao et al.

fabricated MgO nanowhiskers reinforced PLA nanocomposites films for bone repair and fixation [27]. In another study, chitosan was reinforced with spherical MgO nanoparticles for high performance packaging applications [21]. For instance, tensile stress and elastic modulus significantly improved by 86% and 38%, respectively, with the addition of 5% (w/w) of MgO into chitosan matrix.

In this study, MgO nanoparticles reinforced alginate nanofibrous scaffolds were fabricated by electrospinning method for the first time and their mechanical and structural properties were systematically investigated. Spherical MgO nanoparticles were synthesised using a polymer template-based ex situ method. Herein, the influence of MgO nanoparticles on the mechanical, morphological, chemical, and thermal properties of alginate nanofibrous scaffolds was investigated. Furthermore, tensile and structural properties of MgO reinforced alginate scaffolds were compared with those of the glutaraldehyde cross-linked alginate scaffolds. The fabricated MgO reinforced alginate nanofibrous scaffolds exhibited a great potential to be used as an artificial scaffold to substitute extracellular matrices.

2. Materials and Methods

2.1. Materials. Sodium alginate powder and poly(vinyl alcohol) (PVA) (with a M_w of 89,000), acrylic acid (AA) (99% purity), potassium persulfate (99% purity), magnesium chloride hexahydrate (99% purity), and sodium hydroxide (99% purity) were used in this study and purchased from SRL Ltd.

2.2. Synthesis of MgO Nanoparticles. MgO nanoparticles used in this study were synthesised using a method reported by Mantilaka et al. [23] with some modifications. In the current method, poly(acrylic acid) (PAA) was prepared by polymerization of 25 mL of 0.5 M AA using 1 g of $K_2S_2O_8$ initiator in an aqueous medium. PAA was added to 100 mL of 1 M NaOH solution. 25 mL of 0.5 M $MgCl_2$ was added dropwise to the PAA/NaOH mixture while stirring to produce PA^- stabilized $Mg(OH)_2$ precursor nanoparticles. Finally, the precursor was heat-treated at 600°C for 3 h to produce MgO nanoparticles.

2.3. Fabrication of Alginate/MgO Nanofibrous Scaffolds by Electrospinning. Electrospun alginate fibrous mats were prepared using an alginate solution which comprises a secondary polymer, PVA. 2% (w/v) alginate solution was prepared by dissolving alginate in distilled water and corresponding amount of MgO was incorporated under vigorous stirring to prepare 10% (w/w) MgO composition (MgO amount is with respect to the weight of alginate). PVA solution of 10% (w/v) was prepared by dissolving PVA in distilled water at 80°C with continuous stirring for 3-4 h. These alginate/MgO and PVA solutions were mixed together in 3 : 2 weight ratio for 4 h under vigorous stirring and followed by an ultrasound treatment (at an amplitude of 80 Hz for 30 min) to achieve a homogeneous MgO dispersion. The resultant solution was electrospun in a horizontal electrospinning setup. All samples were electrospun with a solution flow rate of 8–10 μL/min, having needle to collector distance of 10 cm, needle diameter

TABLE 1: The optimised values of the operating parameters for spinnability and brief highlights of the morphology of the fibres in the scaffolds.

Fibre composition	Viscosity (P), flow rate (μL/min), and voltage (kV)	Observed morphology
Alginate	68.4 p $\pm$ 0.08, 8–10, and 26	Diameter of 62–180 nm; randomly oriented and continuous; ultrafine, wavy, and smooth surface; beads-free
Alginate/MgO 10% (w/w)	72.2 p $\pm$ 0.0, 8–10, and 26	Diameter of 83–230 nm; randomly oriented and continuous; ultrafine, wavy, and smooth surface; beads-free

of 0.5 mm, and voltage of 26–28 kV (electrospinning parameters are given in Table 1). Alginate/PVA (3 : 2 weight ratio) nanofibres (henceforth, the electrospun alginate/PVA (3 : 2 weight ratio) nanofibres are referred to as alginate nanofibres) without MgO nanoparticles were also fabricated by electrospinning for the comparison purposes. Section S2 (prediction parameters are given in Table S2 in Supplementary Material available online at https://doi.org/10.1155/2017/1391298) describes a numerical method to determine the required MgO loading to reinforce alginate fibres (predetermined amount of 10% (w/w) was selected based on that (Fig. S3)).

Additionally, to compare the effects of particle reinforcement with chemical cross-linking, electrospun alginate fibrous mats were cross-linked by immersing those in 20 ml of 2% (v/v) glutaraldehyde solution for 2 h and samples were dried in vacuum oven at 40°C for 24 h. Over-cross-linking led to disrupted fibre structure (Fig. S2).

2.4. Characterization of MgO Nanoparticles and Alginate/MgO Scaffolds

2.4.1. Morphological Analysis. The morphologies of the electrospun alginate nanofibrous scaffolds as well as the synthesised MgO nanoparticles were examined using a field-emission scanning electron microscope (FE-SEM) (Hitachi SU6600). To prevent electrostatic charging during observation, the samples were coated with a thin layer of gold. The extent of the impregnation of MgO nanoparticles within the nanofibres was determined by carrying out energy-dispersive X-ray (EDX) spectroscopy with a scanning rate of 192000 CPS for 4.5 min. Surface roughness of the fibres was determined by an atomic force microscope (AFM) (Park Systems, XE-100) using the cantilever mode (10 nm tip radius) at 0.5 Hz frequency. Crystallographic structure of synthesised MgO nanoparticles was analysed using X-ray diffractometer (Bruker, Focus D8). The CuKα radiation source was operated at a 40 kV power and 40 mA current and data collected within 20–70° of diffraction angle (2θ).

2.4.2. Fourier Transform Infrared (FTIR) Analysis. FTIR spectroscopy (Bruker Vertex 80) was conducted to identify the presence of polymer phases, filler-matrix interfacial interaction, and chemical homogeneity of the electrospun alginate nanocomposites. The results were also compared with those derived from raw PVA and MgO. All spectra were obtained within 500–4000 cm^{-1} with 32 scans per measurement at 0.4 cm^{-1} resolution.

2.4.3. Tensile Test. Mechanical properties such as tensile strength (σ), elastic modulus (E), and elongation at break (ε) of alginate/MgO nanocomposites were evaluated using an Instron Tensile test rig, following a procedure in accordance with the ASTM D882-02. A strain rate of 5 mm/min was used in this test. The force-displacement data was evaluated to determine the stress-strain data; here stress and strain are defined as the nominal stress and strain. In particular, to determine the nominal cross-sectional area of the specimen, the width and thickness of each scaffold specimen were measured using a digital micrometer screw-gauge (Mitutoyo, 0.001 mm resolution) prior to testing.

One-way analysis of variance (ANOVA) was implemented, complemented by the Tukey Post Hoc test, using commercial software (OriginPro 8) to investigate for significant difference in the respective σ, E, and ε among the three different groups, namely, alginate scaffolds, alginate/MgO scaffolds, and cross-linked alginate scaffolds.

2.4.4. Thermal Properties. The thermal decomposition temperature of electrospun alginate nanocomposites was determined by thermogravimetric analysis (TGA) (STD Q600) from 25 to 800°C at a heating rate of 10°C/min in nitrogen medium.

3. Results and Discussion

3.1. Characteristics of Synthesised MgO Nanoparticles. Figure 1(a) shows the graph of intensity versus angular position to describe the XRD pattern of synthesised MgO nanoparticles. Of note, the peak positions at 2θ = 31.34°, 36.78°, 42.73°, 45.08°, and 62.17° are attributed to the periclase crystalline form of MgO (JCPDS Card Number 75-1525). Any other crystalline phase is not identified in XRD pattern of synthesised MgO nanoparticles. The mean crystallite size of MgO nanoparticles is approximately 23 nm as calculated using the Debye-Scherrer formula. SEM image (Figure 1(b)) of synthesised MgO nanoparticles reveals that the particles are in spherical morphology with an average particle diameter of 45 nm.

3.2. Physicochemical Properties of Electrospun Alginate/MgO Nanocomposite Scaffolds

3.2.1. Morphological Properties. Figures 2(a)–2(d) show SEM images of the electrospun alginate nanofibrous scaffolds, revealing uniform, ultrafine, and randomly oriented alginate

FIGURE 1: Structural analysis of MgO nanoparticles: (a) a graph of intensity versus angular position derived from XRD analysis and (b) a SEM image of MgO nanoparticles.

nanofibres. Of note, the electrospun alginate nanofibrous scaffold appears white (Fig. S1 in supplementary data); all meshes were fabricated to a thickness ranging from 20 to 50 μm. We can estimate the size of the interfibre spacing by examining these SEM images. It is predicted that, to order of magnitude, the pore size ranges from 2 to 50 μm, in good order of magnitude agreement with the results of other types of electrospun scaffold, for example, collagen/PCL/TCP mesh [28]. Insets in Figures 2(b) and 2(d) show the graphs of the number of counts versus energy derived from EDX spectroscopy analysis of the respective alginate scaffolds and alginate/MgO scaffolds. These graphs reveal the presence of Na and Mg peaks corresponding to the alginate and MgO nanoparticles, respectively.

Figures 2(e) and 2(f) show histograms of frequency versus fibre diameters (thickness) derived from a simple image analysis of the electron micrographs of alginate scaffolds and alginate/MgO scaffolds, respectively. This analysis reveals that the alginate-based fibres possess diameters ranging from 62 to 180 nm while alginate/MgO fibres possess diameters of 83–230 nm. Noting that the ranges overlap somewhat, numerically, this suggests that the alginate-based fibres and alginate/MgO fibres do not differ appreciably, valid to order of magnitude (Figure 2(f)). Additionally, it can also be seen that the incorporation of MgO yields no appreciable change in the overall structure of the fibrous scaffolds. The diameter of collagen fibrils ranges from 50 to 350 nm in tendon [29], 20–160 nm in ligament [30], and 30–400 nm in collagen fibril extracts from peristomial membrane of sea urchin (*Paracentrotus lividus*) [31]. Thus, it is seen that the range of values for fibril diameters of the alginate-based/MgO scaffolds overlaps considerably with those in biological tissues. In acellular dermal matrix (ADM) for application as ECM scaffolds, it is found that the diameter of the collagen fibrils ranges from 56.0±8.2 to 60.8±1.9 nm (mean ± SD) from bovine of varying age groups [32]. Thus, it is seen that the lower limit of the fibres diameter of the alginate-based/MgO meshes is in good

order of magnitude agreement with those of acellular dermal matrices.

On the other hand, the cross-linked alginate-based nanofibres are densely packed, fused, and appreciably enlarged (Figures 2(g) and 2(h)). Altogether, these contribute to an appreciable reduction in the interfibre porosity of the mesh. To order of magnitude, the pore size is estimated at around 10 μm or lower. Cross-linking with glutaraldehyde results in acetal bridges, which refer to intramolecular and intermolecular interactions of the hydroxyl groups of PVA with the carbonyl groups of glutaraldehyde [33, 34]. Interfibre porosity is a vital factor in artificial scaffolds; high porosity can facilitate cell adhesion and increase cell proliferation. Hence, the cross-linked alginate scaffold may not be as useful as the alginate/MgO scaffolds for tissue engineering applications.

In all cases, SEM images (Figures 2(a)–2(d)) and AFM images (Figure 3) reveal that the surfaces of the electrospun alginate-based fibres and alginate/MgO fibres are smooth (surface roughness (Ra) is 48 nm) and free of unusual artifacts that might suggest defects. On the other hand, some fillers, such as halloysite and ZnO, have been reported to affect the morphology of electrospun nanofibres—these fibres result in bead formation along the nanofibres [19, 35].

3.2.2. FTIR Analysis. Figure 4 shows the graphs of intensity versus wavenumber derived from FTIR analysis of alginate composite scaffolds. The results from raw alginate and MgO are also presented here for the purpose of comparison. The hydroxyl groups at 3300 cm^{-1}, asymmetric carboxyl at 1600 cm^{-1}, symmetric carboxyl at 1400 cm^{-1}, and carbonyl functional groups at 1015 cm^{-1} [19] that appear in the FTIR spectra are attributed to the major functional groups of sodium alginate of the alginate scaffold (scaffolds contain alginate to PVA weight ratio of 3 : 2). Functional groups of PVA such as O-H stretching at 3300 cm^{-1}, C-H stretching of

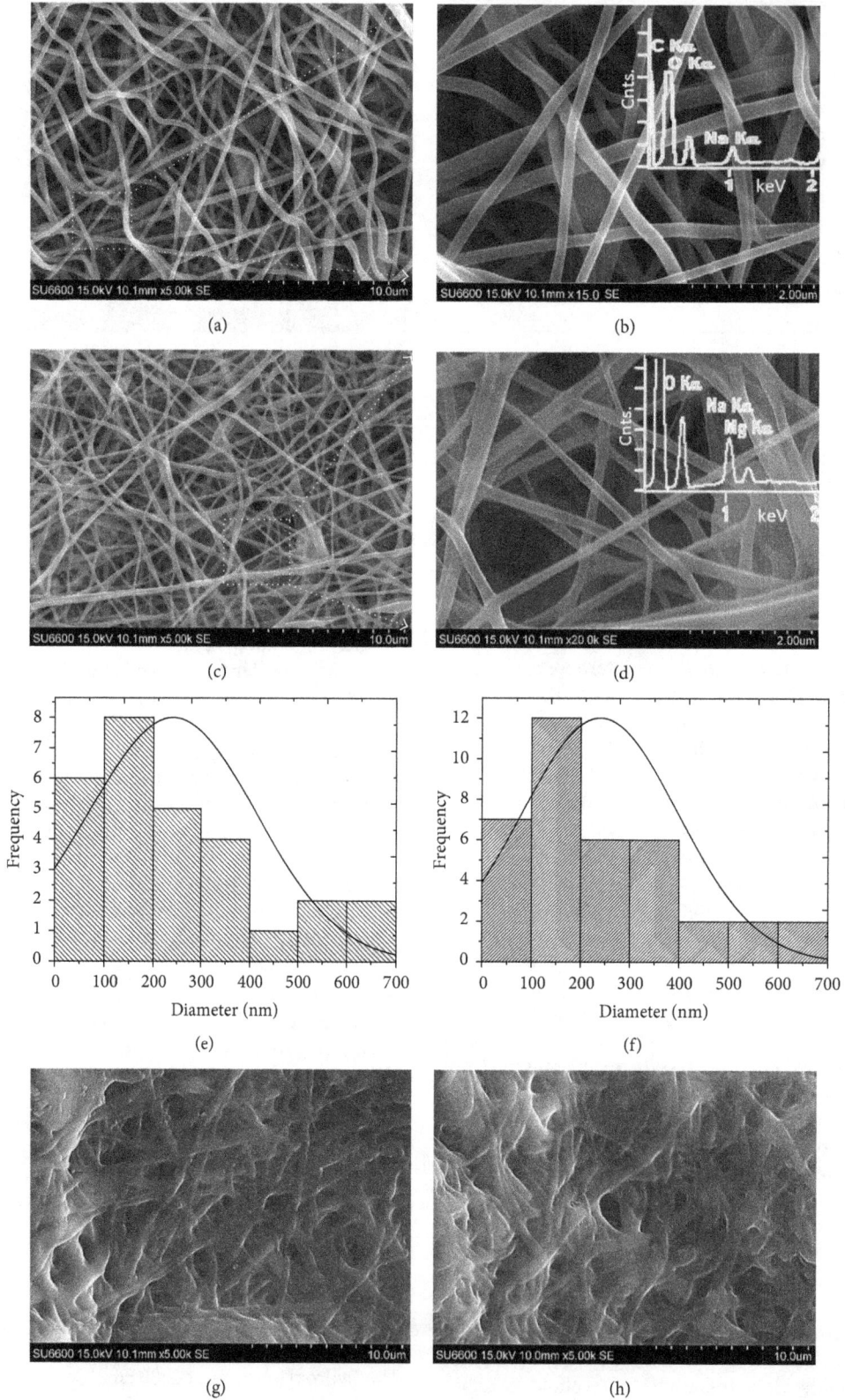

FIGURE 2: Morphology of electrospun alginate-based scaffolds. (a, b) show SEM images of the alginate-based scaffolds. (c, d) show SEM images of the alginate/MgO scaffolds (with 10% (w/w) MgO). Histograms of frequency versus fibre diameter for the (e) alginate-based fibres and (f) alginate/MgO fibres. (g, h) show SEM images of the cross-linked alginate scaffolds. Insets in (b) and (d) are the graphs of the number of counts versus energy derived from EDX analysis.

170202Topography009

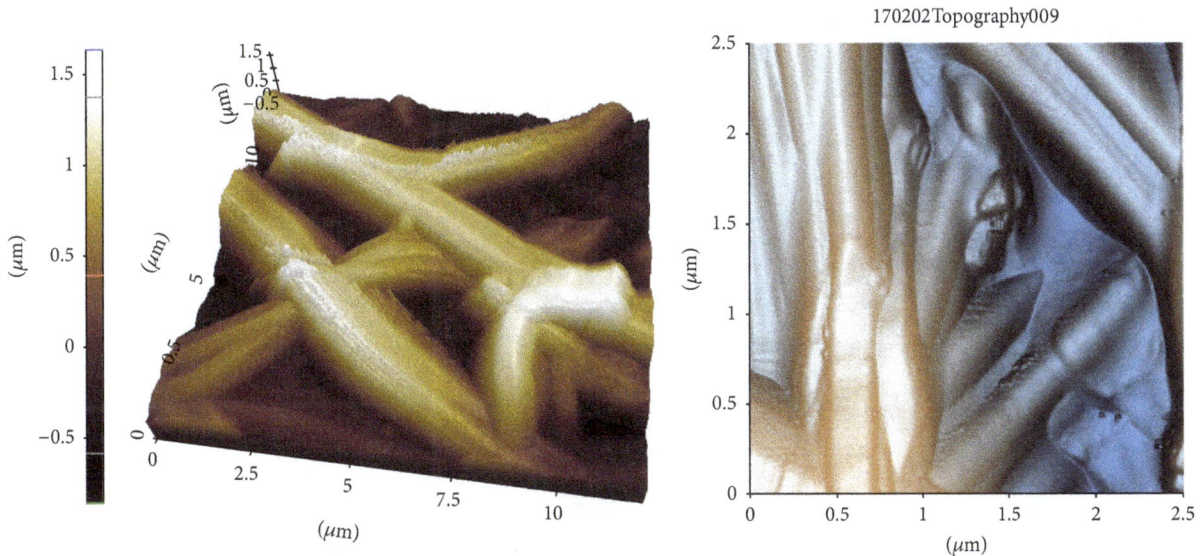

FIGURE 3: 3D and 2D AFM images of alginate nanocomposite scaffolds.

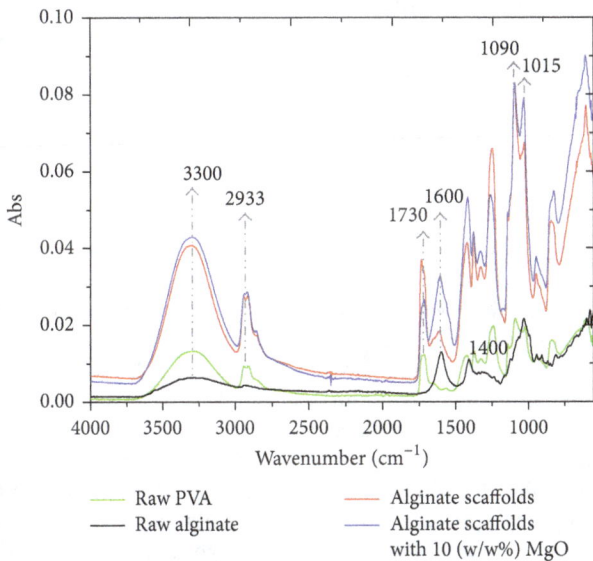

FIGURE 4: FTIR spectrum of raw alginate, raw PVA, and electrospun alginate-based scaffolds and alginate/MgO scaffolds.

alkyl groups at 2933 cm^{-1}, C=O and C-O stretching of acetate groups at 1730 cm^{-1}, and C-C stretching at 1090 cm^{-1} [34] can also be seen in the alginate-based scaffolds although many of these peaks overlap somewhat with the functional groups of alginate. The functional groups of MgO such as Mg-O stretching vibrational band at 540 cm^{-1} could not be isolated in the spectrum of nanocomposite scaffolds since those peaks overlap appreciably with the intense peaks of alginate and PVA. However, the presence of functional groups of alginate and PVA in the electrospun nanocomposite scaffolds confirms that the addition of MgO nanoparticles did not affect the structural integrity of the polymer blend.

3.2.3. Tensile Properties. Figure 5(a) shows the graph of tensile strength (σ_U) and elastic modulus (E) of the respective alginate-based, alginate/MgO, and cross-linked alginate nanofibrous scaffolds. Statistical analysis (ANOVA) reveals that the means of σ_U of the respective scaffolds are significantly different ($P < 0.05$); the Tukey Post Hoc analysis reveals that this occurs between the alginate and alginate/MgO meshes. Similar conclusions have also been observed for E. Thus the addition of MgO nanoparticles results in enhanced mechanical properties. In particular, σ_U of the alginate/MgO mesh (mean σ_U = 4.33 MPa) is approximately three times greater than that of the alginate scaffold (mean σ_U = 1.55 MPa). Similarly, E of the alginate/MgO scaffold (mean E = 0.17 GPa) is three times greater than that of the alginate scaffold (mean E = 0.05 GPa). The increase in σ_U is attributed to the ability to transfer stress from the matrix phase to fillers, facilitated an efficient interaction at the filler-matrix interface, and is directed by hydrogen bonding between the abundant hydroxyl groups of alginate/PVA complex and MgO nanoparticles. Furthermore, these enhanced interfacial interactions between the fillers and matrix also contribute to high structural rigidity (i.e., E), as a result of restriction to the mobility of the polymer chains, when the scaffold deforms under an applied load. Of note, as expected σ_U of cross-linked alginate scaffold is higher than that of the untreated alginate scaffold. However, the cross-linked alginate scaffold is only marginally higher (i.e., 18%) than that of alginate/MgO scaffold. This could be due to the presence of covalent bonds associated with intramolecular and intermolecular interactions of PVA monomers [33]. Although cross-linked alginate scaffolds exhibit better tensile properties than MgO reinforced alginate nanocomposite scaffolds, cross-linking reduces the interfibre porosity of the scaffolds (Section 3.2.1). Consequently, this is unfavourable for promoting cell growth and proliferation as pointed out earlier. Therefore, nanofillers

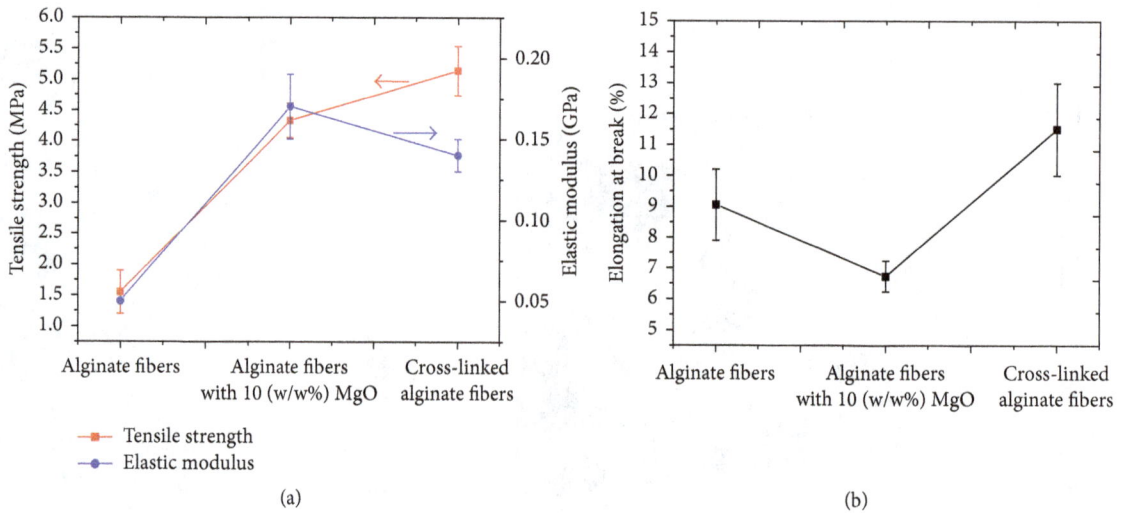

Figure 5: Plots of (a) tensile strength and elastic modulus and (b) elongation at break of electrospun alginate-based, alginate/MgO, and cross-linked alginate scaffolds.

reinforced alginate/PVA scaffolds could be more advantageous than cross-linked alginate/PVA scaffolds for tissue engineering applications.

Figure 5(b) shows a plot of elongation at break (ε_U) versus the respective alginate, alginate/MgO, and cross-linked alginate scaffolds. It is observed that ε_U of alginate/MgO scaffold is smaller than that of alginate scaffold (statistically significant, $P < 0.05$). This suggests that the addition of MgO nanoparticles to alginate-based matrix could contribute to a decrease in ε_U from 9.05% to 6.73%. This could be attributed to the increased rigidity of polymer chains as pointed out in previous paragraphs. However, ε_U of cross-linked alginate scaffold is slightly higher than that of the untreated alginate scaffold. This could be attributed to the formation of plasticizers from the left-over solvent in the scaffold (entrapped solvent in the fibres) after those were cross-linked with 0.5 M glutaraldehyde solution [21].

In order for an implanted scaffold to be able to provide the mechanical (as well as shape, see Section 3.2) stability to the tissue [36], the desired mechanical properties of the structure, that is, the processed scaffold, should be comparable to that of the host tissue. Thus, for the scaffold to be able to take up stress and not fail, that is, rupture, when an external load is acting on the tissue, the mechanical stability is regulated by the strength and extensibility of the scaffold. Table S2 (in supplementary data) lists the mechanical properties, namely, E, σ_U, and ε_U, of some soft connective tissues. With regard to strength, most tissues such as tendons, ligaments, and percardia possess fracture strengths that are at least one order of magnitude higher than what the alginate-based/MgO scaffolds could take, with the exception of the heart valve such as aorta and mitral leaflets. Thus, it would appear that the alginate-based/MgO scaffold is mechanically compatible, from the strength perspective, with the mitral heart valve leaflet as well as the aorta valve. However, it is noted that the extensitivity of these valves is about one order of magnitude

higher than that of the alginate-based/MgO mesh. Nevertheless, as the meshes were tested in dry condition, it may be reasonable to anticipate that the extensibility of the alginate-based/MgO mesh could be appreciably higher when these are tested in wet condition. Of course, the alginate-based scaffolds (albeit a lower strength) may be a possible candidate for ECM-substitute scaffold as its extensibility could match those possessed by the valves.

3.2.4. Thermal Properties. Figure 6 shows plots of mass of the scaffold versus temperature derived from TGA for the respective alginate and alginate/MgO scaffolds. In all cases, initially the mass of the mesh decreases steadily with increasing temperature until at around 50°C. Between 50°C and 250°C, the mass shows no appreciable change with increasing temperature; beyond 250°C the mass decreases drastically. The drastic decrease in mass appears to follow a two-stage process: the first stage corresponds to a rapid decrease in mass while the second stage reveals a less rapid decrease in mass. Thereafter, at about 500°C, the mass of the alginate scaffold is almost zero, implying that the mesh is completely burnt off. On the other hand, the mass of the alginate/MgO scaffold is observed to be equal to 12%, suggesting that some residues including MgO are left behind.

The initial mass loss of about 10–12 wt% (at around 100°C) could be attributed to the removal of entrapped moisture and left-over solvent. The second mass loss at 200–300°C could be due to the thermal decomposition of polymer chains of alginate and PVA. Backbone structure of alginate decomposes at around 250°C due to the degradation of C-H bonds and C-O-C glycoside bonds in the main polysaccharide chain as a result of dehydration of saccharide chains [37]. More interestingly, the temperature of decomposition at 30% mass loss of the alginate/MgO occurs at around 272°C while that of alginate-based scaffold occurs at around 255°C. Thus, the

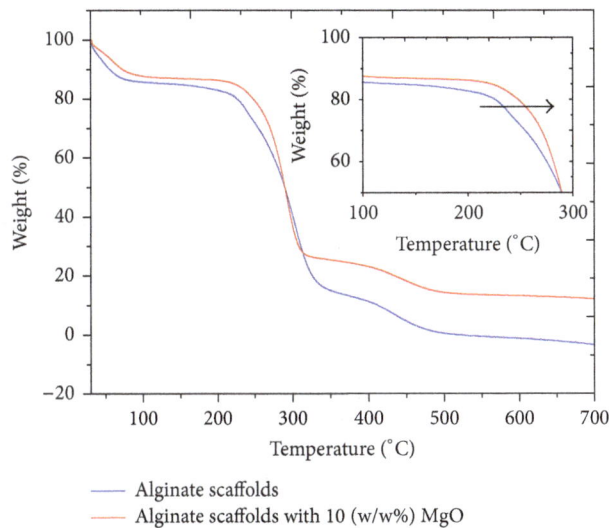

FIGURE 6: Thermogravimetric curves of electrospun alginate and alginate/MgO scaffolds.

thermal stability of the alginate/MgO scaffold is higher than that of alginate-based scaffold, that is, at 30% mass loss. These marginal improvements in thermal stability could be attributed to the high thermal stability of MgO nanoparticles which thermally decompose at around 2800°C.

4. Conclusion

An electrospinning method has been developed to fabricate alginate-based nanofibrous scaffolds, reinforced by MgO nanoparticles (10% (w/w)). The MgO nanoparticles were separately synthesised using a polymer template-based ex situ technique to achieve a near-spherical shape with an average diameter of 45 nm. For the purpose of comparison, the alginate scaffold, as well as glutaraldehyde cross-linked alginate scaffold, was fabricated. The mechanical properties of the alginate/MgO scaffold exhibit the highest tensile strength (σ_U) and elastic modulus (E) among the three different types of scaffolds while retaining the interfibre porosity. The alginate/MgO scaffold yielded randomly oriented, ultrafine, virtually defect-free alginate nanofibres with a diameter ranging from 60 to 250 nm, similar to alginate scaffolds, and pore size is estimated to be 2–50 μm. On the other hand, the cross-linked alginate scaffolds result in densely packed and extensively fused fibres; the resultant scaffolds also exhibit low interfibre porosity (compared with untreated alginate scaffolds). Altogether these suggest that the alginate-based/MgO scaffold is a suitable candidate for further investigation to be utilized as an artificial substitute for extracellular matrix in tissue engineering applications.

Conflicts of Interest

The authors declare that there are no conflicts of interest regarding the publication of this paper.

Acknowledgments

The authors thank Dr. Nicola Gurusinghe for her valuable guidance and contribution.

References

[1] B. Dhandayuthapani, Y. Yoshida, T. Maekawa, and D. S. Kumar, "Polymeric scaffolds in tissue engineering application: a review," *International Journal of Polymer Science*, vol. 2011, Article ID 290602, 19 pages, 2011.

[2] X. Wu, Y. Liu, X. Li et al., "Preparation of aligned porous gelatin scaffolds by unidirectional freeze-drying method," *Acta Biomaterialia*, vol. 6, no. 3, pp. 1167–1177, 2010.

[3] I. A. Rodriguez, P. A. Madurantakam, J. M. McCool et al., "Mineralization potential of electrospun PDO-hydroxyapatite-fibrinogen blended scaffolds," *International Journal of Biomaterials*, vol. 2012, Article ID 159484, 12 pages, 2012.

[4] T. Nie, L. Xue, M. Ge, H. Ma, and J. Zhang, "Fabrication of poly(L-lactic acid) tissue engineering scaffolds with precisely controlled gradient structure," *Materials Letters*, vol. 176, pp. 25–28, 2016.

[5] S. Bose, S. Vahabzadeh, and A. Bandyopadhyay, "Bone tissue engineering using 3D printing," *Materials Today*, vol. 16, no. 12, pp. 496–504, 2013.

[6] Y. He, N. Zhang, Q. Gong et al., "Alginate/graphene oxide fibers with enhanced mechanical strength prepared by wet spinning," *Carbohydrate Polymers*, vol. 88, no. 3, pp. 1100–1108, 2012.

[7] N. Bhattarai, D. Edmondson, O. Veiseh, F. A. Matsen, and M. Zhang, "Electrospun chitosan-based nanofibers and their cellular compatibility," *Biomaterials*, vol. 26, no. 31, pp. 6176–6184, 2005.

[8] G. P. Ma, D. W. Fang, Y. Liu, X. D. Zhu, and J. Nie, "Electrospun sodium alginate/poly(ethylene oxide) core-shell nanofibers scaffolds potential for tissue engineering applications," *Carbohydrate Polymers*, vol. 87, no. 1, pp. 737–743, 2012.

[9] M. Okamoto and B. John, "Synthetic biopolymer nanocomposites for tissue engineering scaffolds," *Progress in Polymer Science*, vol. 38, no. 10-11, pp. 1487–1503, 2013.

[10] C. Zhou, R. Chu, R. Wu, and Q. Wu, "Electrospun polyethylene oxide/cellulose nanocrystal composite nanofibrous mats with homogeneous and heterogeneous microstructures," *Biomacromolecules*, vol. 12, no. 7, pp. 2617–2625, 2011.

[11] J. Venkatesan, I. Bhatnagar, P. Manivasagan, K.-H. Kang, and S.-K. Kim, "Alginate composites for bone tissue engineering: a review," *International Journal of Biological Macromolecules*, vol. 72, pp. 269–281, 2015.

[12] T. Chae, H. Yang, V. Leung, F. Ko, and T. Troczynski, "Novel biomimetic hydroxyapatite/alginate nanocomposite fibrous scaffolds for bone tissue regeneration," *Journal of Materials Science: Materials in Medicine*, vol. 24, no. 8, pp. 1885–1894, 2013.

[13] S. I. Jeong, M. D. Krebs, C. A. Bonino, S. A. Khan, and E. Alsberg, "Electrospun alginate nanofibers with controlled cell adhesion for tissue engineering," *Macromolecular Bioscience*, vol. 10, no. 8, pp. 934–943, 2010.

[14] J.-W. Lu, Y.-L. Zhu, Z.-X. Guo, P. Hu, and J. Yu, "Electrospinning of sodium alginate with poly(ethylene oxide)," *Polymer*, vol. 47, no. 23, pp. 8026–8031, 2006.

[15] K. H. Bouhadir, K. Y. Lee, E. Alsberg, K. L. Damm, K. W. Anderson, and D. J. Mooney, "Degradation of partially oxidized alginate and its potential application for tissue engineering," *Biotechnology Progress*, vol. 17, no. 5, pp. 945–950, 2001.

[16] S. F. Badylak, D. O. Freytes, and T. W. Gilbert, "Extracellular matrix as a biological scaffold material: structure and function," *Acta Biomaterialia*, vol. 5, no. 1, pp. 1–13, 2009.

[17] Y. Ito, H. Hasuda, M. Kamitakahara et al., "A composite of hydroxyapatite with electrospun biodegradable nanofibers as a tissue engineering material," *Journal of Bioscience and Bioengineering*, vol. 100, no. 1, pp. 43–49, 2005.

[18] A. Watthanaphanit, P. Supaphol, H. Tamura, S. Tokura, and R. Rujiravanit, "Fabrication, structure, and properties of chitin whisker-reinforced alginate nanocomposite fibers," *Journal of Applied Polymer Science*, vol. 110, no. 2, pp. 890–899, 2008.

[19] K. T. Shalumon, K. H. Anulekha, S. V. Nair, S. V. Nair, K. P. Chennazhi, and R. Jayakumar, "Sodium alginate/poly(vinyl alcohol)/nano ZnO composite nanofibers for antibacterial wound dressings," *International Journal of Biological Macromolecules*, vol. 49, no. 3, pp. 247–254, 2011.

[20] Y. J. Lee and W. S. Lyoo, "Preparation and characterization of high-molecular-weight atactic poly(vinyl alcohol)/sodium alginate/silver nanocomposite by electrospinning," *Journal of Polymer Science Part B: Polymer Physics*, vol. 47, no. 19, pp. 1916–1926, 2009.

[21] R. T. De Silva, M. M. M. G. P. G. Mantilaka, S. P. Ratnayake, G. A. J. Amaratunga, and K. M. Nalin de Silva, "Nano-MgO reinforced chitosan nanocomposites for high performance packaging applications with improved mechanical, thermal and barrier properties," *Carbohydrate Polymers*, vol. 157, pp. 739–747, 2017.

[22] H. R. Mahmoud, S. M. Ibrahim, and S. A. El-Molla, "Textile dye removal from aqueous solutions using cheap MgO nanomaterials: adsorption kinetics, isotherm studies and thermodynamics," *Advanced Powder Technology*, vol. 27, no. 1, pp. 223–231, 2016.

[23] M. M. M. G. P. G. Mantilaka, H. M. T. G. A. Pitawala, D. G. G. P. Karunaratne, and R. M. G. Rajapakse, "Nanocrystalline magnesium oxide from dolomite via poly(acrylate) stabilized magnesium hydroxide colloids," *Colloids and Surfaces A: Physicochemical and Engineering Aspects*, vol. 443, pp. 201–208, 2014.

[24] N. Salehifar, Z. Zarghami, and M. Ramezani, "A facile, novel and low-temperature synthesis of MgO nanorods via thermal decomposition using new starting reagent and its photocatalytic activity evaluation," *Materials Letters*, vol. 167, pp. 226–229, 2016.

[25] G. Song, X. Zhu, R. Chen, Q. Liao, Y.-D. Ding, and L. Chen, "Influence of the precursor on the porous structure and CO_2 adsorption characteristics of MgO," *RSC Advances*, vol. 6, no. 23, pp. 19069–19077, 2016.

[26] J. Jia, J. Yang, Y. Zhao, H. Liang, and M. Chen, "The crystallization behaviors and mechanical properties of poly(l-lactic acid)/magnesium oxide nanoparticle composites," *RSC Advances*, vol. 6, no. 50, pp. 43855–43863, 2016.

[27] Y. Zhao, B. Liu, C. You, and M. Chen, "Effects of MgO whiskers on mechanical properties and crystallization behavior of PLLA/MgO composites," *Materials and Design*, vol. 89, pp. 573–581, 2016.

[28] Y. B. Kim and G. Kim, "Rapid-prototyped collagen scaffolds reinforced with PCL/β-TCP nanofibres to obtain high cell seeding efficiency and enhanced mechanical properties for bone tissue regeneration," *Journal of Materials Chemistry*, vol. 22, no. 33, pp. 16880–16889, 2012.

[29] K. L. Goh, D. F. Holmes, Y. Lu et al., "Bimodal collagen fibril diameter distributions direct age-related variations in tendon resilience and resistance to rupture," *Journal of Applied Physiology*, vol. 113, no. 6, pp. 878–888, 2012.

[30] K. L. Goh, J. Hiller, J. L. Haston et al., "Analysis of collagen fibril diameter distribution in connective tissues using small-angle X-ray scattering," *Biochimica et Biophysica Acta—General Subjects*, vol. 1722, no. 2, pp. 183–188, 2005.

[31] C. Di Benedetto, A. Barbaglio, T. Martinello et al., "Production, characterization and biocompatibility of marine collagen matrices from an alternative and sustainable source: the sea urchin *Paracentrotus lividus*," *Marine Drugs*, vol. 12, no. 9, pp. 4912–4933, 2014.

[32] H. C. Wells, K. H. Sizeland, H. R. Kayed et al., "Poisson's ratio of collagen fibrils measured by small angle X-ray scattering of strained bovine pericardium," *Journal of Applied Physics*, vol. 117, no. 4, Article ID 044701, 2015.

[33] Y. Wang and Y.-L. Hsieh, "Crosslinking of polyvinyl alcohol (PVA) fibrous membranes with glutaraldehyde and peg diacylchloride," *Journal of Applied Polymer Science*, vol. 116, no. 6, pp. 3249–3255, 2010.

[34] H. S. Mansur, C. M. Sadahira, A. N. Souza, and A. A. P. Mansur, "FTIR spectroscopy characterization of poly (vinyl alcohol) hydrogel with different hydrolysis degree and chemically crosslinked with glutaraldehyde," *Materials Science and Engineering C*, vol. 28, no. 4, pp. 539–548, 2008.

[35] M. Makaremi, R. T. De Silva, and P. Pasbakhsh, "Electrospun nanofibrous membranes of polyacrylonitrile/halloysite with superior water filtration ability," *Journal of Physical Chemistry C*, vol. 119, no. 14, pp. 7949–7958, 2015.

[36] B. P. Chan and K. W. Leong, "Scaffolding in tissue engineering: general approaches and tissue-specific considerations," *European Spine Journal*, vol. 17, no. 4, pp. S467–S479, 2008.

[37] A. Kumar, Y. Lee, D. Kim et al., "Effect of crosslinking functionality on microstructure, mechanical properties, and in vitro cytocompatibility of cellulose nanocrystals reinforced poly (vinyl alcohol)/sodium alginate hybrid scaffolds," *International Journal of Biological Macromolecules*, vol. 95, pp. 962–973, 2017.

Tissue-Engineered Vascular Graft of Small Diameter Based on Electrospun Polylactide Microfibers

P. V. Popryadukhin,[1,2] G. I. Popov,[3] G. Yu. Yukina,[3] I. P. Dobrovolskaya,[1,2] E. M. Ivan'kova,[1,2] V. N. Vavilov,[3] and V. E. Yudin[1,2]

[1]Institute of Macromolecular Compounds, Russian Academy of Sciences, Bolshoy Pr. 31, Saint-Petersburg 199004, Russia
[2]Peter the Great Saint-Petersburg State Polytechnical University, Polytechnicheskaya Str. 29, Saint-Petersburg 194064, Russia
[3]Pavlov First Saint-Petersburg State Medical University, Leo Tolstoy Str. 6-8, Saint-Petersburg 197022, Russia

Correspondence should be addressed to P. V. Popryadukhin; pavel-pn@mail.ru

Academic Editor: Rosalind Labow

Tubular vascular grafts 1.1 mm in diameter based on poly(L-lactide) microfibers were obtained by electrospinning. X-ray diffraction and scanning electron microscopy data demonstrated that the samples treated at $T = 70°C$ for 1 h in the fixed state on a cylindrical mandrel possessed dense fibrous structure; their degree of crystallinity was approximately 44%. Strength and deformation stability of these samples were higher than those of the native blood vessels; thus, it was possible to use them in tissue engineering as bioresorbable vascular grafts. The experiments on including implantation into rat abdominal aorta demonstrated that the obtained vascular grafts did not cause pathological reactions in the rats; in four weeks, inner side of the grafts became completely covered with endothelial cells, and fibroblasts grew throughout the wall. After exposure for 12 weeks, resorption of PLLA fibers started, and this process was completed in 64 weeks. Resorbed synthetic fibers were replaced by collagen and fibroblasts. At that time, the blood vessel was formed; its neointima and neoadventitia were close to those of the native vessel in structure and composition.

1. Introduction

In the modern vascular surgery, the problem of the development of vascular grafts with small diameter still exists. Low patency rates of the synthetic prostheses with diameter less than 5 mm are related, first of all, to development of neointimal hyperplasia at anastomosis sites and the absence of endothelial layer on the inner side of prostheses [1–4]. The problem of using autovenous material is its limited amount and high possibility of pathological changes in autovenous wall after implantation [5, 6]. As for pediatric vascular surgery, it is necessary to repeat reconstructive vascular operations due to the fact that nonresorptive synthetic prostheses cannot grow up and develop with a child organism [7]. Any attempts to create vascular grafts of small diameter by traditional methods were unsuccessful, since thromboses arose inside the grafts over a short period of time [8] because of low blood stream rate in these grafts. Currently, there are several techniques for developing artificial blood vessels

and some of them are now undergoing clinical trials. The main methods are the following: obtaining tissue-engineered vascular grafts (TEVG) by layer-by-layer tissue engineering [9–14]; production of artificial vessels from granulation tissue [15–18]; use of decellularized transplants [19–21]; obtaining artificial vessels based on tubular bioresorbed polymer grafts [22–27].

The problem of thrombosis on early stages of implantation can be solved by using artificial vessels obtained by modern tissue engineering methods with the use of grafts made of biocompatible and bioresorbable polymer. Such artificial blood vessel should imitate structure and functions of the native vessel and be sensitive to neurohumoral action from recipient organism. The method consists in cultivating cells on bioresorbable graft in the bioreactor which imitates biological and mechanical factors providing proliferation and differentiation of the cells. It is expected that after implantation of this TEVG into recipient organism, biodegradation of polymer structures will be accompanied by the formation of a new vascular wall [28]. Another approach to development of artificial vessel includes implantation of a polymer graft into living organism where the cells from the surrounding tissues migrate to the graft and fill it forming a TEVG. Thus, functional tissues are formed and in parallel with this process; resorption of a polymer graft takes place under the action of active biological medium. When resorption is completed, a new blood vessel should be formed on the place of the graft. New artificial vessel should meet the following requirements: (1) to be biocompatible and infection-resistant; (2) to be hermetical and resistant to thrombosis (thus, the inner surface of the artificial vessel should be covered with endothelium); (3) to possess mechanical characteristics which allow carrying out surgical manipulations and also endure prolonged hydrodynamic loadings; (4) to possess vasoactive physiological properties (including ability to undergo spasm or dilatation as a response to nervous or chemical stimuli). Besides, it is necessary to have possibility of producing vascular grafts with various characteristics in sufficient amounts for solving any clinical problems [29]. Recently, there are a number of publications which describe attempts to omit the stage of cell cultivation on grafts *in vitro* and thus to simplify the technique and approximate it to clinical trials [30].

One of the promising methods to produce polymer vascular grafts is electrospinning. The method allows obtaining materials based on nano- and microfibers which demonstrate high porosity and specific surface area (the latter features are necessary for migration and proliferation of cells in graft volume) and simultaneously keep tightness with respect to blood [31–33]. The vascular grafts obtained by electrospinning possess the necessary mechanical characteristics. They are able to be integrated quickly into living organism, and their inner surface is covered with endothelium, which significantly reduces risk of thrombosis [34, 35].

The aim of the present work was development of a method for producing vascular grafts with small diameter based on poly(L-lactide) (PLLA) microfibers, studies of their structure, strength and deformation properties, *in vivo* investigation of biological tissue formation on the obtained grafts, and analysis of bioresorption mechanism.

2. Materials and Methods

2.1. Materials. The objects of the study were tubular grafts based on microfibers which were produced from PLLA «Purasorb PL-10» (Corbion Purac, Netherlands). Chloroform (Sigma-Aldrich, USA) was used as a solvent. All materials were of purissimus grade.

2.2. Processing of Nanofibers by Electrospinning. The microfibers were produced by electrospinning from the solution of PLLA in chloroform using a laboratory-scale Nanon-01A instrument (Japan). PLLA solution with a concentration of 15 wt.% was pumped through the die in the electrical field ($V = 16$ kV); the distance between the electrodes was 0.15 m; fiber deposition occurred on cylindrical electrodes. The rotational speed of cylindrical electrode (having a diameter of 1.1 mm) was 1500 rpm. Tubular samples with an inner

diameter of 1.1 mm and wall thickness of 320 μm were produced.

2.3. Investigation of Mechanical Properties. Mechanical properties of the grafts were studied using a universal Instron 5943 setup (United Kingdom) in the uniaxial tension mode at a rate of 10 mm/min. The Young modulus, tensile strength, and tensile deformation were determined for tubular samples based on PLLA with diameters of 1.1 mm and 850 μm and wall thicknesses of 370, 320, and 250 μm as well as for samples of native rat aorta. The length of test section of all tested samples was 20 mm.

2.4. Contact Angle Measurements. Wetting angle was measured on the surface of PLLA films and on the inner side of tubular samples from PLLA microfibers with the help of a DSA 30 instrument (KRUSS, Germany).

2.5. Investigation of Sample Structure. Structure of the tubular samples based on the PLLA microfibers was investigated using scanning electron microscope Supra 55 VP (Carl Zeiss, Germany). The sample surface was previously sputtered by platinum. Measuring of the wall thickness of tubular specimens, fiber diameter, and pore size (interfiber space) was performed using analysis of microphotographs.

Mean pore diameters and diameter of fibers were determined by image analysis (Scandium, ©OLYMPUS Soft Imaging Solutions) of SEM micrographs of tubular specimens. For this purpose SEM images of several areas perpendicular to the pore direction were captured. At least 20 pores were analyzed from different locations of the same sample.

X-ray diffraction (WAXD, Bruker D8 DISCOVER, Germany) was used for studying fine structure of the grafts.

Glass transition temperature, melting point, and crystallization temperature of PLLA were determined by differential scanning calorimetry (DSC) using a DSC 204 F1 Phoenix instrument (NETZSCH, Germany) in argon atmosphere.

2.6. Implantation Technique. Female rats (weight of 350 g) of a single genetic line were operated with inhalation anesthesia (1.5% isoflurane, induction of anesthesia, 3% isoflurane). After Y-shaped laparotomy, mobilization of an infrarenal aorta with ligation of lumbar arteries and the right renal artery clipping were made. The infrarenal aorta was replaced with the graft made of crystallized PLLA with an inner diameter of 1.1 mm and a length of 5 mm. An operating microscope, special tools, and an atraumatic needle (Prolene 9-0) were used. The mean diameter of the rat abdominal aorta was 1.0 mm. Anticoagulants were not applied. Surgical interventions were made in 42 female albino Wistar rats. Animals received free access to water and standard diet, which included PC-120-1 all-mash. The evaluations were made in 1, 4, 12, 24, 48, 56, and 64 weeks after operation. The conclusions were drawn from the experiments with 6 animals.

Ischemia time of the lower part of the body, the hind limbs, and the right kidney was about 50 minutes. In all trials, the hemorrhage through the anastomosis was not significant. Patency was assessed by classical technique immediately and 30 minutes after operation [36]. The abdominal wall was sutured layer by layer. The animals were kept in individual cages in a vivarium with free access to food and water. Postoperatively, the color and temperature of the skin of the hind limbs of the animals and their physical activity were examined.

All animals were treated appropriately according to Order number 1179 of Ministry of Health Care (10.10.1983); Order number 267 of Ministry of Health Care (19.06.2003); "Rules of Using Experimental Animals"; the principles of the European Convention (Strasbourg, 1986); and the Helsinki Declaration of the World Medical Association about humane treatment of animals (1996).

2.7. Angiographic Study. Angiographic studies were carried out with the aid of a Philips Allura Xper FD 20 setup (Netherlands) after open puncture of rats abdominal aorta in the area above prosthesis zone. The contrast substance Omnipaque (300 mg/mL) was used.

2.8. Microscopic Examination. In order to carry out histological examinations, the prosthesis and the surrounding portion of the native vessel were fixed in 10% neutral formalin phosphate buffer (pH 7.4) for 24 hours. Special histological technique with isopropyl alcohol or petroleum was used for embedding material into paraffin blocks. Paraffin sections having thickness of 5 μm were cut and stained by hematoxylin and eosin. The Mallory and the Weigert-Van Gieson methods were used (Mallory trichrome and Weigert-Van Gieson, Bio-Optica, Italy) for visualization of connective tissue and elastic fibers.

To identify the endothelium, an immunohistochemical method was used and the immunoreactivity of the PECAM-1/CD31 protein was evaluated. After the standard dewaxing and rehydration procedures, the antigen was thermally smeared in pH 6.0 buffer (Diagnostic BioSystems, USA) for 60 min. The incubation was carried out for 30 minutes at room temperature. At the first stage, incubation of sections with polyclonal goat antibodies to PECAM-1/CD31 (Santa Cruz Biotechnology, USA) was carried out. Antibodies were diluted as 1 : 100 on a blocking solution (Diagnostic BioSystems, USA). To reveal the antigen-antibody complex, a set of reagents of Super Sensitive Polymer-HRP IHC Detection System (BioGenex, USA) was used. The preparations were additionally stained with Gill's hematoxylin (Bio-Optica, Italy).

Photographs were taken using a Leica DM750 optical microscope and an ICC50 camera (Leica, Germany). The tissue sections were observed under the microscope using the eyepiece N 10, and lenses N 4, 10, and 40. Morphometric analysis was performed using the "Videotest" system; eyepieces with magnification ×10 and lens N 40 were used.

3. Results and Discussion

According to the SEM data, the grafts obtained from PLLA solution by electrospinning consist of microfibers 1.5–4 μm in diameter.

Inner diameter of the grafts is 1.1 mm; wall thickness is 320 μm. The microfibers have no preferential orientation and pores between them have average sizes of tens and hundreds of micrometers (Figures 1(a) and 1(b)). The X-ray structural analysis data demonstrated that the microfibers have amorphous structure.

After implantation of these samples into rats abdominal aorta, the prosthesis was deformed, and the wall integrity was affected during contact with sutures at anastomosis zones. Besides, a portion of fibers on the inner side of a tube was deformed during surgical manipulations; vascular lumen became smaller, and thus thromboses in early postoperative period were provoked. Implantation of these prostheses was difficult and inefficient.

It is well known that PLLA is a crystallizable polymer. It was revealed by DSC examination that PLLA is able to form crystalline phase inside the graft at temperatures close to its glass transition temperature (53–59°C). In this connection, there is a risk of spontaneous crystallization of the polymer at body temperature and under the action of biological media. Crystallization would be accompanied by shrinkage of microfibers and, correspondingly, by decrease in graft diameter; these processes could also lead to increasing risk of thromboses.

In order to improve mechanical properties and stability in biological medium, PLLA grafts were subjected to thermal treatment in free and fixed states at $T = 70°C$ for 1 h. The DSC and X-ray diffraction data revealed that after thermal treatment at temperature above the glass transition point partial crystallization of the polymer occurred; degree of crystallinity of the treated PLLA samples was about 44%.

In the case of grafts treated in free state (as seen in Figures 1(c) and 1(d)), wall thickness increased from 320 μm (for amorphous sample) to 370 μm, and inner diameter decreased from 1.1 mm to 0.85 mm; the structure of graft wall became more loose and fibrous.

Thermal treatment of amorphous sample in the fixed state was carried out on a cylindrical mandrel 1.1 mm in diameter that led to decrease in graft wall thickness down to 250 μm (Figures 1(e) and 1(f)) without decrease in the inner diameter; more dense fibrous structure was formed.

Mechanical characteristics of the thermally treated grafts were considerably better than those of the initial amorphous samples and exceeded the corresponding values for the native vessel (Table 1).

An important characteristic of biomaterials for tissue engineering is their wettability by water and other liquid media of an organism. On the one hand, high hydrophobicity of these materials prevents adhesion of cells on their surface and thus affects tissue formation. On the other hand, high hydrophilicity and, consequently, swelling and strong interaction between a graft and molecules of liquid environment can cause formation of thrombi [37]. The present studies demonstrated that the wetting contact angle of PLLA film after 1 h heat treatment was 80°. This value is typical of materials with rather low hydrophobicity. However, the contact angle of inner surface of the graft based on PLLA microfibers was 120°. Such value is characteristic of materials with high hydrophobicity and was reached due to surface relief. It is expected to reduce the risk of thrombosis during blood stream through the graft. However, it does not prevent

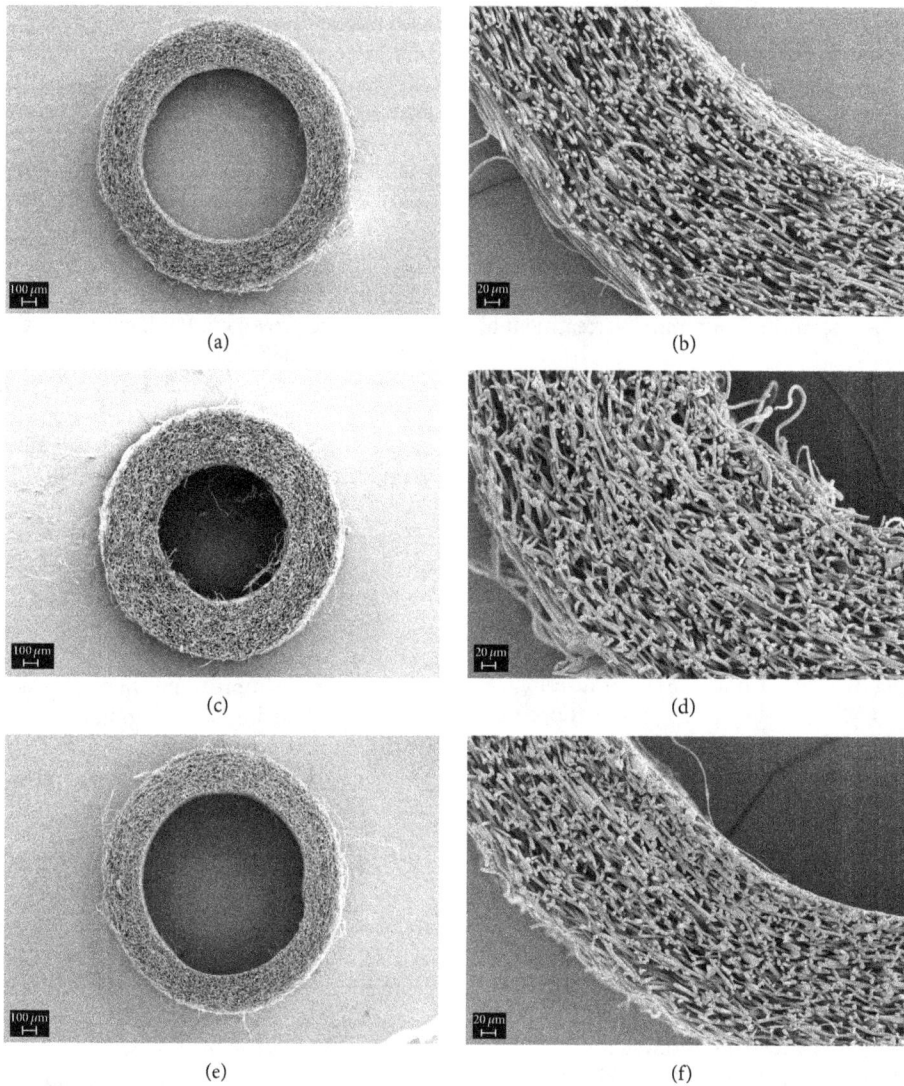

FIGURE 1: Cross sections of the grafts based on PLLA microfibers: the initial sample (a, b) and the sample crystallized in the free (c, d) and fixed (e, f) states.

TABLE 1: Mechanical characteristics of the grafts based on PLLA microfibers.

Medium	Wall thickness, μm	Tensile strength, MPa	Young's modulus, MPa	Elongation, %
Native rat aorta	150	2.27 ± 0.56	17.3 ± 5.1	139 ± 32
The initial PLLA graft	320	2.64 ± 0.29	27.0 ± 2.8	258 ± 24
PLLA graft thermally treated in the fixed state	250	3.63 ± 0.41	145 ± 11	175 ± 18
PLLA graft thermally treated in the free state	370	2.61 ± 0.36	109 ± 13	96.3 ± 21

further cell adhesion and endothelization of the surface graft. The most stable values of hydrophobicity were observed on samples crystallized in a fixed state. Besides, the increased initial mechanical characteristics for bioresorbable vascular grafts as compared to the native vessel are some advantage because the bioresorption process can go nonuniformly and not always qualitatively. Thus, a margin of safety is quite necessary. The best mechanical characteristics were obtained on samples crystallized in a fixed state (Table 1).

On the basis of the above data of mechanical characteristics, hydrophobicity parameters, stability of internal diameter, and thickness and homogeneity of the wall, the grafts of 5 mm in length thermally treated in the fixed state were selected for in vivo experiments (Figure 2).

The experiments involving implantation of the grafts into rat abdominal aorta demonstrated that in 1 week all grafts remained passable; aorta was adherent to the graft; removal of sutures did not cause destruction of anastomosis;

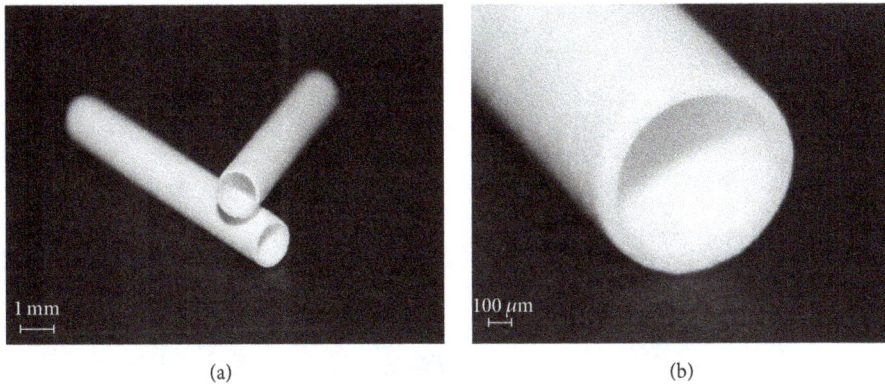

(a)   (b)

FIGURE 2: The grafts produced from the crystallized PLLA microfibers by electrospinning and photographed at low (a) and high (b) magnification.

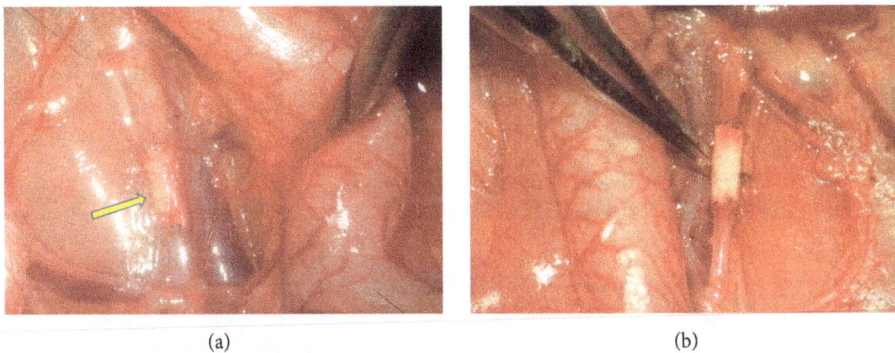

(a)   (b)

FIGURE 3: The graft four weeks after implantation covered with tissue (a) and separated (b). The arrow indicates the implanted graft.

no pathological influence on the surrounding tissue was revealed. Morphological analysis of the material showed that endothelial layer was formed on the inner side of the graft starting from distal and proximal anastomoses; the central part was covered with inhomogeneous network layer of fibrin. Between PLLA fibers, mainly from the adventitia side, fibroblast nuclei were found; thin collagen fibers started to appear. Giant polynuclear cells of foreign bodies were located on the outer side of the graft. The presence of these cells is typical of the organism's reaction to a foreign body, since they are actively involved in its destruction. However, in this case the cell sizes do not allow them to penetrate through the pores into the interior of the graft. In four weeks, all grafts remained passable; they were covered with connective tissue on the outside and adhered to connective tissue (Figure 3).

Histological studies showed that the graft was completely covered with endothelium. Subendothelial layer was formed along the whole length of the graft and contained thin network of the collagen fibers. The thickest neointima was formed in anastomosis zones ($55 \mu m$), and its thickness gradually decreased towards the center of graft (down to $10 \mu m$). There were no signs of hyperplasia of the neointima. The whole width of the graft was filled with fibroblasts and pierced with thin collagen fibers. From the side of the adventitia, the collagen fibers were thicker, and the outer side of the graft was

surrounded by a capsule formed from the connective tissue (Figure 4).

No immune response of an organism towards the graft was revealed. Bioresorption of the polymer was not observed either; structure of microfibers was retained.

Two of six grafts became thrombosed in 12 weeks. The grafts retained their shape and structure. The signs of beginning bioresorption were found; cross section of the fibers demonstrated low porosity.

Morphological studies revealed that endothelium and subendothelium containing collagen fibers covered the graft completely and neointima was formed. Within the thickness of the graft, between PLLA fibers, fibroblasts and the newly formed collagen fibers were located. Numerous giant polynuclear cells of foreign bodies can be seen from the adventitia side.

In 24 weeks after implantation, all grafts remained passable, numerous cross cracks were observed on microfibers, and a portion of fibers was fragmented (Figure 5(a)). Cross sections of the microfibers demonstrate pronounced porosity (Figure 5(b)) which was indicative of active bioresorption.

Histological analysis of the grafts performed after 24 weeks (similarly to the data obtained in the early postoperational period) showed completely formed endothelium and subendothelium consisting of loose collagen fibers. In one graft, completely formed plethoric blood vessels were

(a)

(b)

(c)

(d)

(e)

(f)

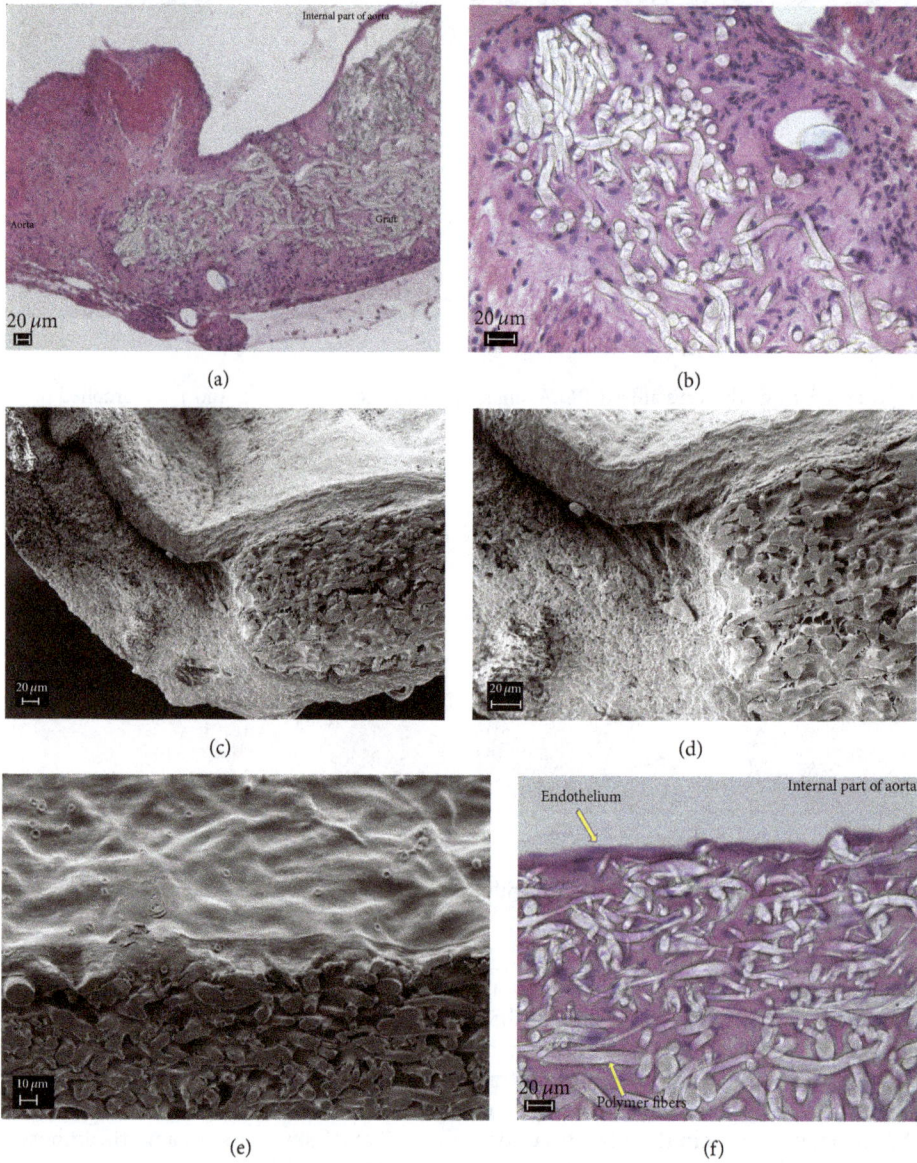

FIGURE 4: The results of histological and SEM studies of the vascular grafts after exposure for 4 weeks; anastomosis (a, b, c, d); fragment of the wall with neointima (e, f).

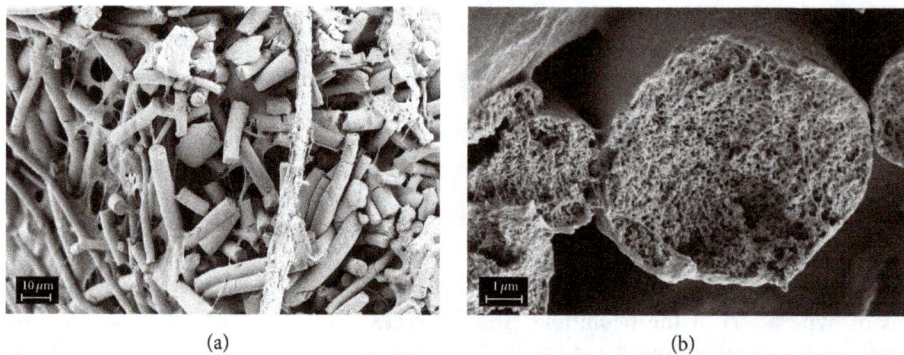

(a)

(b)

FIGURE 5: Fragment of the prosthesis wall in 24 weeks after implantation (a); fragment of the microfiber (b).

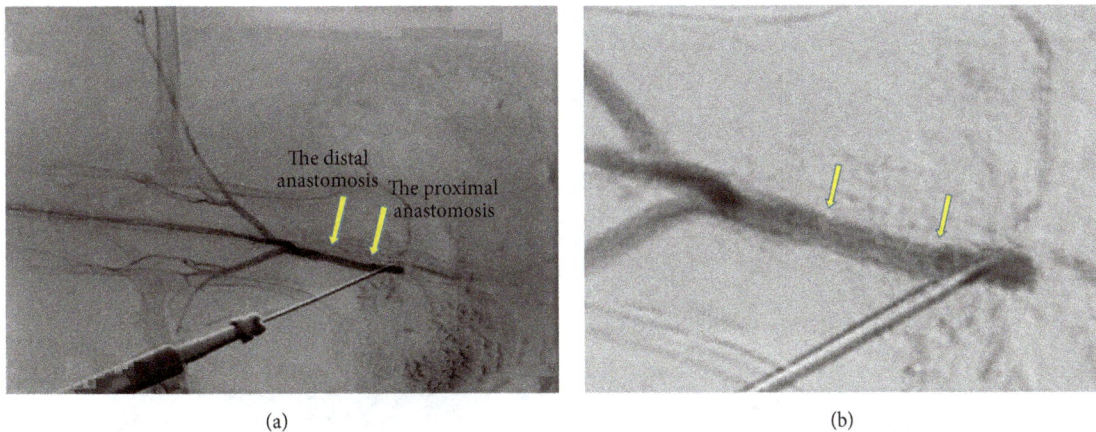

(a)   (b)

FIGURE 6: Intravital angiographic study performed in 56 weeks after implantation (a); the same sample shown at higher magnification (b). Arrows indicate the boundaries of the implanted graft.

revealed in the subendothelial layer. The grafts were completely occupied with fibroblasts and pierced with the collagen fibers. Over the whole volume of the grafts, numerous round cavities with diameter ranging from 1.750 to 7.320 μm filled with weakly oxyphilic contents were observed between PLLA microfibers. These cavities were located chaotically and not surrounded by capsules; they destroy the structure of the wall of the new formed vessel that reduces its mechanical properties. Perhaps the appearance of cavities is associated with the resorption of polymer fibers [38]. In this regard, to exclude the risk of destruction of the vascular wall and the appearance of life-threatening bleeding, additional reliable strengthening of the vascular wall is necessary. On the outside of the grafts, the capsule of connective tissue containing giant polynuclear cells of foreign bodies was found.

In 48 weeks, one of six grafts was impassable and contained a thrombus which ran the entire length of the prosthesis. In passable grafts, microfibers were strongly fragmented and partially resorbed.

Morphological studies demonstrated that subendothelial layer was wide and consisted of densely packed collagen fibers and individual elastic fibers. In the subendothelial layer, calcium salts were deposited. Neomedia consisted of thick bundles of the collagen fibers which were randomly arranged. Fragments of PLLA fibers could be seen and mainly fibroblasts were located between them. Round cavities with weakly oxyphilic contents were observed over the whole graft volume; their area varied from 6.590 μm^2 to 11.500 μm^2. Adventitia consisted of loose fibrous connective tissue which contained vessels and giant polynuclear cells of foreign bodies.

In 56 weeks, all grafts of this group were still passable; in one sample, uniform hyperplasia of intima was revealed throughout the graft; its diameter decreased by 50%. Insignificant aneurisms were observed in 2 grafts.

According to the results of intravital angiographic study, the grafts were passable and conducted a pulse wave without difficulty; the prosthesis boundaries were barely distinguishable; there were no signs of stenosis or dilatation (Figure 6).

Resorption of polylactide fibers was found to continue; longitudinal sizes of visible fragments of the microfibers varied from 1 to 10 μm (Figures 7(a) and 7(b)).

Histological sections of the studied samples were similar to those obtained earlier. Endothelium and subendothelium consisting of collagen and the elastic fibers were revealed throughout the graft. However, neointima thickness was not the same in various parts. Neomedia consisted of disordered bundles of the collagen fibers, fibroblasts, and fragments of the PLLA fibers. The area of newly formed cavities between the microfibers ranged from 3.360 μm^2 to 7.040 μm^2. Vessels and giant polynuclear cells of foreign bodies were revealed in adventitia consisting of the connective tissue along the whole surface of the graft.

In 64 weeks, all grafts remained passable and consisted completely of newly formed tissues with the wall width varying from 100 to 160 μm (Figures 7(c) and 7(d)). All newly formed vessels demonstrated aneurysms of various degrees (Figure 8). The presence of an aneurysm creates a risk of thrombosis, thromboembolism, and rupture of an aneurysm. These complications can threaten the patient's life and require urgent treatment [38, 39].

Neointima consisted of the endothelium and subendothelial layer which included collagen and elastic fibers. Neomedia consisted mainly of the fibroblasts. Collagen fibers formed disordered bundles. No PLLA fiber fragments were revealed. The area of cavities in the newly formed neomedia reached the values from 2.690 μm^2 to 21.700 μm^2 (Figures 7(e) and 7(f)). Neoadventitia consisted of loose fibrous connective tissue with vessels. Giant polynuclear cells of foreign bodies were observed in neoadventitia.

4. Conclusions

The studies of the vascular grafts consisting of PLLA microfibers by electrospinning technique demonstrated that the most promising mechanical and performance characteristics were inherent in the grafts with partially crystalline

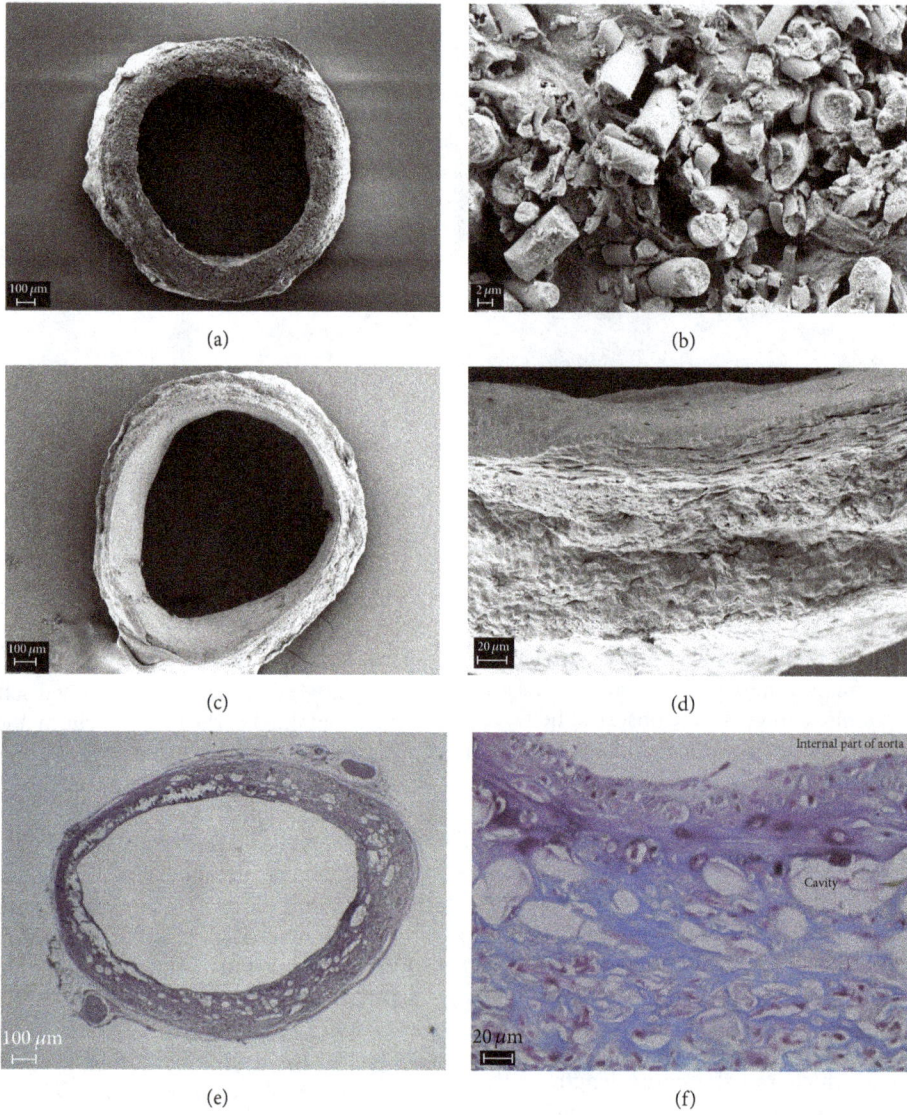

(a)

(b)

(c)

(d)

(e)

(f)

FIGURE 7: The graft in 56 weeks after implantation (a); fragment of the graft wall (b). The graft in 64 weeks after implantation (c, e); fragment of the graft wall (d, f). Histological sections (e, f) were stained according to the Mallory method.

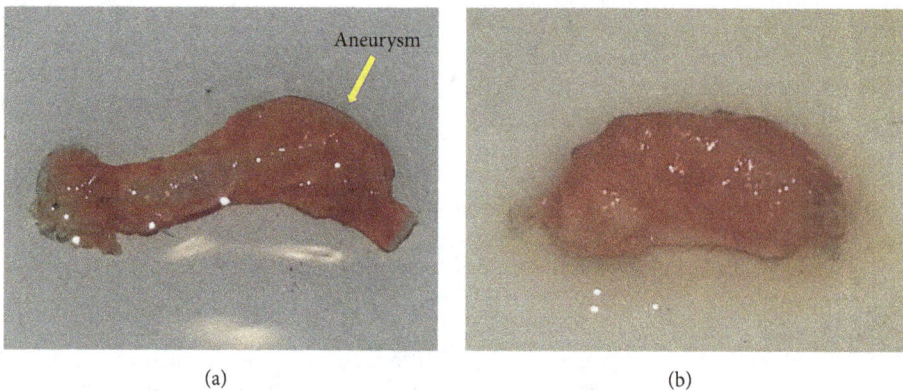

(a)

(b)

FIGURE 8: The graft in 64 weeks after implantation; aneurysms of various degrees: minimal (a) and maximal (b).

microfibers; this structure was obtained after thermal treatment of the initial tubular sample in the fixed state.

The grafts implanted into the rat's abdominal aorta did not cause pathological reactions in the rats; in 4 weeks, their inner side became completely covered with endothelium, and the whole wall was occupied by fibroblasts. These facts indicated good integration between the graft and the body of the rats. The presence of the endothelium layer in early postoperational period facilitated high stability of the graft against thrombosis.

First signs of bioresorption of microfibers were observed in 12 weeks after implantation; bioresorption continued until the 56th week and did not cause visible pathological reactions. Resorbed fibers were replaced by collagen fibers and fibroblasts. It is important to note that resorption rate of polymer graft did not exceed the rate of growth of cells and tissues inside it.

The blood vessels had been formed in 64 weeks after implantation; these vessels had intima and adventitia almost similar to those of the native vessel. It should be noted that neomedia consisted of disordered bundles of collagen fibers, fibroblasts, and oxyphilic cavities. The formation of this partially completed structure led to reduction of mechanical characteristics of vessels and appearance of aneurysms; at the same time, very high total permeability of the grafts was observed (93%).

One can conclude from the obtained results that PLLA grafts produced by electrospinning are promising for clinical uses, although some methods for strengthening walls of the newly formed blood vessels should be developed.

Conflicts of Interest

The authors declare that there are no conflicts of interest with any organization regarding the material discussed in this manuscript.

Acknowledgments

The authors are grateful to the Russian Science Foundation for financial support (Grant no. 14-33-00003).

References

[1] R. Y. Kannan, H. J. Salacinski, P. E. Butler, G. Hamilton, and A. M. Seifalian, "Current status of prosthetic bypass grafts: a review," Journal of Biomedical Materials Research Part B: Applied Biomaterials, vol. 74, no. 1, pp. 570–581, 2005.

[2] J. D. Kakisis, C. D. Liapis, C. Breuer, and B. E. Sumpio, "Artificial blood vessel: the Holy Grail of peripheral vascular surgery," Journal of Vascular Surgery, vol. 41, no. 2, pp. 349–354, 2005.

[3] G. R. Campbell and J. H. Campbell, "Development of tissue engineered vascular grafts," Current Pharmaceutical Biotechnology, vol. 8, no. 1, pp. 43–50, 2007.

[4] D. G. Seifu, A. Purnama, K. Mequanint, and D. Mantovani, "Small-diameter vascular tissue engineering," Nature Reviews Cardiology, vol. 10, no. 7, pp. 410–421, 2013.

[5] R. E. Harskamp, R. D. Lopes, C. E. Baisden, R. J. De Winter, and J. H. Alexander, "Saphenous vein graft failure after coronary artery bypass surgery: pathophysiology, management, and future directions," Annals of Surgery, vol. 257, no. 5, pp. 824–833, 2013.

[6] T. Athanasiou, S. Saso, C. Rao et al., "Radial artery versus saphenous vein conduits for coronary artery bypass surgery: Forty years of competition - which conduit offers better patency? A systematic review and meta-analysis," European Journal of Cardio-Thoracic Surgery, vol. 40, no. 1, pp. 208–220, 2011.

[7] N. Hibino, E. McGillicuddy, G. Matsumura et al., "Late-term results of tissue-engineered vascular grafts in humans," The Journal of Thoracic and Cardiovascular Surgery, vol. 139, no. 2, pp. 431–436, 2010.

[8] P. Klinkert, P. N. Post, P. J. Breslau, and J. H. van Bockel, "Saphenous vein versus PTFE for above-knee femoropopliteal bypass. A review of the literature," European Journal of Vascular and Endovascular Surgery, vol. 27, no. 4, pp. 357–362, 2004.

[9] N. L'Heureux, N. Dusserre, G. Konig et al., "Human tissue-engineered blood vessels for adult arterial revascularization," Nature Medicine, vol. 12, no. 3, pp. 361–365, 2006.

[10] N. L'Heureux, L. Germain, R. Labbé, and F. A. Auger, "In vitro construction of a human blood vessel from cultured vascular cells: A morphologic study," Journal of Vascular Surgery, vol. 17, no. 3, pp. 499–509, 1993.

[11] N. L'Heureux, S. Pâquet, R. Labbé, L. Germain, and F. A. Auger, "A completely biological tissue-engineered human blood vessel," The FASEB Journal, vol. 12, no. 1, pp. 47–56, 1998.

[12] N. L'Heureux, N. Dusserre, A. Marini, S. Garrido, L. de la Fuente, and T. McAllister, "Technology Insight: The evolution of tissue-engineered vascular grafts - From research to clinical practice," Nature Clinical Practice Cardiovascular Medicine, vol. 4, no. 7, pp. 389–395, 2007.

[13] N. L'Heureux, S. Paquet, R. Labbe, L. Germain, and F. A. Auger, "Tissue-engineered blood vessel for adult arterial revascularization," The New England Journal of Medicine, vol. 357, pp. 1451–1453, 2007.

[14] T. N. McAllister, N. Dusserre, M. Maruszewski, and N. L'Heureux, "Cell-based therapeutics from an economic perspective: primed for a commercial success or a research sinkhole?" Journal of Regenerative Medicine, vol. 3, no. 6, pp. 925–937, 2008.

[15] W.-L. Chue, G. R. Campbell, N. Caplice et al., "Dog peritoneal and pleural cavities as bioreactors to grow autologous vascular grafts," Journal of Vascular Surgery, vol. 39, no. 4, pp. 859–867, 2004.

[16] S. Ravi and E. L. Chaikof, "Biomaterials for vascular tissue engineering," Journal of Regenerative Medicine, vol. 5, no. 1, pp. 107–120, 2010.

[17] J. H. Campbell, J. L. Efendy, and G. R. Campbell, "Novel vascular graft grown within recipient's own peritoneal cavity," Circulation Research, vol. 85, no. 12, pp. 1173–1178, 1999.

[18] T. W. Gilbert, T. L. Sellaro, and S. F. Badylak, "Decellularization of tissues and organs," Biomaterials, vol. 27, no. 19, pp. 3675–3683, 2006.

[19] A. Bader, G. Steinhoff, and A. Haverich, "Tissue engineering of vascular grafts: Human cell seeding of decellularised

porcine matrix," *European Journal of Vascular and Endovascular Surgery*, vol. 19, no. 4, pp. 381–386, 2000.

[20] P. J. Schaner, N. D. Martin, T. N. Tulenko et al., "Decellularized vein as a potential scaffold for vascular tissue engineering," *Journal of Vascular Surgery*, vol. 40, no. 1, pp. 146–153, 2004.

[21] K. Shimizu, A. Ito, M. Arinobe et al., "Effective cell-seeding technique using magnetite nanoparticles and magnetic force onto decellularized blood vessels for vascular tissue engineering," *Journal of Bioscience and Bioengineering*, vol. 103, no. 5, pp. 472–478, 2007.

[22] L. E. Niklason, J. Gao, W. M. Abbott et al., "Functional arteries grown in vitro," *Science*, vol. 284, no. 5413, pp. 489–493, 1999.

[23] J. D. Berglund, M. M. Mohseni, R. M. Nerem, and A. Sambanis, "A biological hybrid model for collagen-based tissue engineered vascular constructs," *Biomaterials*, vol. 24, no. 7, pp. 1241–1254, 2003.

[24] S. Ravi, Z. Qu, and E. L. Chaikof, "Polymeric materials for tissue engineering of arterial substitutes," *Vascular*, vol. 17, supplement 1, pp. S45–S54, 2009.

[25] B.-S. Kim, I.-K. Park, T. Hoshiba et al., "Design of artificial extracellular matrices for tissue engineering," *Progress in Polymer Science*, vol. 36, no. 2, pp. 238–268, 2011.

[26] D. Shum-Tim, U. Stock, J. Hrkach et al., "Tissue engineering of autologous aorta using a new biodegradable polymer," *The Annals of Thoracic Surgery*, vol. 68, no. 6, pp. 2298–2305, 1999.

[27] T. Shin'oka, G. Matsumura, N. Hibino et al., "Midterm clinical result of tissue – engineered vascular autografts seeded with autologous bone marrow cells," *Journal of Thoracic and Cardiovascular Surgery*, vol. 129, no. 6, pp. 1330–1338, 2005.

[28] Q.-Z. Chen, S. E. Harding, N. N. Ali, A. R. Lyon, and A. R. Boccaccini, "Biomaterials in cardiac tissue engineering: ten years of research survey," *Materials Science and Engineering: R: Reports*, vol. 59, no. 1–6, pp. 1–37, 2008.

[29] A. C. Thomas, G. R. Campbell, and J. H. Campbell, "Advances in vascular tissue engineering," *Cardiovascular Pathology*, vol. 12, no. 5, pp. 271–276, 2003.

[30] S. L. Dahl, A. P. Kypson, and J. H. Lawson, "Readily available tissue-engineered vascular grafts," *Science Translational Medicine*, vol. 3, no. 68, Article ID 68ra9, 2011.

[31] S. L. M. Dahl, J. Koh, V. Prabhakar, and L. E. Niklason, "Decellularized native and engineered arterial scaffolds for transplantation," *Cell Transplantation*, vol. 12, no. 6, pp. 659–666, 2003.

[32] Y. Naito, T. Shinoka, D. Duncan et al., "Vascular tissue engineering: Towards the next generation vascular grafts," *Advanced Drug Delivery Reviews*, vol. 63, no. 4, pp. 312–323, 2011.

[33] Y. Naito, K. Rocco, H. Kurobe, M. Maxfield, C. Breuer, and T. Shinoka, "Tissue engineering in the vasculature," *Anatomical Record*, vol. 297, no. 1, pp. 83–97, 2014.

[34] I. P. Dobrovolskaya, P. V. Popryadukhin, V. E. Yudin et al., "Structure and properties of porous films based on aliphatic copolyamide developed for cellular technologies," *Journal of Materials Science: Materials in Medicine*, vol. 26, no. 1, p. 5381, 2015.

[35] P. V. Popryadukhin, G. I. Popov, I. P. Dobrovolskaya et al., "Vascular Prostheses Based on Nanofibers from Aliphatic Copolyamide," *Cardiovascular Engineering and Technology*, vol. 7, no. 1, pp. 78–86, 2016.

[36] M. C. Bernard and O. Brien, *Microvascular Reconstructive Surgery*, Churchill Livingston, Edinburgh, UK, 1977.

[37] S. Sarkar, K. M. Sales, G. Hamilton, and A. M. Seifalian, "Addressing thrombogenicity in vascular graft construction," *Journal of Biomedical Materials Research Part B: Applied Biomaterials*, vol. 82, no. 1, pp. 100–108, 2007.

[38] S. Tara, H. Kurobe, M. W. Maxfield et al., "Evaluation of remodeling process in small-diameter cell-free tissue-engineered arterial graft," *Journal of Vascular Surgery*, vol. 62, no. 3, pp. 734–743, 2015.

[39] L. M. Harris, G. L. Faggioli, R. Fiedler, G. R. Curl, and J. J. Ricotta, "Ruptured abdominal aortic aneurysms: Factors affecting mortality rates," *Journal of Vascular Surgery*, vol. 14, no. 6, pp. 812–820, 1991.

13

Silicone Substrate with Collagen and Carbon Nanotubes Exposed to Pulsed Current for MSC Osteodifferentiation

Daniyal Jamal and Roche C. de Guzman

Bioengineering Program, Department of Engineering, Hofstra University, Hempstead, NY 11549, USA

Correspondence should be addressed to Roche C. de Guzman; roche.c.deguzman@hofstra.edu

Academic Editor: Feng-Huei Lin

Autologous human adipose tissue-derived mesenchymal stem cells (MSCs) have the potential for clinical translation through their induction into osteoblasts for regeneration. Bone healing can be driven by biophysical stimulation using electricity for activating quiescent adult stem cells. It is hypothesized that application of electric current will enhance their osteogenic differentiation, and addition of conductive carbon nanotubes (CNTs) to the cell substrate will provide increased efficiency in current transmission. Cultured MSCs were seeded and grown onto fabricated silicone-based composites containing collagen and CNT fibers. Chemical inducers, namely, glycerol phosphate, dexamethasone, and vitamin C, were then added to the medium, and pulsatile submilliampere electrical currents (about half mA for 5 cycles at 4 mHz, twice a week) were applied for two weeks. Calcium deposition indicative of MSC differentiation and osteoblastic activity was quantified through Alizarin Red S and spectroscopy. It was found that pulsed current significantly increased osteodifferentiation on silicone-collagen films without CNTs. Under no external current, the presence of 10% (m/m) CNTs led to a significant and almost triple upregulation of calcium deposition. Both CNTs and current parameters did not appear to be synergistic. These conditions of enhanced osteoblastic activities may further be explored ultimately towards future therapeutic use of MSCs.

1. Introduction

Stem cells have the potential to revolutionize contemporary medicine and therapy methods. They enable us to screen new drugs, investigate the causes of birth defects, and understand the development of complex organisms from a single cell [1, 2]. More importantly, their greatest potential lies in cell-based and tissue engineering therapies to combat a range of diseases [3]. The idea of growing functional tissues and organs is no longer farfetched. Hematopoietic stem cells from bone marrow are one of the few clinically approved stem cells which are proven to increase the survival rate for thousands of patients [4]. Bone marrow-derived mesenchymal stem cells, or the blood stem cells that are nonhematopoietic, are currently being used in clinical trials and show promise in improving recovery based on early indicators [5]. Mesenchymal stem cells (MSCs) are multipotent stromal cells in connective tissues capable of differentiating into a variety of specialized and functional cells, including osteoblasts, chondroblasts, myocytes, and adipocytes. MSCs are easy to isolate, propagate to sufficient numbers, can be differentiated using simple chemicals, and have been proven safe [6]. They can be utilized as an autologous graft (from the same individual) or an allograft (from another human source) which, when introduced into the patient, leads to less likelihood of immune rejection, thus making them ideal for transplants and in vivo applications [7, 8]. Gaining control of MSCs differentiation pathways in vitro and within their local environment is a crucial feat for an efficient clinical approach. For autologous human adipose tissue-derived MSCs, enhanced differentiation into osteoblastic phenotype can be advantageous and may be used with matrix biomaterials for regeneration of the patient's own damaged bones due to fractures, osteoporosis, and deformities. The development of new technologies such as this tissue engineering approach is very important since millions of people [9] suffer from bone injuries on a yearly basis. A possible differentiation enhancer is the application of an external stimulus to stem cells in the form of electricity [10–12].

Bones contain a matrix of collagen proteins, which are piezoelectric materials that accumulate small electrical charges when subjected to mechanical stresses, leading to stimulation and deposition of intracellular calcium during fracture healing [13]. Biophysical stimulation methods [9] such as pulsed electric currents and electromagnetic fields have been clinically shown to significantly improve bone fracture repair—even the healing of nonunions, which is an advanced stage of fracture where healing is ceased by the body [14–16]. Rat tibial osteoporosis was reversed back to nonosteoporotic condition through capacitively coupled electric currents [17]. These definitive bone healing activities imply that quiescent local MSCs are activated for osteogenesis directly or through indirect effects of biological and chemical processes influenced by bursts of electricity [18] with varying strengths and frequencies. To provide a more efficient medium of electrical conduction, covalently linked elemental carbon in the form of carbon nanotubes (CNTs) can be introduced into the stem cell-adhesion substrates. CNTs are hexagonally ordered carbon atoms that form a lattice in the shape of hollow cylindrical tubes [19]. A single tube's outer diameter typically ranges from 10 nm to 50 nm. CNTs are primarily available in two forms: single-walled and multiwalled nanotubes, which vary in conductivity, fiber strength, and applications. They have exceptional mechanical properties, reported to be one of the strongest materials to be discovered. CNTs can also be chemically modified allowing for the attachment of functional carboxyl, amine, and hydroxyl groups within their structure for further functionalization. They are highly electroconductive; hence, they can enhance the flow of electric current.

Accordingly, in this study, the osteogenic differentiation effects of low intensity pulsed electric currents on cultured and induced MSCs using chemical factors were tested. Additionally, the presence of electrically conductive carbon nanotubes in cell-attachment silicone-collagen composite films was investigated for their MSC differentiation ability.

2. Materials and Methods

2.1. Materials.
Sylgard® 184 Silicone Elastomer Kit was purchased from Dow Corning (Greensboro, NC) for silicone matrix polymerization. Composite discontinuous phase components: type I collagen was prepared in the lab from rat tails generously given by Dr. Mark Van Dyke's lab (Virginia Tech, Blacksburg, VA, from IACUC-approved animal protocol), while carboxylic acid- (COOH-) functionalized graphitized multiwalled carbon nanotubes (CNTs with outside diameter = 20 to 30 nm and length = 10 to 30 μm) were obtained from Cheap Tubes (Cambridgeport, VT). Common lab reagents as well as N-(3-dimethylaminopropyl)-N′-ethylcarbodiimide hydrochloride (EDAC), N-hydroxysulfosuccinimide sodium salt (sulfo-NHS), MES buffer, poly(dimethylsiloxane-co-(3-aminopropyl)methylsiloxane), poly(dimethylsiloxane), bis(3-aminopropyl) terminated, 10% neutral-buffed formalin (NBF), Alizarin Red

S, and dimethyl sulfoxide (DMSO) were purchased from Sigma-Aldrich (St. Louis, MO). Deionized water was used as a solvent, unless stated otherwise. Human adipose tissue-derived mesenchymal stem cells (MSCs) were acquired from ATCC ([PCS-500-011] Manassas, VA). These cells were reported by ATCC to be positive (90% to 100%) for MSC markers, CD29, CD44, CD73, CD90, CD105, and CD166, and negative (<3%) for surface markers (non-MSCs like hematopoietic cells, blood cells, and endothelial cells), CD14, CD31, CD34, and CD45. ATCC, as well as our lab, has tested their in vitro differentiation into adipocytes and osteoblast (Figure 1) on tissue culture-treated plates and on scaffolds with collagen. Maintenance culture medium (CM) for these primary cells utilized Dulbecco's Modified Eagle's medium (DMEM) high glucose (4.5 mg/L), 10% fetal bovine serum, and Gibco® Antibiotic-Antimycotic Reagent (Thermo Fisher Scientific, Waltham, MA). Osteogenic culture medium (OM) for their differentiation into osteoblast-like cells contained CM with 10 mM β-glycerol phosphate, 100 nM dexamethasone, and 200 μM ascorbic acid 2-phosphate.

2.2. Collagen Extraction.
Type I collagen protein fibers were isolated from rat tail tendons based on modified Bornstein's methods [20]. Briefly, tendon bundles were dissected out, dissolved in acetic acid solution, precipitated with high salt, and centrifuged, and precipitates containing collagen were redissolved in acid. The solutions were then dialyzed for solvent exchange into dilute hydrochloric acid, freeze-dried, and stored at 4°C. The collagen yield was determined to be 175 mg per tail. A 1% (10 mg/mL ≈ 10 mg/g) collagen solution was made by dissolving the cotton-like powder in 1 M HCl.

2.3. Composite Film Preparation.
Substrates composed of polydimethylsiloxane (PDMS; silicone) continuous phase with collagen and CNTs discontinuous phases were fabricated as follows. Sylgard base and curing reagents, collagen solution (1% in 1 M HCl), and CNTs were weighed, according to the values in Table 1, into wells of 6-well tissue culture-treated plates (Thermo Fisher Scientific) and then slowly mixed to make a homogenous solution. Samples were placed overnight in a vacuum oven at 70°C to remove air bubbles and to cure the film substrates. Three groups of composites, namely, high CNTs with ~10% (m/m) CNTs, low CNTs with ~5% CNTs, and no CNTs (negative control; without CNTs) were made.

Sylgard reagents were also mixed with different ratios of amine-linked PDMS: poly(dimethylsiloxane-co-(3-aminopropyl)methylsiloxane) and poly(dimethylsiloxane), bis(3-aminopropyl) terminated prior to curing to potentially introduce amine groups to the silicone matrix.

To ensure surface adhesion of cells, a collagen coating was applied onto the silicone-based substrates. Collagen solution (60 μM in 1 mM HCl) was mixed with a carbodiimide crosslinker EDAC (150 μM) and sulfo-NHS (75 nM) in MES buffer, pH 6.0, for 15 min at room temperature (RT) to activate collagen. A 1.5 mL aliquot of this solution was added to the surface of the cured films and reacted for 3 hours at RT. Complete reaction expected a yield of 36 mg of collagen (for

TABLE 1: Composition of experimental silicone-based films.

C_i	Components	High CNTs		Low CNTs		No CNTs	
		m	C_f	m	C_f	m	C_f
100%	Sylgard base	0.75	68.81%	0.75	72.82%	0.75	76.53%
100%	Sylgard curing	0.2	18.35%	0.2	19.42%	0.2	20.41%
1%	Collagen	0.03	0.03%	0.03	0.03%	0.03	0.03%
	HCl		0.05%		0.05%		0.05%
	Water		2.68%		2.83%		2.98%
100%	CNTs	0.11	10.09%	0.05	4.85%	0	0%
	Total	*1.09*	*100%*	*1.03*	*100%*	*0.98*	*100%*

m = mass (g), C_i = initial concentration (m/m), and C_f = final concentration (m/m).

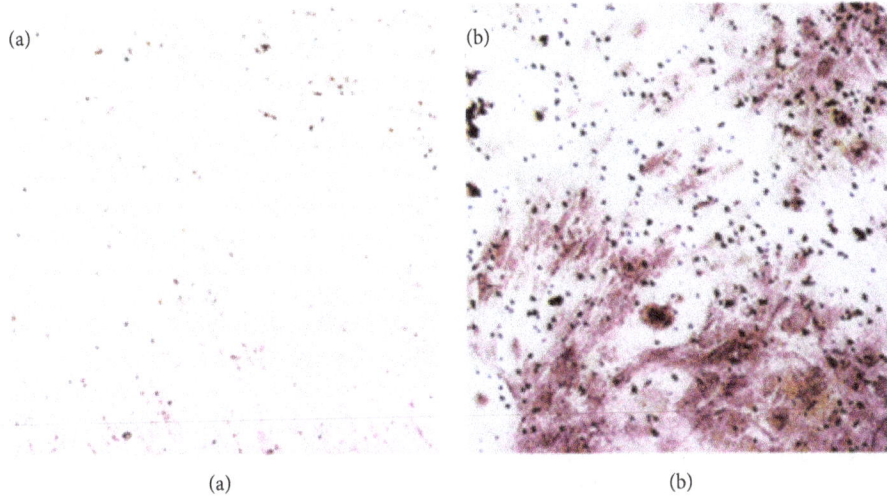

(a)   (b)

FIGURE 1: MSCs grown on plasma-treated polystyrene surface (a) left uninduced and (b) induced with osteogenic culture medium and then stained with Alizarin Red S. Induced cultures showed more rounded morphological features and a 21-fold increase in red dye absorption.

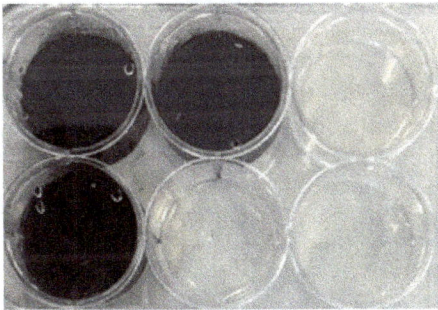

FIGURE 2: Plates containing surface-crosslinked collagen on films with no CNTs (clear) and high CNTs (black).

MW ~ 400 kDa) or a surface area density of 3.8 mg/cm^2 for an area of 9.5 cm^2. Collagen was expected to crosslink to the carboxylic acid groups of the functionalized CNTs and the collagen present within the material. After the reaction, films were washed with phosphate-buffed saline (PBS) thrice to remove uncrosslinked reactants and products (Figure 2).

2.4. Measurement of Electrical Properties. Film substrate groups (high CNTs, low CNTs, and no CNTs) in wells of

6-well plates were washed thrice and then soaked in 1.5 mL of CM. Positive and negative electrodes were immersed into the liquid culture medium and penetrating the substrates at 2 cm apart ($d = 0.02$ m). An electrical circuit (Figure 3(a)) was constructed connecting the electrodes and powered by a 5 V source with a 220 Ω resistor on a PB-505 circuit design trainer (Global Specialties, Yorba Linda, CA). A multimeter (ammeter function) was used to monitor the direct electrical current (I). Current values were measured at 1 min intervals for 10 min. Electrical conductivity (σ) was computed based on the equation

$$\sigma = \frac{I}{Vd},\qquad(1)$$

where I is average stable electric current in A, V is electric potential in V, and d is distance between electrodes in m.

2.5. Mesenchymal Stem Cell Culture on Silicone-Based Films. MSCs were revived, proliferated, and cultured up to sufficient numbers in T-75 tissue culture flasks (Thermo Fisher Scientific) with culture medium, CM, at 5% CO_2, > 90% relative humidity, and 37°C (in a mammalian cell incubator). CM was replaced twice a week. Trypsinized and CM-resuspended

(a)

(b)   (c)

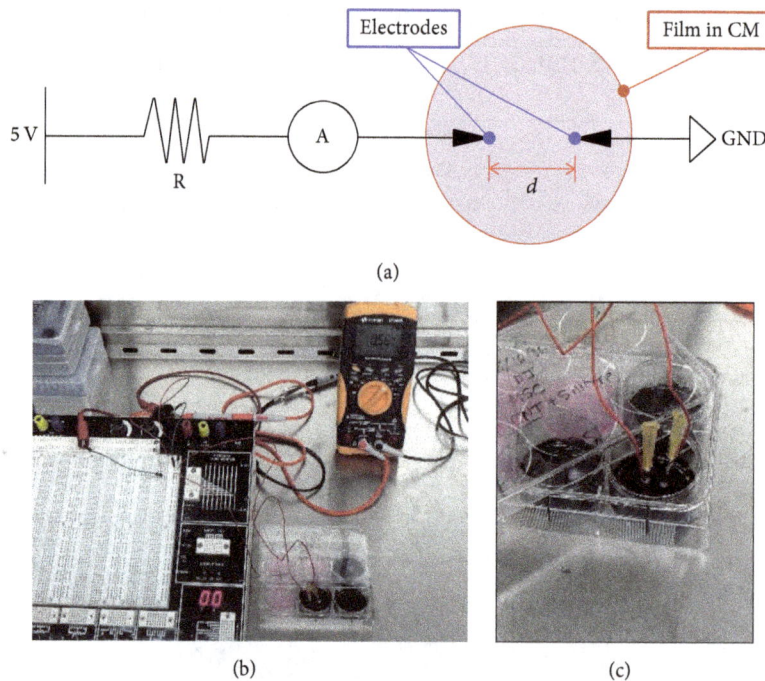

FIGURE 3: (a) Circuit diagram and (b) actual experimental setup for application of electric current to the composite films soaked in culture medium. (c) Close-up image of the electrodes penetrating the silicone substrate with collagen and carbon nanotube fibers. R = resistor, A = ammeter, GND = ground, and d = distance separating the two electrodes.

MSCs were seeded at 9×10^3 cells/well onto the test substrates: high CNTs and no CNTs then incubated for 24 hours at 37°C. The next day, nonadherent cells (cells in suspension) were counted using a hemocytometer. Cell attachment was reported as

$$\text{cell attachment} = \frac{(\text{initial cell count} - \text{cells in suspension})}{\text{initial cell count}}. \tag{2}$$

Adherent cells growing on films were allowed to proliferate for 7 days in CM.

2.6. Osteogenic Differentiation with Pulsed Current.
The culture medium of proliferating MSCs on composite substrates was replaced with osteogenic medium, OM, for 2 weeks to induce differentiation of mesenchymal stem cells into the osteoblastic phenotype. For each of the two groups, high CNTs and no CNTs, two subgroups were created: with current (experimental) and no current (control). For the "with current group," after 3 days in OM, a total of 6 wells were subjected to pulsed electrical currents (Figures 3(b)-3(c)), 3 for high CNTs and 3 for no CNTs. Low milliampere-level currents, 0.5 to 0.6 mA (at 5 to 6 mS/m conductivity), were applied for a 2 min duration followed by a 2 min duration of no current for a total of 20 minutes (5 cycles at $240\,s^{-1}$ ~ 4 mHz frequency). Ohm's Law was utilized to calculate the total circuit resistance, which included the resistor, liquid medium, and composite film resistance. The total resistance was in the range of 8.3 to 10 kΩ. Pulsed current application

was performed in a Biosafety Level 2 (BSL-2) cabinet twice a week prior to the culture medium change.

At the 2-week endpoint after OM induction, media were removed from the samples; 10% NBF was added for 15 min to fix cells. Representative sample regions for the "no current" groups were saved for elemental analysis. The remaining samples were then washed thrice with PBS and twice with water and stained with 250 μL per well of 2% Alizarin Red S, pH 4.2, for 5 min for deposited calcium staining. The stained composite films with MSC cells were destained with water thrice and solubilized with 750 μL DMSO per well for 10 min. Aliquots (200 μL) of liquid samples were transferred into wells of a 96-well plate and their absorbances were read in an iMark™ spectrometer (Bio-Rad, Hercules, CA) at a wavelength of 490 nm. DMSO solvent was used as a blank. Colorimetric readings were normalized on sample surface areas. Films before and after DMSO treatment looked similar, indicating that DMSO only dissolved cells and Alizarin Red S-calcium deposits, making the colorimetric test reliable for calcium precipitate detection.

2.7. Elemental Analysis of Osteogenic-Induced Cells.
Sampled spots from the no current groups (high CNTs and no CNTs) were washed in water, air-dried, and then loaded into the sample holder of Q250 SEM scanning electron microscope (FEI, Hillsboro, OR) with UltraDry™ (energy-dispersive X-ray spectroscopy) EDS detector (Oxford Instruments, Oxfordshire, England, UK). SEM was operated at 30 kV and $1\,mm^3$ areas were selected for EDS analysis. Elemental carbon (C) and calcium (Ca) mass fractions were obtained and Ca/C ratios were reported.

(a)   (b)

FIGURE 4: Fabricated silicone-based substrates with high CNTs (a) and no CNTs (b).

2.8. Statistical Analysis of Data. All experiments were performed with sample sizes (replicates) of $n \geq 3$. Computed values and graph points were reported as average ±1 standard deviation. Excel (Microsoft, Redmond, WA) and Prism (GraphPad, San Diego, CA) software were used for generating bar graphs and scatter plots. Student's *t*-test and one-way analysis of variance (ANOVA) with Tukey's post hoc multiple comparison analyses were made with Prism at 95% confidence intervals and 5% probability of type I error.

3. Results and Discussion

3.1. Fabricated Composite Substrates. Silicone-based composite films were synthesized (Figure 4) containing surface-crosslinked collagen for cell attachment and CNTs for promotion of increased electrical conductivity. The elemental carbon tubes were evenly dispersed creating a uniform black layer. Multiple washes of PBS demonstrated stability of the CNTs within the films since no loose particles were detected from the waste solution. Immobilization and crosslinking into the silicone matrix with collagen thereby prevented the possible cell internalization and cytotoxicity of free fibers of CNTs in suspension [21].

In groups with no CNTs, addition of 0.03% (m/m) = 0.3 mg/g collagen (Table 1) and surface collagen coating did not affect the visible appearance and transparency of the films relative to PDMS films alone. The amine-linked PDMS additives tested, poly(dimethylsiloxane-co-(3-aminopropyl)methylsiloxane), and poly(dimethylsiloxane), bis(3-aminopropyl) terminated, were not used as matrix precursors since they did not allow for the polymerization and complete curing of the silicone-based films.

3.2. Conductivity of Films in Culture Medium. Electrical current (direct current, DC) passing between electrodes placed 2 cm apart along a film, immersed in culture medium at 5 V electric potential connected to a 220 Ω resistor, was found to increase from 0 to about 7 mA over the 10 min time period (Figure 5(a)). At the 6 to 10 min marks, the current flattened indicating stability. ANOVA and Tukey's test showed that the mean values of the current transmission of films with 10% CNTs (high CNTs) were significantly greater ($p < 0.05$) than those with 5% CNTs (low CNTs) and substrates without CNTs (no CNTs). Low CNTs generated significantly higher DC current ($p < 0.05$) than those of negative control films. After stabilization, the high CNTs group allowed 11.1% more electrical current than the no CNTs group and 5.1% more than the low CNTs group. The total system resistance from the resistor and substrate and liquid medium resistance were reported as 827 Ω (no CNTs), 782 Ω (low CNTs), and 760 Ω (high CNTs). Accordingly, the corresponding electrical conductivities (Figure 5(b)), computed in mS/m at $d = 0.02$ m, were determined to be highest (although not statistically significant) in films with high CNTs (at 69.4 mS/m) compared to the other two groups (low CNTs and no CNTs). The conductivity values obtained were close to that of drinking water and within the expected range of silicone polymers [22, 23]. Increasing amounts of CNTs led to increased conductivity of the composite silicone-based material; hence, the 10% CNT formulation was utilized in subsequent cell culture and induction experiments for potentially higher electrical responses.

3.3. Adhesion of Mesenchymal Stem Cells onto the Substrates. Seeding MSCs on composite films after 24-hour incubation resulted in significantly more attached cells ($p < 0.05$) on the silicone-collagen composites with 10% CNTs (high CNTs) compared to those without (0% CNTs; no CNTs) (Figure 6). The presence of CNTs led to 13% more initial cell adhesion on the film surface (87% on high CNTs versus 74% on no CNTs). This result implies that carbon nanotube fibers aid in cell attachment, possibly due to the presence of charged functional carboxylic acid groups promoting extracellular matrix adsorption, which in turn allows cell binding through their integrin receptors. MSCs are adhesion-dependent cells and can only survive and differentiate after successful substrate attachment. Collagen added as coatings (≤36 mg) and mixed in bulk films (~0.3 mg) successfully provided the initial

(a)

(b)

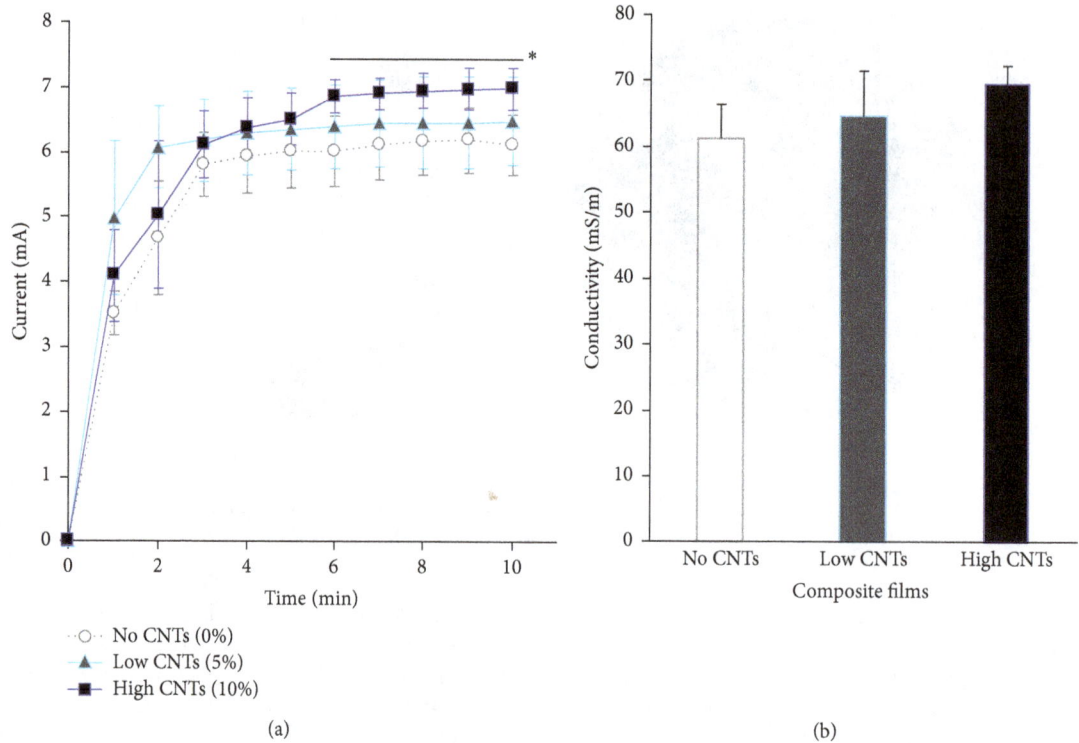

FIGURE 5: (a) Measured direct electrical currents (in mA) at constant voltage in composite films (no CNTs, low CNTs, and high CNTs groups) immersed in culture medium. $^*p < 0.05$ for high CNTs versus low CNTs and no CNTs at currents from 6 to 10 min time points (horizontal line). (b) Electrical conductivity (in mS/m) derived from 1 per system resistance and electrodes distance.

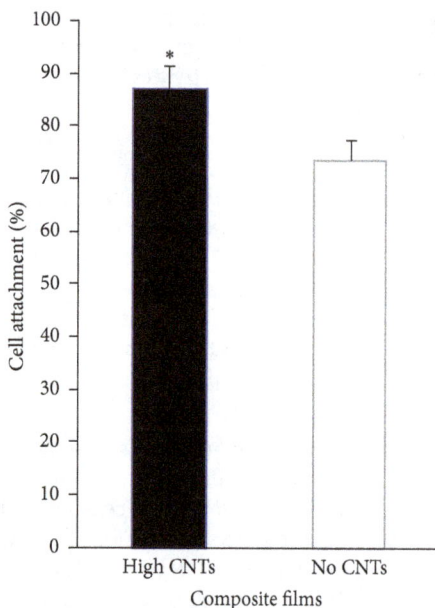

FIGURE 6: Adhesion of mesenchymal stem cells on silicone-collagen substrates with high CNTs and no CNTs. $^*p < 0.05$ for high CNTs versus no CNTs.

substrate adhesion of these cells. A previous study showed similar enhancement of stem cell adhesion due to reinforcing CNTs within the substrate material [24].

3.4. Pulsed Current Exposure during Cell Differentiation. The electrical current applied to MSCs on films was in the submilliampere level, specifically in the range of 0.5 to 0.6 mA, and delivered at interrupted regular intervals (i.e., noncontinuous and pulsatile). These exposure conditions enable current to flow through cells without killing them [15, 25]. After the 2-week time point, where MSCs were subjected to induction by chemical factors, namely, glycerol phosphate, dexamethasone, and ascorbic acid (vitamin C), as well as with pulsed electrical current, cells on the surface of translucent no CNTs substrates showed dark red staining with Alizarin Red S (Figure 7(a)), indicating calcium deposition and osteoblastic lineage cell differentiation. Contrastingly, those grown on black films with high CNTs (Figure 4) were not observed using light microscopy because of black-body light absorption. Alizarin Red S-calcium precipitates were quantified by DMSO solubilization and subsequent spectrophotometric analysis at 490 nm (red signal). ANOVA and Tukey's multiple comparison tests showed that the presence of CNTs alone (high CNTs versus no CNTs at no current) led to significantly more production of deposited calcium ($p < 0.05$; 2.9-fold increase) based on Alizarin Red S staining (Figure 7(b) right). Furthermore, elemental analysis of film surfaces demonstrated significantly elevated ($p = 0.0044$), 6-fold calcium relative to carbon level increase (Figure 8). These results indicated that carbon nanotubes induced more osteoblast differentiation, which may primarily be due to more cells adhering onto the high CNTs composites (Figure 6), leading to high survival of cells and increased osteogenic response.

(a)

(b)

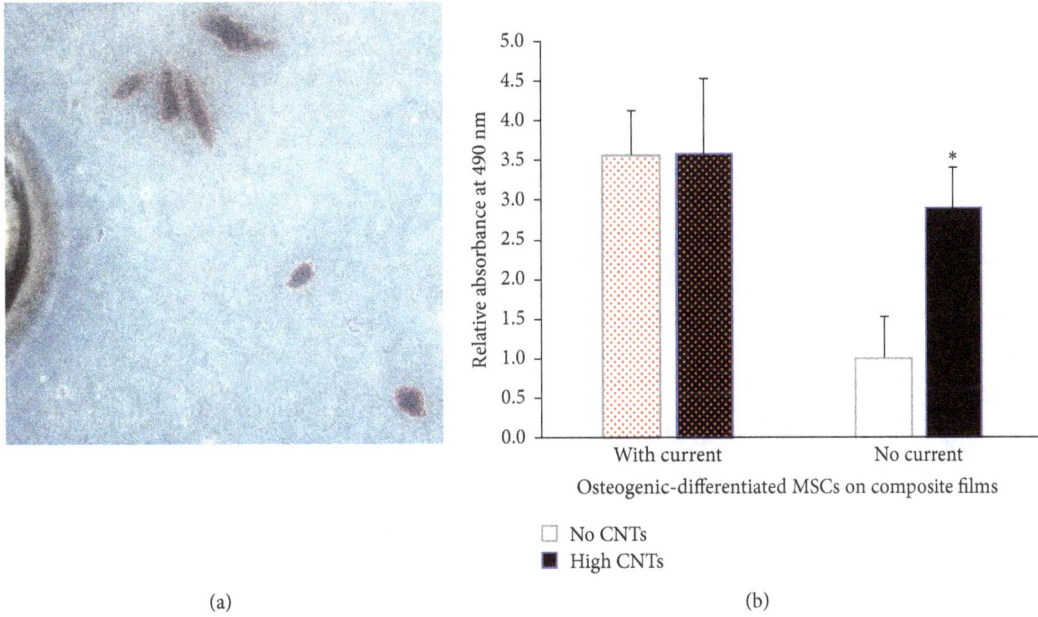

FIGURE 7: (a) A representative image of Alizarin Red S-stained cell clusters on the surface of no CNTs silicone-collagen film. (b) Absorbance spectroscopy at 490 nm (red intensity) of DMSO-solubilized Alizarin Red S-calcium crystals on composites with high CNTs and no CNTs, with applied electrical stimulation (with current) and without it (no current control). $^*p < 0.05$ for high CNTs versus no CNTs at a no current condition.

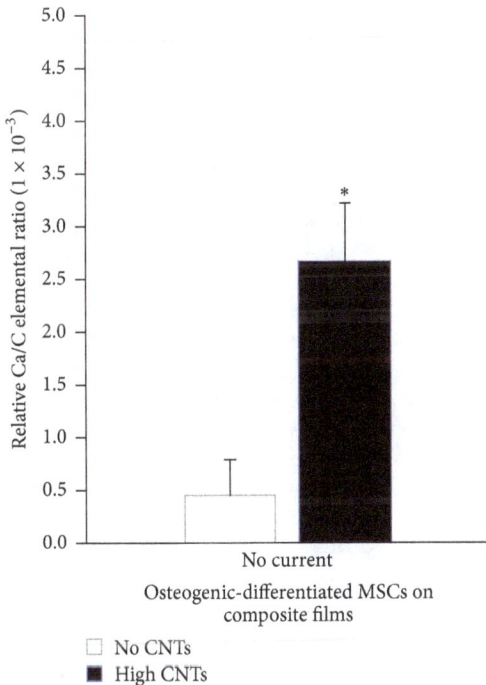

FIGURE 8: Relative amounts of Ca to C on induced MSCs on silicone-collagen substrates with high CNTs and no CNTs. $^*p < 0.05$ for high CNTs versus no CNTs.

The application of pulsed current in silicone-collagen groups only (with current versus no current at no CNTs) led to significantly more Alizarin Red S staining ($p < 0.01$), suggesting that pulsed current was beneficial towards induction of more MSCs, directly by stimulating few cells to enable calcium deposition or indirectly by initially enabling few cells to proliferate prior to differentiation. The higher the cell number, the greater the relative amount of calcium phosphate that precipitated. Other researchers demonstrated similar enhanced differentiation into osteoblast-like cells after biophysical electrical stimulation, like pulsed electromagnetic field and alternating currents [15, 26]. Conflicting studies with negative impacts in stem cell growth and differentiation [27, 28], however, suggest the need for current delivery with optimal properties including the right frequency, magnitude, and duration of cell exposure; otherwise, unwanted effects are observed.

Groups containing high CNTs (with current versus no current at high CNTs) showed elevated response with application of pulsatile current (1.8-fold increase), although nonsignificant ($p > 0.05$; Figure 7(b) black bars). This could indicate that longer duration of current exposure or longer culture time may be needed to observe a significant effect of pulsed current in composite films with 100 mg/g of carbon tubes. Alternatively, the amount of cells on composite films may have reached saturation; hence, any increased response was not sensitive for detection. Further testing is recommended.

The current film substrate composition containing carbon nanotubes (high CNTs) can be utilized as an in vitro preconditioning culture system for production of osteoblasts that can ultimately be transplanted into hosts for bone regeneration therapy. Since free CNTs are cytotoxic [21] at high concentrations in the body, they can be used as a substrate additive just for cell culture but not for direct implantation into patients.

Pulsed current did not affect the osteogenic calcium activity of differentiated MSCs on silicone composite films (high CNTs versus no CNTs at with current group). The high CNTs and no CNTs groups demonstrated similar levels of absorbance ($p > 0.05$; Figure 7(b) left). Submilliampere-level electric currents were able to transmit across the media containing the composite films, cells, and liquid culture medium; thus, the presence of CNTs did not matter. The average relative absorbance of groups with current was found to be 3.572 ± 0.002, compared to the 1.942 ± 1.332 value for no current. Application of electric current led to 1.8 times higher osteoblastic calcium deposition activity.

Osteogenic induction based on calcium deposition staining showed relatively high activities on groups with pulsatile current (Figure 7(b) left). To further improve MSCs differentiation, groups with longer exposure to electric current (i.e., more cycles: 10 and 20 cycles; longer differentiation period: 3 weeks and 4 weeks) will be added in the follow-up study. Results from these additional experimental groups will enable us to determine the temporal conditions that generate the greatest osteodifferentiation using the current system.

4. Conclusions

In summary, this study showed that silicone (PDMS) composites with cell-adhesive collagen and stable reinforcing CNTs were successfully made into films. The immobilized CNTs provided increased electrical conductivity to the substrate, enabled more MSC attachment, and allowed for increased surface calcium deposition when chemical inducers, namely, glycerol phosphate, dexamethasone, and vitamin C, were added to the cell culture media. Application of pulsatile, low intensity, and low frequency external DC electrical current significantly enhanced osteoblastic response on silicone-collagen films without CNTs. The combined differentiation effects of CNTs and pulsed current were found to be nonsynergistic or nonadditive in the tested system. These silicone-collagen substrates with CNTs may further be developed for enhancement of mesenchymal differentiation into osteoblasts which can be used in the clinic for treatment of patients with bone injuries. Other potential applications of CNTs in bioengineering include nanotechnology, microelectronics, and medical devices. On the other hand, in composites devoid of CNTs, delivery of pulsed electric current can also be employed towards MSC differentiation in vitro or in vivo in combination with piezoelectric, degradable, implantable, and biocompatible scaffolds as tissue-engineered constructs for bone repair and regeneration.

Conflicts of Interest

The authors declare that there are no conflicts of interest regarding the publication of this paper.

Acknowledgments

The authors would like to thank the members of the Bioengineering Materials Lab including Emily Diaz, Jennifer Miller, Liam Lang, Reid Wasserman, Kimberly Lewis, Horacio Reyna, Bradley Sheen, and Hazel Consunji for technical assistance, manuscript proofreading, and discussion, Jacqueline Scarola and Lori Castoria for help with reagents purchasing, and Sina Rabbany for his continuing support and collaboration in conducting biomaterials research at Hofstra SEAS. Support for this study was provided by Hofstra University internal research funds.

References

[1] A. Bajek, N. Gurtowska, J. Olkowska, L. Kazmierski, M. Maj, and T. Drewa, "Adipose-derived stem cells as a tool in cell-based therapies," *Archivum Immunologiae et Therapiae Experimentalis*, vol. 64, no. 6, pp. 443–454, 2016.

[2] K. A. Hasty and H. Cho, "Stem cell considerations for the clinician," *Physical Medicine and Rehabilitation Clinics of North America*, vol. 27, no. 4, pp. 855–870, 2016.

[3] T. Almela, I. M. Brook, and K. Moharamzadeh, "The significance of cell-related challenges in the clinical application of tissue engineering," *Journal of Biomedical Materials Research - Part A*, vol. 104, no. 12, pp. 3157–3163, 2016.

[4] T. Hahn, P. L. McCarthy Jr., A. Hassebroek et al., "Significant improvement in survival after allogeneic hematopoietic cell transplantation during a period of significantly increased use, older recipient age, and use of unrelated donors," *Journal of Clinical Oncology: Official Journal of the American Society of Clinical Oncology*, vol. 31, no. 19, pp. 2437–2449, 2013.

[5] D. S. Shenaq, F. Rastegar, D. Petkovic et al., "Mesenchymal progenitor cells and their orthopedic applications: forging a path towards clinical trials," *Stem Cells International*, Article ID 519028, 2010.

[6] K. E. Hammerick, A. W. James, Z. Huang, F. B. Prinz, and M. T. Longaker, "Pulsed direct current electric fields enhance osteogenesis in adipose-derived stromal cells," *Tissue Engineering Part A*, vol. 16, no. 3, pp. 917–931, 2010.

[7] B. Larijani, E. N. Esfahani, P. Amini et al., "Stem cell therapy in treatment of different diseases," *Acta Medica Iranica*, vol. 50, no. 2, pp. 79–96, 2012.

[8] S. E. New, A. Ibrahim, L. Guasti et al., "Towards reconstruction of epithelialized cartilages from autologous adipose tissue-derived stem cells," *Journal of Tissue Engineering and Regenerative Medicine*, 2016.

[9] "The frequency of bone disease," in *Bone Health and Osteoporosis: A Report of the Surgeon General*, Office of the Surgeon General (US), Ed., Rockville, MD, 2004.

[10] J. Zhang, M. Li, T. Kang, and K. G. Neoh, "Electrical stimulation of adipose-derived mesenchymal stem cells in conductive scaffolds and the roles of voltage-gated ion channels," *Acta Biomaterialia*, vol. 32, pp. 46–56, 2016.

[11] R. Mishra, D. B. Raina, M. Pelkonen, L. Lidgren, M. Tagil, and A. Kumar, "Study of in vitro and in vivo bone formation in composite cryogels and the influence of electrical stimulation," *International Journal of Biological Sciences*, vol. 11, no. 11, pp. 1325–1336, 2015.

[12] J. G. Hardy, M. K. Villancio-Wolter, R. C. Sukhavasi et al., "Electrical stimulation of human mesenchymal stem cells on conductive nanofibers enhances their differentiation toward osteogenic outcomes," *Macromolecular Rapid Communications*, vol. 36, no. 21, pp. 1884–1890, 2015.

[13] C. Frias, J. Reis, F. Capela e Silva, J. Potes, J. Simoes, and A. T. Marques, "Polymeric piezoelectric actuator substrate for osteoblast mechanical stimulation," *Journal of Biomechanics*, vol. 43, no. 6, pp. 1061–1066, 2010.

[14] W. G. de Haas, J. Watson, and D. M. Morrison, "Non-invasive treatment of ununited fractures of the tibia using electrical stimulation," *Journal of Bone and Joint Surgery - Series B*, vol. 62, no. 4, pp. 465–470, 1980.

[15] M. Hronik-Tupaj, W. L. Rice, M. Cronin-Golomb, D. L. Kaplan, and I. Georgakoudi, "Osteoblastic differentiation and stress response of human mesenchymal stem cells exposed to alternating current electric fields," *BioMedical Engineering Online*, vol. 10, article 9, 2011.

[16] J. D. Heckman, A. J. Ingram, R. D. Loyd, J. V. Luck Jr., and P. W. Mayer, "Nonunion treatment with pulsed electromagnetic fields," *Clinical Orthopaedics and Related Research*, vol. 161, pp. 58–66, 1981.

[17] C. T. Brighton, G. T. Tadduni, and S. R. Pollack, "Treatment of sciatic denervation disuse osteoporosis in the rat tibia with capacitively coupled electrical stimulation. dose response and duty cycle," *Journal of Bone and Joint Surgery - Series A*, vol. 67, no. 7, pp. 1022–1028, 1985.

[18] N. Selvamurugan, S. Kwok, A. Vasilov, S. C. Jefcoat, and N. C. Partridge, "Effects of BMP-2 and pulsed electromagnetic field (PEMF) on rat primary osteoblastic cell proliferation and gene expression," *Journal of Orthopaedic Research*, vol. 25, no. 9, pp. 1213–1220, 2007.

[19] C. J. Serpell, K. Kostarelos, and B. G. Davis, "Can carbon nanotubes deliver on their promise in biology? harnessing unique properties for unparalleled applications," *ACS Central Science*, vol. 2, no. 4, pp. 190–200, 2016.

[20] M. B. Bornstein, "Reconstituted rattail collagen used as substrate for tissue cultures on coverslips in Maximow slides and roller tubes," *Laboratory Investigation*, vol. 7, no. 2, pp. 134–137, 1958.

[21] S. Y. Madani, A. Mandel, and A. M. Seifalian, "A concise review of carbon nanotube's toxicology," *Nano Reviews*, vol. 4, no. 1, Article ID 21521, 2013.

[22] S. Ramirez-Garcia, S. Alegret, F. Cespedes, and R. J. Forster, "Carbon composite electrodes: surface and electrochemical properties," *The Analyst*, vol. 127, no. 11, pp. 1512–1519, 2002.

[23] Y. H. Yu, S. H. Chen, C. L. Chang, C. T. Lin, W. D. Hairston, and R. A. Mrozek, "New flexible silicone-based EEG dry sensor material compositions exhibiting improvements in lifespan, conductivity, and reliability," *Sensors (Basel)*, vol. 16, no. 11, article 1826, 2016.

[24] D. Lahiri, A. P. Benaduce, L. Kos, and A. Agarwal, "Quantification of carbon nanotube induced adhesion of osteoblast on hydroxyapatite using nano-scratch technique," *Nanotechnology*, vol. 22, no. 35, Article ID 355703, 2011.

[25] K. A. Chang, J. W. Kim, J. A. Kim et al., "Biphasic electrical currents stimulation promotes both proliferation and differentiation of fetal neural stem cells," *PLoS ONE*, vol. 6, no. 4, Article ID e18738, 2011.

[26] C. H. Lohmann, Z. Schwartz, Y. Liu et al., "Pulsed electromagnetic field stimulation of MG63 osteoblast-like cells affects differentiation and local factor production," *Journal of Orthopaedic Research*, vol. 18, no. 4, pp. 637–646, 2000.

[27] L. A. Norton, D. W. Witt, and L. A. Rovetti, "Pulsed electromagnetic fields alter phenotypic expression in chondroblasts in tissue culture," *Journal of Orthopaedic Research*, vol. 6, no. 5, pp. 685–689, 1988.

[28] L. A. Norton, "Effects of a pulsed electromagnetic field on a mixed chondroblastic tissue culture," *Clinical Orthopaedics and Related Research*, vol. 167, pp. 280–290, 1982.

14

Graphene Family Nanomaterials: Properties and Potential Applications in Dentistry

Ziyu Ge,[1] Luming Yang,[2] Fang Xiao,[2] Yani Wu,[1] Tingting Yu,[2] Jing Chen (ID),[2] Jiexin Lin,[2] and Yanzhen Zhang (ID)[1]

[1]Department of General Dentistry, The Second Affiliated Hospital, Zhejiang University School of Medicine, 310052, China
[2]Zhejiang University, 310058, China

Correspondence should be addressed to Yanzhen Zhang; 2191004@zju.edu.cn

Academic Editor: Carlo Galli

Graphene family nanomaterials, with superior mechanical, chemical, and biological properties, have grabbed appreciable attention on the path of researches seeking new materials for future biomedical applications. Although potential applications of graphene had been highly reviewed in other fields of medicine, especially for their antibacterial properties and tissue regenerative capacities, *in vivo* and *in vitro* studies related to dentistry are very limited. Therefore, based on current knowledge and latest progress, this article aimed to present the recent achievements and provide a comprehensive literature review on potential applications of graphene that could be translated into clinical reality in dentistry.

1. Introduction

Oral cavity is an extremely demanding setting. Dental materials when placed within oral cavity are fully contacted with saliva, gingival crevicular fluid, and water. At the same time, it is exposed to high temperature, masticatory forces, and variety of abrasion causing mechanical failures and overtime requiring restoration replacement with extra cost. Furthermore, most dental materials are in intimate contact with oral tissue for a long time; they must be noncytotoxic and biocompatible for them to have a harmonious interaction with host while performing desired functions. Therefore, there is always a huge interest and strong trend in continuous development of dental materials with improving properties.

Nanotechnology, "the manufacturing technology of the 21st century", is an art of manipulating matter on a scale of less than 100nm to create numerous materials with various properties and functions. Over the past decades, with the discovery of fullerene in 1985 and carbon nanotubes in 1991, carbon based nanomaterials have been merited on the scientific stage (see Figure 1). Graphene is a 2D single layer of sp^2 hybridized carbon atoms with hexagonal packed configuration (see Figure 2). The in-depth investigation of graphene conducted by Andre Geim and Konstantin Novoselov in 2004 has proven that graphene was the building block for all graphitic carbon materials such as graphite, diamond, nanoribbons, CNTs, and fullerenes. Moreover, it possesses exceptional physicochemical, optical, and mechanical properties. Since then, research efforts have been focused on excavating its potential applications including various biomedical applications such as drug delivery carriers [1], imaging agents [2], biosensors [3], bimolecular analysis, and tissue engineering scaffolds [4].

Graphene family nanomaterials (GFNs) include ultrathin graphite, few-layer graphene (FLG), graphene oxide (GO; from monolayer to few layers), reduced graphene oxide (rGO), and graphene nanosheets (GNS) [5]. They differ from each other in terms of surface properties, number of layers, and size [6]. Among other members of graphene family nanomaterial, graphene oxide (GO) is one of the most important chemical graphene derivatives which could be produced through energetic oxidation of graphite through Hummers method using oxidative agents. GO possessed a variety of chemically reactive functional groups on its surface, which facilitate connection with various materials including polymers, biomolecules, DNA, and proteins [7]. The large

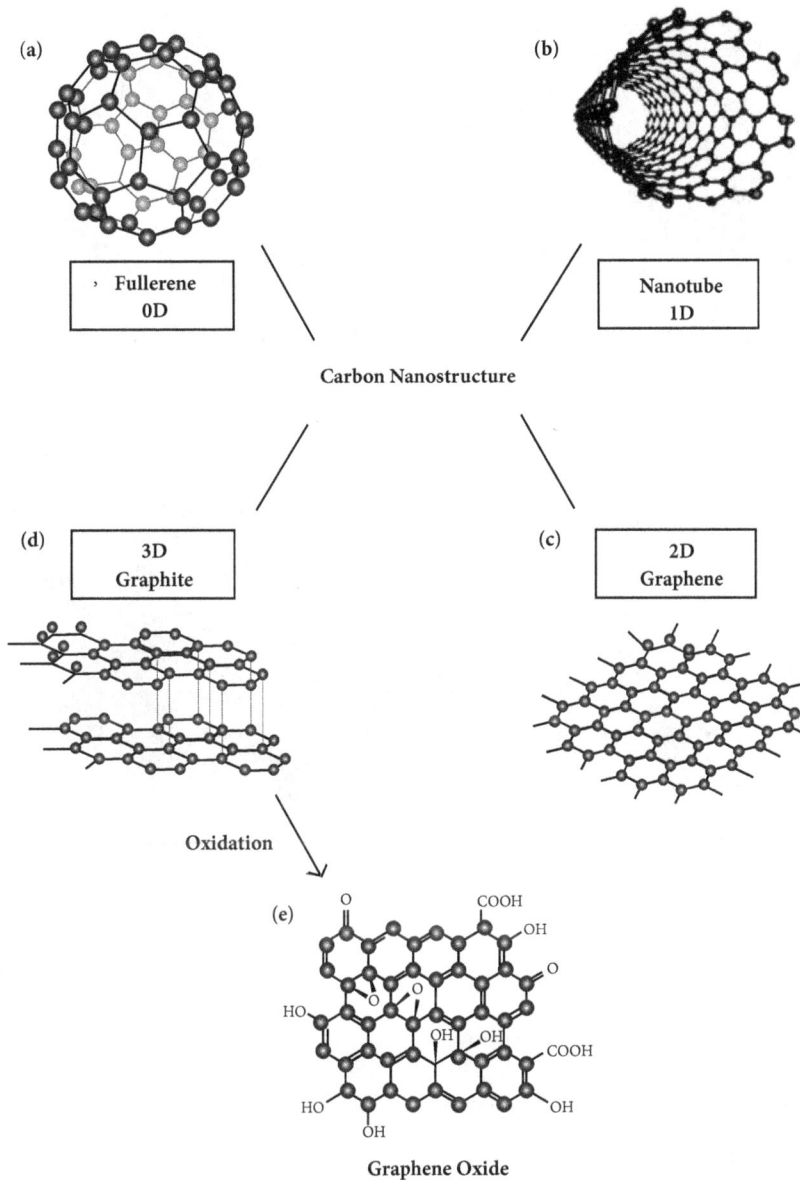

FIGURE 1: Different allotropes of carbon nanostructure: (a) 0D Fullerenes; (b) 1D Carbon Nanotubes; (c) 2D Graphene; (d) 3D Graphite. (e) Graphene Oxide can be synthesized through oxidation of graphite, with common method called Hummers method.

interactive aromatic surface area of GO is at least an order of magnitude higher compared with other nanomaterials endows it with high drug loading capacity [8]. Reduced graphene oxide (rGO) can be obtained by chemically, thermally, or electrochemically reducing graphene oxide, which possesses heterogeneous electron-transfer properties [9]. Fluorinated graphene (FG) is an uprising member in the graphene family. FG has favorable biocompatibility, exhibiting a neuroinductive effect via spontaneous cell polarization and enhancing adhesion and proliferation of mesenchymal cells providing scaffold for their growth [1].

Although the developments and researches of graphene-based biomaterials related to dentistry are still at infancy, their unique properties and their abilities to functionalize

alone or combined with biomaterials offer several opportunities in possible clinical applications. In this review, we intended to provide readers with an overview of the potential applications of graphene correlated to dentistry. Their biocompatibility aspect and antibiotic properties were briefly discussed. Perspectives related to graphene-based technologies aimed at oral care are presented and organized by different fields of dentistry.

2. Biocompatibility

The first aspect to consider in the introduction of a new biomedical material is its biocompatibility. For a safer development of graphene-based nanomaterial, it is necessary to

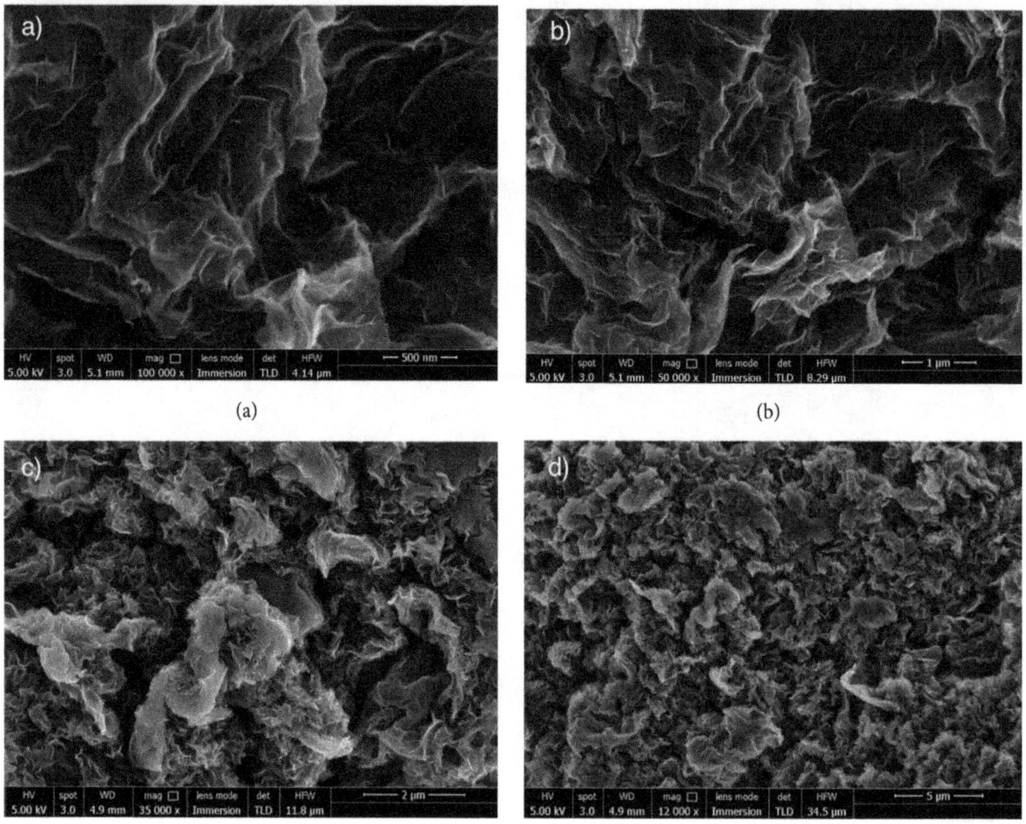

FIGURE 2: Graphene under scanning electron microscope (SEM) at (a) 100000× magnification, (b) 50000× magnification, (c) 35000× magnification, and (d) 12000× magnification.

understand the interaction of graphene and their derivatives with living systems and their toxicity *in vivo* and *in vitro* [10].

Accumulating evidences have suggested that cytotoxicity of GFNs can not be generalized as it depends on various factors including their morphology (size, shape, and sharp edges), surface charge, surface functionalization, dispensability, state of aggregation, number of layers, purity, and methods of synthesis [11]. This is because different morphology, shape, and size of GFNs could influence their cellular uptake characteristics; moreover, distinctive functional groups on the surface can alter their interactions with proteins, biomolecules, and micronutrients. In relation to the concentration of GFNs on toxicity, many studies suggested that 50ug/ ml might be a toxicity threshold for GO on normal mammalian cells. Concentration higher than 50ug/ml might harm human fibroblast cells and T lymphocytes [7, 12]. Lateral size of GFN may determine the location of accumulation and amount of toxicity uptake by different organs in human body as nanoparticles with sizes <100nm can enter cell, <40nm can enter nucleus, and <35nm can cross blood brain barrier [13]. Surface structure, such as wettability, also plays a role in GFN's toxicity. Graphene is recognized as a hydrophobic material while GO is slightly hydrophilic due to the presence of oxygen containing functional groups on its basal plane [14]. In general, compared to hydrophobic ones,

hydrophilic (more oxidized) graphene nanoparticles may be more cytocompatible as they tend to form a stable colloid dispersion and avoid aggregation, and, therefore, could be easier internalized and excreted from the application site [15]. GFNs' purity also deserves attention because sometimes contaminating metal may account for toxicity reaction in cells rather than GFNs themselves. Traditionally prepared GO often contains high levels of Mn^{2+} and Fe^{2+}, which are highly mutagenic to cells leading to high levels of cytotoxicity and random scission of DNA [16]. Greener exfoliating methods suggested by Peng et al. were able to produce a high purity GO containing much lower Mn^{2+} and Fe^{2+} and consequently significantly lowered the cytotoxicity of GFNs [17]. Therefore, choosing a high purity GFNs synthesis method is a vital step toward safer bioapplications.

The exact mechanism underlying the toxicity of GFNs remains obscure; several possible toxicity mechanisms of GFNs were proposed. Oxidative stress is suggested to be the main cause as the elevated ROS level may oxidize various molecules including DNA, lipids, and proteins inducing apoptosis or necrosis [18]. GFNs may also cause cell necrosis by directly influencing cell mitochondrial activity through induction of mitochondrial membrane potential dissipation, which subsequently increases the generation of intracellular ROS and triggers apoptosis by activating the mitochondrial pathway [19]. For instance, GO with lower degree of oxidation possesses more free electrons facilitates more OH production from H_2O_2. The formation of OH and the cytochrome c/H_2O_2 electron-transfer system could enhance oxidative and thermal stress to impair the mitochondrial respiration system and eventually impose stronger oxidative damage on normal cells displaying stronger toxicity [20].

Studies focused on the cytotoxicity of GFNs in oral setting are very limited. A study conducted by Olteanu et al. assessed the cytotoxic potential of GO, thermally reduced graphene oxide (TRGO) and Nitrogen doped graphene (N–Gr), on human dental follicle stem cells. The result showed that GFNs, especially GO, increased the intracellular ROS generation in a concentration and time-dependent manner. At high concentration (40 ug /mL), cells viability was reduced and mitochondria membrane potential was altered. While, at low concentrations (4 ug/mL), they exhibited a good safety profile providing high antioxidant defense. Their authors concluded that, among these three investigated graphenes, GO exhibited the lowest levels of cytotoxicity and induced least amount of damage to human dental follicle stem cells [21].

Nevertheless, lack of consensus is reflected in *in vivo* studies as concentration and variations of GFNs tested can significantly change the toxicity outcome, not a conclusive answer can suit them all [22]. Thus, biosafety constraints, especially targeting dental tissue, should be solved to translate GFNs onto clinical applications.

3. Antibacterial Effect of Graphene

Interestingly, because of GNFs' versatility, their usage as potential antimicrobial agents has gained substantial interest in the field of nanomedicine [23–26]. However, the antibacterial effect of GFNs have reported to be controversial as the effect is also highly determined by size, shape, stability, and distribution [27] and the underlying experimental designs remained inconsistent [28].

A thorough understanding of the antimicrobial mechanism is still in its infancy, but, with an increasing number of investigations on the antimicrobial activities of GRNs, three predominate mechanisms were proposed. Physical damage is induced by blade like graphene materials piercing through the microbial cellular membrane causing leakage of intracellular substance leading to cell death. Wrapping and photothermal ablation mechanism could also provoke bacterial cell damage by enclosing the bacterial cells, providing an unique flexible barrier to isolate bacteria growth medium, inhibiting bacteria proliferation, and decreasing microbial metabolic activity and cell viability. Chemical effect is primary oxidative stress mediated with production of ROS as excessive intracellular ROS accumulation could cause intracellular protein inactivation, lipid peroxidation, and dysfunction of the mitochondria, which lead to gradual disintegration of cell membrane and eventual cell death [29]. There had also been researches theorizing that the antibacterial activity of graphene on metal substrate involved electron transfer from the bacterial membrane, producing ROS independent oxidative stress to the microbial membrane, interrupting electron transport in respiratory chain,

and leading to destruction of microbial integrity and cell death [30]. A study conducted by Dellieu et al. tested the growth of *Escherichia coli* and *Streptococcus mutans* on gold and copper substrate in contact with CVD graphene and the result showed that the facile transfer of electrons from microbial membranes to graphene played no role on bacterial viability, denying the influence of metal substrates' conductive character on the antibacterial activity of CVD graphene [31].

Oral cavity is a complex ecosystem, forming structurally and functionally organized dental biofilm embedded in a matrix of polymers of host and bacterial origins. Of clinical relevance is the fact that imbalanced microbial homeostasis in dental biofilms is associated with etiology of dental diseases; for this reason, research efforts have been made to target putative pathogens by antimicrobial or antiadhesive strategies to prevent disease initiation and progression [32]. Although studies have established the antimicrobial activity of GNF against several bacteria such as *Escherichia coli*, *Streptococcus aureus*, *Klebsiella* sp., and *Pseudomonas aeruginosa*, only few studies focused on oral pathogens. *Streptococcus mutans* is the primary gram-positive facultative anaerobic bacteria involved in caries formation while *Porphyromonas gingivalis* and *Fusobacterium nucleatum* are Gram-negative anaerobic bacteria associated with periodontitis and root canal infection.

A study conducted by He et al. evaluated the antibacterial activity of GO nanosheets against these three common types of bacteria and found that GO nanosheets were highly effective in inhibiting the growth of dental pathogens. At GO concentration 40 μg/mL, bacterial growth of *P. gingivalis* and *F. nucleatum* was inhibited, while, at concentration 80 μg/mL, GO absolutely killed all *S. mutans* [33]. This could be explained by the difference of resistance toward oxidative stress generated by GO between anaerobic and facultative anaerobic bacteria, in which GO nanosheets were more bactericidal against obligate anaerobic bacteria. It was also observed in TEM images that GO nanosheets can insert or cut through the cell membranes of bacteria and extract large amounts of phospholipids, mechanically destruct cell membrane integrity, and cause leakage of intracellular substances.

Another graphene derivative, graphene nanoplatelet, produced via thermal exfoliation of graphite intercalation compound, was also investigated for their antibacterial properties against *S. mutans*. Scanning electron microscopy analysis revealed that there is a strong mechanical interaction between cells and GNPs, firstly involving cell trapping and consequent shrinking and secondly involving piercing through soft cell wall with sharp edges of GNP flakes which eventually killed the planktonic form of *S. mutans* [34]. CVA-grown graphene also showed disruption of proliferation and formation of biofilm formed by *S. mutans* and *E. faecalis*, which infer to be due to the surface properties not electron diffusion of graphene material [35]. These research findings suggested that GNPs can be an effective dental material for controlling *S. mutans* and, consequently, caries, further proving the potential graphene hold for biomedical applications

Graphene, when combined with other compounds, showed improved synergistic antimicrobial, antibiofilm, and antiadherence activities against oral pathogens. Zinc oxide nanoparticles have been widely used in biomedical applications for its superior bactericidal effect, but its aggregation properties producing toxicity to mammalian cells have been hindering its use. When graphene material is synthesized with zinc oxide producing graphene/zinc oxide (GZNC) nanocomposite, it not only causes a much lower toxicity, but also forms a unique nanointerface to interact with microbes as compared to ZnO alone. It was observed that GZNC could decrease the synthesis of EPS, one of the key virulence factors of cardiogenicity, and reduce the amount of insoluble glucans, which influences biofilm formation by disturbing its physical integrity and stability, significantly reducing the biofilm and cariogenic properties *S. mutans* [36].

These antibacterial properties of GFNs could be very beneficial when integrating into biomaterials for potential clinical application. For instance, surgical sutures may be one of the most widely used medical adjunct nowadays and a good suture material should not only possess good mechanical properties, but also the antimicrobial ability to prevent breeding of bacteria. It was recently suggested that compared with conventional polyvinyl alcohol (PVA) fiber which has no antimicrobial property, PVA matrix dispersed with mechanically exfoliated graphene (MEG) showed high antibacterial effect toward gram-positive bacteria, thereby efficaciously accelerated the healing of wound, making PVA/MEG nanocomposite fiber a promising new candidate for surgical suture [37].

Studies on graphene-based materials for *in vivo* bactericidal applications are still at initial stage; whether graphene has long-term and broad-spectrum antibacterial effects stay debatable. A recent study suggested that GO is not an intrinsic antimicrobial material but a general growth enhancer that can act as a biofilm that allows bacterial attachment and proliferation [38]. Nevertheless, GFNs are still a potential antibacterial agent that are easy to obtain, cheaper, and capable of providing support to disperse and stabilize various nanomaterials synergistically yielding high antibacterial activities [39]. Additional in-depth experimental studies, especially toward dental pathogen, need to be introduced to further analyze the interaction between GFNs and biosystems for future clinical applications.

4. Potential Applications of Graphene in Tissue Engineering

Tissue loss due to trauma, disease, or congenital abnormality is a major healthcare concern worldwide. The ongoing challenge of dental treatment is to restore missing teeth with their periodontal structure [40]. There has been an evolution from the use of materials to simply replace diseased tissue, to that of utilizing specific biomaterials, which will nurture and regenerate a functionally and structurally acceptable tissue [41]. It is inevitable that tissue regeneration research topic is growing quickly in the clinical fields. Experimental development of stem cells together with their supporting biomaterials is a fundamental component of tissue engineering research [42].

Scaffolds play a significant role in tissue engineering. Either in the absence or presence of chemical inducers and growth factors, it should be optimally designed to provide a biocompatible three-dimensional environment that can not only mechanically 'support' and 'guide' bone regeneration, but also 'stimulate' proliferation and differentiation of stem cells into their specific tissue lineage [43]. So far, most artificial biomaterials in the markets lack tissue inductive activities, which means that fast healings and functional reconstructions can not be satisfied especially in patients with infections and weak healing ability [38]. On the way searching for new material strategies that can overcome these limitations, a few studies have shown that graphene, without signs of cytotoxicity, accelerated the proliferation and differentiation of human mesenchymal stem cells (hMSCs) into bone cells with a rate that is comparable to the one achieved with common growth factors [44]. Starting from these results, the possible roles of graphene in enhancing osteogenesis have been extensively investigated.

Among graphene derivatives, GO, with many functional groups, has outstanding surface activities which can exert adsorptive capability to drugs, growth factors, and other biomolecules. Several in vitro experiments have demonstrated that pristine GO may upregulate β-catenin protein expression and activate catenin/Wnt signaling pathway, markedly increasing the degree of proliferation and differentiation of cultured cells, and led to acceleration of bone formation [45, 46]. A study conducted by Nishida et al. evaluated the tissue proliferative behaviors in relation to GO scaffold in the tooth extraction socket of dogs. It was observed that the bone formation in GO scaffold was fivefold more than collagen scaffold, which further confirmed the high bone-forming capability of GO scaffolds [47].

Besides being used in their pristine form, they can be combined with different biomaterials. The resultant graphene modified scaffold presented enhanced bioactivity. Addition of GO to chitosan 3D scaffold's composition stimulated the interconnected pore structure, improved mechanical properties, and enhanced the bioactivity of the scaffold materials for osteogenesis [48]. β-tricalcium phosphate scaffold modified with GO significantly stimulated the proliferation and osteogenic differentiation of human bone marrow stromal osteoprogenitor cells and, importantly, accelerated new bone formation *in vivo* [16]. GO application to 3D collagen scaffolds stimulated tissue ingrowth behaviors improved their physical properties, enzyme resistance, and Ca and proteins absorption [45].

Dental pulp stem cells (DPSC) are self-renewing and multipotent cells which contain mesenchymal stem cells that can relatively easily obtained from the extracted teeth without esthetic damage [49]. They may undergo both osteoblastic and odontoblastic differentiation making it an interesting model for tissue regeneration. Rosa et al. first confirmed the capability of GO to allow DPSC attachment and proliferation [50]. Xie et al. then tested the potential ability of graphene to induce DPSC's odontogenic and osteogenic differentiation. Without stimuli from bioactive factors, graphene produced by the chemical vapor deposition (CVD) method downregulated the expression of odontoblastic genes (MSX-1, PAX, and DMP), which implied that CVD grown graphene may not serve as a platform for endodontic and pulp regeneration.

However, osteogenic genes and proteins including RUNX2, COL, and OCN were significantly upregulated on graphene. This is very likely that the osteogenic differentiation of stem cells presented higher potential as the rigidity of substrate increased; conversely, the odontogenic differentiation of DPSC is better achieved on soft substrate. Nevertheless, graphene could be a potential material to be used for bone tissue engineering and regeneration.

Another stem cell from dental tissue that can be influenced by graphene scaffold is periodontal ligament stem cells (PDLSCs). They could be obtained from dental tissue and have shown multiple differentiation capacity and immunomodulation, capable of differentiating into both cementoblasts and collagen forming cells [51]. A study conducted by Xie et al. investigated the expression profile of PDLSCs for osteogenic differentiation on two-dimensional graphene (2DGp) and three-dimensional graphene scaffold (3DGp) [52]. The result showed that, in both arrangements, without use of osteogenic medium, there was an increase in bone related genes RUNX2 and OCN. Remarkably, an upregulation of MYH10 and MYH10-V2 was observed, especially in 3DGp. MYH10 and MYH10-V2 are cytoskeletal proteins associated with mitochondrial DNA which increase with substrate rigidity. Their upregulation implied that substrate stiffness may play a role in regulating PDLSCs cell lineage specification and promoting differentiation. In fact, it was previously established by other researches that compared to two-dimensional structure; three-dimensional scaffold allowed stem cells to respond better to hormones and exhibited lower requirements for growth factor, thereby improving stem cells' viability, degree of efficiency, consistency, and predictability [53, 54]. Nevertheless, this study suggested that both chemical characteristics and intrinsic physical properties of graphene take part in osteoblastic differentiation of PDLSCs [52].

Research has also been done to analyze the performance of Silk fibroin (SF) and GO in cell proliferation and mesenchymal phenotype expression of PDLSCs [55]. Although SF was previously proved to allow optimal adhesion of mesenchymal stem cell, PDLSCs showed limited attachment on it. When GO was added to SF, layered molecular structures of fibroin and graphene reinforced each other producing an unusually high robust construction, very suitable to serve as a cellular environment where mechanical resistance is required. As it was shown under MTT assay and cytometry, without interfering the mesenchymal phenotype of PDLSCs, the initial adhesion of PDLSCs was significantly improved and rapid spreading was observed. This fact opens a big step for usage of graphene in regenerative dentistry. Another study conducted by Sanchez et al. tested the capability of PDLSCs cultured on GO/SF or rGO/rSF composite to initiate cementoblast differentiation. Remarkably, together with enhanced level of RUNX2, ALP, and OSX, there was an overexpression of CEMP1, which is a novel cementum component exclusively expressing for cementoblasts and their progenitors. This finding may suggest the presence of advanced spontaneous cementoblast differentiation with moderate rate of proliferation [56]. This is very appreciated in cell regeneration as most artificial materials require multiple growth factors to promote HSCs differentiation, whereas GO/SF may provide a new stage for cementoblast differentiation in the absence of biochemical factor [57].

Only part of recent studies was included in this section; nevertheless, it could be speculated that graphene is one of the most promising biocompatible scaffolds for MSCs adhesion, proliferation, and differentiation, particularly toward the osteogenic lineage. Graphene alone with optimal concentration may spontaneously drive stem cells' osteogenesis, but this effect could be further promoted with presence of growth factors. For future engineering of graphene-based substrates for targeted biomedical applications, a deeper understanding of the ability of graphene to improve the biological properties of different scaffold materials will be essential for future biomedical applications.

5. Potential Applications of Graphene in Dental Implants

After dental implantation, fibroosseous integration took place between host biological system and dental implant. At the hard tissue interface, osteogenic properties of implant material are essential for osseointegration while at the soft tissue interface, to ensure a tight epithelial seal preventing bacterial invasion is obligatory [58]. Leak of seal at either interface may result in bacterial contamination and colonization, which may eventually impair osteogenesis and induce bone loss [59]. Therefore, the challenge today is to outrace bacterial colonization over tissue integration, which may be achieved by either inhibiting microorganism colonization or accelerating tissue healing and osseointegration [60].

Even though the alternative materials for dental implants have gained increasing interests, titanium or titanium alloy still remained popular to be the material of choice [57]. Recent nanotechnology researches have been reporting that modifications on titanium implant features such as surface composition, hydrophilicity, surface roughness topography, and geometry can affect the rate and quality of osseointegration [61]. Due to graphene's potential osteogenic and antibacterial ability, it appeared to be an excellent implant coating material to favor better osseointegration.

When graphene is coated on titanium substrate, the hydrophobic character of graphene film exerted self-cleaning effect on its surfaces decreasing the adhesion of microorganism including S. sanguinis and S. mutans. Additionally, compared to titanium alone, graphene possesses osteogenic property enhancing the expression of osteogenic related genes RUNX2, COL-I, and ALP, boosting osteocalcin gene and protein expression, and consequently increasing the deposition of mineralized matrix [35]. Attempts had also been made to coat GO on titanium substrate as a cell culture platform for PDLSCs differentiation. GO-Ti substrate provided a suitable environment for the attachment, proliferation, and differentiation of PDLSCs. When compared with Na-Ti substrate, expression level of osteogenesis-related markers of COL-I, ALP, BSP, RUNX2, and OCN was higher [62]. These findings confirmed that coating of titanium with graphene could be a promising strategy to improve osseointegration and prevent biofilm formation on implants and devices.

Multiphase nanocomposite could also be a promising biomedical material to prevent implant-associated infection. As discussed in the previous session, the antibacterial properties of GO had been contradicting; hence a study conducted by Jin et al. added an antimicrobial nanomaterial, silver nanoparticle, to the GO-Ti composite. GO, with many carboxyl, hydroxyl, and carbonyl on its surfaces, are negatively charged, which could readily combine with positively charged Ag ions in the aqueous solution. Loading GO thin film and silver nanoparticles onto titanium exhibited excellent antiadherence and antimicrobial ability, especially toward *S. mutans* and *P. gingivalis* [63].

Hydroxyapatite (HA) coating is widely applied as an osteoinductive modification of titanium implant; however its slow biological interaction mechanism and low mechanical strength restricted its application [64]. Graphene oxide/chitosan/hydroxyapatite-titanium (GO/CS/HA-Ti) is produced by incorporating GO and chitosan (CS) into hydroxyapatite-titanium substrate through electrophoretic deposition method. It showed better bioactivity by improving the adhesion, proliferation, and differentiation of BMSC cells *in vivo* and possessed superior osseointegration *in vitro*. Furthermore, bonding strength between composite coating and titanium substrate of GO/CS/HA composite coating were enhanced compared with HA, GO/HA, and CS/HA coatings [65].

The direct fabrication of graphene on NiTi based dental implant using chemical vapor deposition technique upregulated the expression level of osteogenic related genes (OCN, OPN, BMP-2, and RUNX2) and promoted expression of integrin β1 and initial adhesion of MSCs, indicating that graphene-functionalized NiTi allows better MSCs cytoskeleton development and spontaneous osteogenic differentiation [66].

Aside from that, GO can be used as a carrier for BMP-2, an osteoinductive, and Substance P, a MSC recruitment agent. Application of GO on titanium allowed dual delivery of SP and BMP-2, showing the greatest new bone formation on Ti implant in the mouse calvaria [67]. However, BMP-2 has short half-life hindering its long-term release at therapeutic dose. Ren et al. further incorporated GO and rGO into a hydrothermally prepared porous titanate scaffold on Ti implants, constructing a delivery vehicle for dexamethasone, an osteoinductive synthetic glucocorticoid. Their study results showed that both DEX-GO-Ti and DEX-rGO-Ti enhanced ALP activity of rBMSCs and upregulated the gene expression levels of OPN and OCN, confirming their osteopromotive ability to promote proliferation and to accelerate osteogenic differentiation of rBMSCs [68].

Despite all the excitement, developing a universally accepted transfer route of graphene to titanium implants remains to be a challenge as the deposition technique should ensure transfer effectiveness without compromising the surface properties graphene possess. Several transferring methods have been developed in the market such as wet method, electrochemical delimitation, thermal release tape, hot press transfer, and roll to roll; these methods may allow large scale transfer of graphene onto planar substrate, but are poorly suited for transferring onto 3D objects such as dental implant. Recently, Morin et al. described a vacuum-assisted dry transfer method which is to coat graphene onto the target object and maintained by a mound to sustain conformation. Using a pressure differential provided via partial vacuum, uniform force is applied to the surface facilitating successful coating of graphene onto the object without changing graphene characteristics [69].

Graphene and its derivatives when coated on titanium implant have remarkable abilities to improve properties of titanium, enabling binding of biomolecules, and induce osseointegration. These characteristics place them under the spotlight for improvement and modification of implant materials.

6. Potential Applications of Graphene in Endodontics

The purpose of root canal treatment is to biomechanical clean an infected root canal system to destroy intracanal pathogens, decontaminate residually infected tissue, and avoid further intraoperative/postoperative infection. The main cause of endodontic failure is persisting infection in the root canal [70]. Recently developed photodynamic therapy has gained attention for effective canal disinfection while preserving dentin structures. One of the material that plays a key role in this technique is nontoxic photosensor, such as indocyanine green (ICG), but its poor stability and concentration-dependent aggregation have been concerning [71]. Modifying ICG with GO not only significantly reduced number of *E. faecalis* and *S. mutans*, but also improved the stability and bioavailability of ICG, preventing its degradation and aggregation. [72].

For many years, sodium hypochlorite has been used as the most common intracanal irrigants for its strong antibacterial and tissue-dissolving abilities. However, sodium hypochlorite extrusion during root canal treatment causes acute immediate symptoms and serious potential sequelae including rapid hemolysis and ulceration of surrounding tissues, destruction of endothelial, and fibroblast cells [73]. Incorporating graphene into silver nanoparticles showed strong antibacterial property, as efficacy as 3% sodium hypochlorite in canal disinfection, but with less cytotoxic effect to bone and soft tissues [74].

Bioactive cements have been widely used in endodontics for management of perforation, retrograde root filling, and pulp capping. Among them, Biodentine (BIO) and Endocem-Zr (ECZ) are considered as the safest cements that exhibits the least discoloration and calcification of tooth, but with shortcomings such as high pull-out bond strength, long setting time, and modest mechanical properties. With 3 wt % addition of graphene nanosheets, the setting time of both cements significantly decreased, which could be explained by the role of carbon based materials to act as a matrix for the development of C–S–H and calcium hydroxide, thereby reducing the induction period and accelerating the hydration process [75]. However, a decrease in push out strength of ECZ was observed which requires further studies for clinical use.

Another cement, calcium silicate (CS), well known as mineral trioxide aggregate (MTA), has become popular in endodontic treatment in recent years. It is a hydration product of Portland cement, which has the capacity to upregulate differentiation of odontoblasts and promote calcium phosphate deposition [76]. However, its brittle nature, low fracture toughness, and poor wear resistance, because of the presence of relatively large pores that are potential in initiating macrocracks, have been hindering its use in clinical application. Thus, attempt was made to use GFNs as reinforcement to CS to enhance their mechanical properties. With incorporation of GFNs, significant grain size refinement occurs as GFNs may wrap around grains and inhibit their growth. Aside from decrease in grain size, through crack bridging, pullout GFNs, crack branching, and crack deflection mechanism, there was also an increase in indentation fracture toughness and brittle index. Moreover, incorporation of GFNs into CS has a beneficial effect on proliferation of human osteoblastic cells (hFOB), indicating their biocompatibility suitability for cell proliferation [77].

Incorporating GO into Portland cement also exhibited positive influence on the workability of the cement. It enhanced the degree of hydration by increasing the nonevaporable water content and calcium hydroxide hydroxide content, contributing to the refinement of pore structures, and subsequently increasing the compressive strengths and suppressing crack propagation in the matrix at nanoscale. Introduction of 0.03% by weight GO into cement paste can increase their compressive strength and tensile strength by more than 40% [78].

In conclusion, addition of graphene into dental cements leads to refinement of pore structure which not only strengthens the cement material but also blocks the entry for possible bacterial invasion. It could be promising reinforcing cementitious material for future dental applications.

7. Potential Applications of Graphene in Restorative Dentistry

Glass ionomer (GI), with favorable coefficient of linear thermal expansion, ability to chemically bond to sound tooth structure, and dynamic fluoride release, had been utilized in a wide range of clinical application. However, its poor physiomechanical properties remained to be a concern despite the developments in GI constituents with addition of various filler types including fibers, metallic powders, and hydroxyapatite powders [79]. In recent years, attempts had also been made to incorporate graphene derived nanomaterial into commercially available glass ionomer for reinforcement. Graphene, when combined with glass ionomer prepared with poly(acrylic acid), has significantly enhanced physiomechanical properties of GIs [11]. Fluoride graphene when prepared by hydrothermal reaction of graphene oxide and mechanically blend with glass ionomer could produce a GICs/FG composites matrix, which could significantly enhance the mechanical, tribological, and antibacterial properties of glass ionomer [80]. With the increase of FG content in glass ionomer, there is a decrease of pores and microcracks in the internal structure of material and an increase in antibacterial

ability making it less susceptible to erosion disintegration and microbial invasion. Reinforcing resin polymer matrices with graphene gold nanoparticles as fillers showed improvement of degree of conversion and surface properties, offering a good solution to improve physicochemical properties of dental nanocomposite [81].

Due to the presence of microcavities between the healthy tissue and dental restoration causing bacteria invasion, there has been an increasing interest in development of antibiofilm adhesives. One concern with dental adhesive monomers is its excessive ROS production which not only cause oxidative stress associated toxicity toward fibroblasts and pulp cell, but also affect saliva redox equilibrium and decreasing natural oral immune system defenses [82]. Due to graphene's antibacterial properties, Bregnocchi et al. proposed using GFNs as an antimicrobial and antibiofilm filler for dental adhesive. The study result showed that GFNs modified dental adhesive significantly inhibited the adhesion and growth of S. mutans without interfering its original mechanical performances and without producing a surplus of ROS [83].

8. Potential Applications of Graphene in Periodontology

In treatment for periodontal bone defects using guided tissue regeneration (GTR) and guided bone regeneration (GBR), barrier membranes have been a crucial biomaterials, which is to create a secluded space between soft connective tissue and regenerating bone, for formation of unimpeded bone promoting faster differentiation of mesenchymal cells into odontoblast/osteoblast [84]. Attempts to modify barrier membrane have been made to improve its biocompatibility. Radunovic et al. investigated the effect of collagen membrane coated with GO (10 μg/mL) on the viability and metabolic activity of dental pulp mesenchymal cells. The result showed that GO coating at the higher concentration induces PGE2 secretion, controls inflammation, and promotes DPSCs differentiation which is probably due to GO's large reactive surface area providing idea platform for biofunctionalization and concentrating chemical, proteins, and growth factors for faster differentiation [85]. An animal study conducted by Kawamoto evaluated the periodontal wound healing capability of GO scaffold on dogs with class II furcation defect. The result showed that GO scaffold succeeded in periodontal ligament like and cementum like tissue formation, followed by alveolar bone formation. The study also speculated that GO application to collagen scaffold stimulated scaffold degradation and replaced them with newly regenerated periodontal tissue [86].

9. Conclusion

Researches toward graphene in dental materials mainly focused on two ways: one is to prepare new dental materials as GFNs alone, and the other is to modify the common dental materials by transferring appropriate GFNs onto different substrates. As clearly highlighted in this paper, in either way, GFNs improved the physical, chemical, and mechanical properties of biomaterials, holding enormous potential in

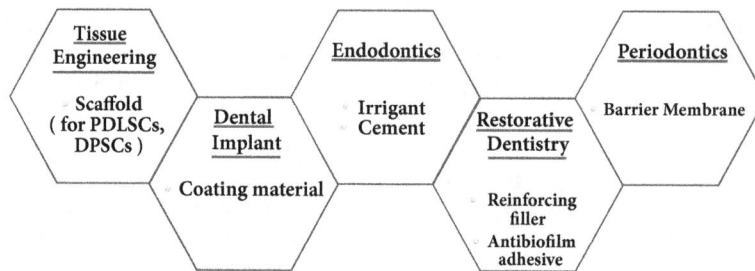

FIGURE 3: Currently, no studies correlating dentistry and GFNs have been done on human subjects. Based on the properties presented in *in vivo* and *in vitro* studies, the potential applications of GFNs that could be translate to clinical reality in dentistry were summarized by different dental disciplines.

new therapeutic strategies in dental field. In this review, we firstly discussed the toxicity and antibacterial properties of GFNs on cells *in vitro* and *in vivo*. Then, we presented a comprehensive summary of the latest research progress related to the potential applications of GFNs in dentistry (see Figure 3).

Safety and potential risks of GFNs should be emphasized and research efforts should be made to ensure harmless use of graphene in oral environment. Additionally, some very promising properties of GFNs have been extensively investigated in other organs of human body *in vitro* and *in vivo*, but no studies have been done on human subjects and in-depth studies are still scarce in oral settings. GFNs is a rather fascinating material worthy of in-depth investigation; further exploration on their underlying mechanism of interaction toward oral tissues in terms of cell-signaling, metabolic pathway, osteogenesis, and antibacterial effects is needed. We hope that this review article could provide some valuable elicitation for the future scientific and technological innovations of graphene in dentistry.

Conflicts of Interest

The authors declare that there are no conflicts of interest regarding the publication of this paper.

References

[1] L. Feng, L. Wu, and X. Qu, "New horizons for diagnostics and therapeutic applications of graphene and graphene oxide," *Adv. Mater*, vol. 25, pp. 168–186, 2013.

[2] V. Biju, "Chemical modifications and bioconjugate reactions of nanomaterials for sensing, imaging, drug delivery and therapy," *Chemical Society Reviews*, vol. 43, no. 3, pp. 744–764, 2014.

[3] L. C. Geraldine et al., "A graphene-based physiometer array for the analysis of single biological cells," *Sci. Rep*, vol. 4, p. 6865, 2014.

[4] Y. Zhang, T. R. Nayak, H. Hong, and W. Cai, "Graphene: a versatile nanoplatform for biomedical applications," *Nanoscale*, vol. 4, no. 13, pp. 3833–3842, 2012.

[5] N. Chatterjee, H.-J. Eom, and J. Choi, "A systems toxicology approach to the surface functionality control of graphene-cell interactions," *Biomaterials*, vol. 35, no. 4, pp. 1109–1127, 2014.

[6] S. Y. Wu, S. S. A. An, and J. Hulme, "Current applications of graphene oxide in nanomedicine," *International journal of nanomedicine*, vol. 10, no. 9, 2015.

[7] H. Wang, W. Gu, N. Xiao, L. Ye, and Q. Xu, "Chlorotoxin-conjugated graphene oxide for targeted delivery of an anti-cancer drug," *International Journal of Nanomedicine*, vol. 9, no. 1, pp. 1433–1442, 2014.

[8] S. Goenka, V. Sant, and S. Sant, "Graphene-based nanomaterials for drug delivery and tissue engineering," *Journal of Controlled Release*, vol. 173, no. 1, pp. 75–88, 2014.

[9] B. Gadgil, P. Damlin, and C. Kvarnström, "Graphene vs. reduced graphene oxide: A comparative study of graphene-based nanoplatforms on electrochromic switching kinetics," *Carbon*, vol. 96, pp. 377–381, 2016.

[10] X. Guo and N. Mei, "Assessment of the toxic potential of graphene family nanomaterials," *Journal of Food and Drug Analysis*, vol. 22, no. 1, pp. 105–115, 2014.

[11] S. Malik, F. M. Ruddock, A. H. Dowling et al., "Graphene composites with dental and biomedical applicability," *Beilstein Journal of Nanotechnology*, vol. 9, no. 1, pp. 801–808, 2018.

[12] X. Ding, H. Liu, and Y. Fan, "Graphene-Based Materials in Regenerative Medicine," *Advanced Healthcare Materials*, vol. 4, no. 10, pp. 1451–1468, 2015.

[13] M. Jennifer and W. Maciej, "Nanoparticle technology as a double-edged sword: cytotoxic, genotoxic and epigenetic effects on living cells," *Journal of Biomaterials and Nanobiotechnology*, vol. 4, pp. 53–63, 2013.

[14] N. Wei, C. Lv, and Z. Xu, "Wetting of graphene oxide: A molecular dynamics study," *Langmuir*, vol. 30, no. 12, pp. 3572–3578, 2014.

[15] A. B. Seabra, A. J. Paula, R. De Lima, O. L. Alves, and N. Durán, "Nanotoxicity of graphene and graphene oxide," *Chemical Research in Toxicology*, vol. 27, no. 2, pp. 159–168, 2014.

[16] C. Wu, L. Xia, P. Han et al., "Graphene-oxide-modified β-tricalcium phosphate bioceramics stimulate in vitro and in vivo osteogenesis," *Carbon*, vol. 93, pp. 116–129, 2015.

[17] L. Peng, Z. Xu, Z. Liu et al., "An iron-based green approach to 1-h production of single-layer graphene oxide," *Nature communications*, vol. 6, p. 5716, 2015.

[18] A. Nel, T. Xia, L. Mädler, and N. Li, "Toxic potential of materials at the nanolevel," *Science*, vol. 311, no. 5761, pp. 622–627, 2006.

[19] S. Gurunathan, J. W. Han, V. Eppakayala, and J.-H. Kim, "Green synthesis of graphene and its cytotoxic effects in human breast cancer cells," *International Journal of Nanomedicine*, vol. 8, pp. 1015–1027, 2013.

[20] A. Jarosz, M. Skoda, and I. Dudek, "Oxidative stress and mitochondrial activation as the main mechanisms underlying graphene toxicity against human cancer cells," *Oxidative Medicine and Cellular Longevity*, vol. 2016, Article ID 5851035, 14 pages, 2016.

[21] D. Olteanu, A. Filip, C. Socaci et al., "Cytotoxicity assessment of graphene-based nanomaterials on human dental follicle stem cells," *Colloids and Surfaces B: Biointerfaces*, vol. 136, pp. 791–798, 2015.

[22] H. Xie, T. Cao, F. J. Rodríguez-Lozano, E. K. Luong-Van, and V. Rosa, "Graphene for the development of the next-generation of biocomposites for dental and medical applications," *Dental Materials*, vol. 33, no. 7, pp. 765–774, 2017.

[23] A. F. De Faria, D. S. T. Martinez, S. M. M. Meira et al., "Anti-adhesion and antibacterial activity of silver nanoparticles supported on graphene oxide sheets," *Colloids and Surfaces B: Biointerfaces*, vol. 113, pp. 115–124, 2014.

[24] J. Ma, J. Zhang, Z. Xiong, Y. Yong, and X. S. Zhao, "Preparation, characterization and antibacterial properties of silver-modified graphene oxide," *Journal of Materials Chemistry*, vol. 21, no. 10, pp. 3350–3352, 2011.

[25] J. Tang, Q. Chen, L. Xu et al., "Graphene oxide-silver nanocomposite as a highly effective antibacterial agent with species-specific mechanisms," *ACS Applied Materials & Interfaces*, vol. 5, no. 9, pp. 3867–3874, 2013.

[26] W.-P. Xu, L.-C. Zhang, J.-P. Li et al., "Facile synthesis of silver@graphene oxide nanocomposites and their enhanced antibacterial properties," *Journal of Materials Chemistry*, vol. 21, no. 12, pp. 4593–4597, 2011.

[27] M. Moritz and M. Geszke-Moritz, "The newest achievements in synthesis, immobilization and practical applications of antibacterial nanoparticles," *Chemical Engineering Journal*, vol. 228, pp. 596–613, 2013.

[28] F. Perreault, A. F. De Faria, S. Nejati, and M. Elimelech, "Antimicrobial Properties of Graphene Oxide Nanosheets: Why Size Matters," *ACS Nano*, vol. 9, no. 7, pp. 7226–7236, 2015.

[29] X. Zou, L. Zhang, Z. Wang, and Y. Luo, "Mechanisms of the Antimicrobial Activities of Graphene Materials," *Journal of the American Chemical Society*, vol. 138, no. 7, pp. 2064–2077, 2016.

[30] S. Panda, T. K. Rout, A. D. Prusty, P. M. Ajayan, and S. Nayak, "Electron Transfer Directed Antibacterial Properties of Graphene Oxide on Metals," *Advanced Materials*, vol. 30, no. 7, 2018.

[31] L. Dellieu, E. Lawarée, N. Reckinger et al., "Do CVD grown graphene films have antibacterial activity on metallic substrates?" *Carbon*, vol. 84, no. 1, pp. 310–316, 2015.

[32] P. D. Marsh, "Dental plaque as a biofilm and a microbial community – implications for health and disease," *BMC Oral Health*, vol. 6, no. Suppl 1, p. S14.

[33] J. He, X. Zhu, Z. Qi et al., "Killing dental pathogens using antibacterial graphene oxide," *ACS Applied Materials & Interfaces*, vol. 7, no. 9, pp. 5605–5611, 2015.

[34] I. Rago, A. Bregnocchi, E. Zanni et al., "Antimicrobial activity of graphene nanoplatelets against Streptococcus mutans," in *Proceedings of the 15th IEEE International Conference on Nanotechnology (IEEE-NANO '15)*, pp. 9–12, July 2015.

[35] N. Dubey, K. Ellepola, F. E. D. Decroix et al., "Graphene onto medical grade titanium: an atom-thick multimodal coating that promotes osteoblast maturation and inhibits biofilm formation from distinct species," *Nanotoxicology*, vol. 12, no. 4, pp. 274–289, 2018.

[36] S. Kulshrestha, S. Khan, R. Meena, B. R. Singh, and A. U. Khan, "A graphene/zinc oxide nanocomposite film protects dental implant surfaces against cariogenic Streptococcus mutans," *Biofouling*, vol. 30, no. 10, pp. 1281–1294, 2014.

[37] Y. Ma, D. Bai, X. Hu et al., "Robust and Antibacterial Polymer/Mechanically Exfoliated Graphene Nanocomposite Fibers for Biomedical Applications," *ACS Applied Materials & Interfaces*, vol. 10, no. 3, pp. 3002–3010, 2018.

[38] P. C. Wu, H. H. Chen, S. Y. Chen et al., "Graphene oxide conjugated with polymers: a study of culture condition to determine whether a bacterial growth stimulant or an antimicrobial agent?," *Journal of nanobiotechnology*, vol. 16, no. 1, 2018.

[39] H. Ji, H. Sun, and X. Qu, "Antibacterial applications of graphene-based nanomaterials: Recent achievements and challenges," *Advanced Drug Delivery Reviews*, vol. 105, pp. 176–189, 2016.

[40] K. Niibe, F. Suehiro, M. Oshima, M. Nishimura, T. Kuboki, and H. Egusa, "Challenges for stem cell-based "regenerative prosthodontics"," *Journal of Prosthodontic Research*, vol. 61, no. 1, pp. 3–5, 2017.

[41] E. A. Abou Neel, W. Chrzanowski, V. M. Salih, H.-W. Kim, and J. C. Knowles, "Tissue engineering in dentistry," *Journal of Dentistry*, vol. 42, no. 8, pp. 915–928, 2014.

[42] G. Sammartino, D. M. D. Ehrenfest, J. A. Shibli, and P. Galindo-Moreno, "Tissue Engineering and Dental Implantology: Biomaterials, New Technologies, and Stem Cells," *BioMed Research International*, vol. 2016, Article ID 5713168, 3 pages, 2016.

[43] A. A. Zadpoor, "Bone tissue regeneration: The role of scaffold geometry," *Biomaterials Science*, vol. 3, no. 2, pp. 231–245, 2015.

[44] T. R. Nayak, H. Andersen, V. S. Makam et al., "Graphene for controlled and accelerated osteogenic differentiation of human mesenchymal stem cells," *ACS Nano*, vol. 5, no. 6, pp. 4670–4678, 2011.

[45] E. Nishida, H. Miyaji, H. Takita et al., "Graphene oxide coating facilitates the bioactivity of scaffold material for tissue engineering," *Japanese Journal of Applied Physics*, vol. 53, no. 6S, p. 06JD04, 2014.

[46] C. Wei, Z. Liu, F. Jiang, B. Zeng, M. Huang, and D. Yu, "Cellular behaviours of bone marrow-derived mesenchymal stem cells towards pristine graphene oxide nanosheets," *Cell Proliferation*, vol. 50, no. 5, 2017.

[47] E. Nishida, H. Miyaji, A. Kato et al., "Graphene oxide scaffold accelerates cellular proliferative response and alveolar bone healing of tooth extraction socket," *International Journal of Nanomedicine*, vol. 11, pp. 2265–2277, 2016.

[48] S. Dinescu, M. Ionita, and A. M. Pandele, "In vitro cytocompatibility evaluation of chitosan/graphene oxide 3D scaffold composites designed for bone tissue engineering," *Biomed Mater Eng*, vol. 24, no. 6, pp. 2249–2256, 2014.

[49] B. Mead, A. Logan, M. Berry, W. Leadbeater, and B. A. Scheven, "Concise Review: Dental Pulp Stem Cells: A Novel Cell Therapy for Retinal and Central Nervous System Repair," *Stem Cells*, vol. 35, no. 1, pp. 61–67, 2017.

[50] V. Rosa, H. Xie, N. Dubey et al., "Graphene oxide-based substrate: physical and surface characterization, cytocompatibility and differentiation potential of dental pulp stem cells," *Dental Materials*, vol. 32, no. 8, pp. 1019–1025, 2016.

[51] L. Guo, Y. Hou, L. Song, S. Zhu, F. Lin, and Y. Bai, "D-Mannose Enhanced Immunomodulation of Periodontal Ligament Stem Cells via Inhibiting IL-6 Secretion," *Stem Cells International*, 2018.

[52] H. Xie, T. Cao, J. V. Gomes, A. H. Castro Neto, and V. Rosa, "Two and three-dimensional graphene substrates to magnify osteogenic differentiation of periodontal ligament stem cells," *Carbon*, vol. 93, pp. 266–275, 2015.

[53] X. Meng, P. Leslie, Y. Zhang, and J. Dong, "Stem cells in a three-dimensional scaffold environment," *Springerplus*, vol. 3, no. 1, p. 80, 2014.

[54] B. J. Lawrence and S. V. Madihally, "Cell colonization in degradable 3D porous matrices.," *Cell adhesion & migration*, vol. 2, no. 1, pp. 9–16, 2008.

[55] F. J. Rodríguez-Lozano, D. García-Bernal, S. Aznar-Cervantes et al., "Effects of composite films of silk fibroin and graphene oxide on the proliferation, cell viability and mesenchymal phenotype of periodontal ligament stem cells," *Journal of Materials Science: Materials in Medicine*, vol. 25, no. 12, pp. 2731–2741, 2014.

[56] M. Vera-Sánchez, S. Aznar-Cervantes, E. Jover et al., "Silk-fibroin and graphene oxide composites promote human periodontal ligament stem cell spontaneous differentiation into osteo/cementoblast-like cells," *Stem Cells and Development*, vol. 25, no. 22, pp. 1742–1754, 2016.

[57] D. Torii, T. W. Tsutsui, N. Watanabe, and K. Konishi, "Bone morphogenetic protein 7 induces cementogenic differentiation of human periodontal ligament-derived mesenchymal stem cells," *Odontology*, vol. 104, no. 1, pp. 1–9, 2016.

[58] F. Rupp, L. Liang, J. Geis-Gerstorfer, L. Scheideler, and F. Hüttig, "Surface characteristics of dental implants: A review," *Dental Materials*, vol. 34, no. 1, pp. 40–57, 2018.

[59] N. Robitaille, D. N. Reed, J. D. Walters, and P. S. Kumar, "Periodontal and peri-implant diseases: identical or fraternal infections?" *Molecular Oral Microbiology*, vol. 31, no. 4, pp. 285–301, 2016.

[60] X. Wu, S.-J. Ding, K. Lin, and J. Su, "A review on the biocompatibility and potential applications of graphene in inducing cell differentiation and tissue regeneration," *Journal of Materials Chemistry B*, vol. 5, no. 17, pp. 3084–3102, 2017.

[61] R. Rasouli, A. Barhoum, and H. Uludag, "A review of nanostructured surfaces and materials for dental implants: surface coating, patterning and functionalization for improved performance," *Biomaterials Science*, 2018.

[62] Q. Zhou, P. Yang, X. Li, H. Liu, and S. Ge, "Bioactivity of periodontal ligament stem cells on sodium titanate coated with graphene oxide," *Scientific Reports*, vol. 6, p. 19343, 2016.

[63] J. Jin, L. Zhang, M. Shi, Y. Zhang, and Q. Wang, "Ti-GO-Ag nanocomposite: The effect of content level on the antimicrobial activity and cytotoxicity," *International Journal of Nanomedicine*, vol. 12, pp. 4209–4224, 2017.

[64] D. Gopi, N. Murugan, S. Ramya, and L. Kavitha, "Electrode-position of a porous strontium-substituted hydroxyapatite/zinc oxide duplex layer on AZ91 magnesium alloy for orthopedic applications," *Journal of Materials Chemistry B*, vol. 2, no. 34, pp. 5531–5540, 2014.

[65] L. Suo, N. Jiang, Y. Wang et al., "The enhancement of osseointegration using a graphene oxide/chitosan/hydroxyapatite composite coating on titanium fabricated by electrophoretic deposition," *Journal of Biomedical Materials Research Part B: Applied Biomaterials*, 2018.

[66] J. Li, G. Wang, H. Geng et al., "CVD growth of graphene on NiTi alloy for enhanced biological activity," *ACS Applied Materials & Interfaces*, vol. 7, no. 36, pp. 19876–19881, 2015.

[67] W. G. La, M. Jin, S. Park et al., "Delivery of bone morphogenetic protein-2 and substance P using graphene oxide for bone regeneration," *International journal of nanomedicine*, vol. 9, supplement 1, p. 107, 2014.

[68] N. Ren, J. Li, J. Qiu et al., "Growth and accelerated differentiation of mesenchymal stem cells on graphene-oxide-coated titanate with dexamethasone on surface of titanium implants," *Dental Materials*, vol. 33, no. 5, pp. 525–535, 2017.

[69] J. L. P. Morin, N. Dubey, F. E. D. Decroix, E. K. Luong-Van, A. C. Neto, and V. Rosa, "Graphene transfer to 3-dimensional surfaces: a vacuum-assisted dry transfer method," *2D Materials*, vol. 4, no. 2, Article ID 025060, 2017.

[70] M. Yamaguchi, Y. Noiri, Y. Itoh et al., "Factors that cause endodontic failures in general practices in Japan," *BMC oral health*, vol. 18, no. 1, p. 70, 2018.

[71] V. Chrepa, G. A. Kotsakis, T. C. Pagonis, and K. M. Hargreaves, "The effect of photodynamic therapy in root canal disinfection: a systematic review," *Journal of Endodontics*, vol. 40, no. 7, pp. 891–898, 2014.

[72] T. Akbari, M. Pourhajibagher, F. Hosseini et al., "The effect of indocyanine green loaded on a novel nano-graphene oxide for high performance of photodynamic therapy against Enterococcus faecalis," *Photodiagnosis and Photodynamic Therapy*, vol. 20, pp. 148–153, 2017.

[73] S. A. Farook, V. Shah, D. Lenouvel, O. Sheikh, Z. Sadiq, and L. Cascarini, "Guidelines for management of sodium hypochlorite extrusion injuries," *British Dental Journal*, vol. 217, no. 12, pp. 679–684, 2014.

[74] D. K. Sharma, M. Bhat, V. Kumar, D. Mazumder, S. V. Singh, and M. Bansal, "Evaluation of Antimicrobial Efficacy of Graphene Silver Composite Nanoparticles against E. faecalis as Root Canal Irrigant: An ex-vivo study," *Int. J. Pharm. Med. Res*, vol. 3, no. 5, pp. 267–272, 2015.

[75] N. Dubey, S. S. Rajan, Y. D. Bello, K.-S. Min, and V. Rosa, "Graphene nanosheets to improve physico-mechanical properties of bioactive calcium silicate cements," *Materials*, vol. 10, no. 6, 2017.

[76] C. Prati and M. G. Gandolfi, "Calcium silicate bioactive cements: biological perspectives and clinical applications," *Dental Materials*, vol. 31, no. 4, pp. 351–370, 2015.

[77] M. Mehrali, E. Moghaddam, S. F. S. Shirazi et al., "Mechanical and in vitro biological performance of graphene nanoplatelets reinforced calcium silicate composite," *PLoS ONE*, vol. 9, no. 9, 2014.

[78] K. Gong, Z. Pan, A. H. Korayem et al., "Reinforcing effects of graphene oxide on portland cement paste," *Journal of Materials in Civil Engineering*, vol. 27, no. 2, Article ID A4014010, 2014.

[79] M. S. Baig and G. J. P. Fleming, "Conventional glass-ionomer materials: a review of the developments in glass powder, polyacid liquid and the strategies of reinforcement," *Journal of Dentistry*, vol. 43, no. 8, pp. 897–912, 2015.

[80] L. Sun, Z. Yan, Y. Duan, J. Zhang, and B. Liu, "Improvement of the mechanical, tribological and antibacterial properties of glass ionomer cements by fluorinated graphene," *Dental Materials*, vol. 34, no. 6, pp. e115–e127, 2018.

[81] C. Sarosi, A. R. Biris, A. Antoniac et al., "The nanofiller effect on properties of experimental graphene dental nanocomposites," *Journal of Adhesion Science and Technology*, vol. 30, no. 16, pp. 1779–1794, 2016.

[82] M. Battino, M. Greabu, and B. Calenic, "Oxidative stress in oral cavity: interplay between reactive oxygen species and antioxidants in health, inflammation, and cancer," in *Textbook of oxidative stress and antioxidant protection: the science of free*

radical biology and disease, D. Armstrong and R. D. Stratton, Eds., pp. 155–166, Wiley Blackwell, London, 2016.

[83] A. Bregnocchi, E. Zanni, D. Uccelletti et al., "Graphene-based dental adhesive with anti-biofilm activity," *Journal of Nanobiotechnology*, vol. 15, no. 1, 2017.

[84] S. Pajoumshariati, H. Shirali, S. K. Yavari et al., "GBR membrane of novel poly (butylene succinate-co-glycolate) co-polyester co-polymer for periodontal application," *Scientific Reports*, vol. 8, no. 1, p. 7513, 2018.

[85] M. Radunovic, M. De Colli, P. De Marco et al., "Graphene oxide enrichment of collagen membranes improves DPSCs differentiation and controls inflammation occurrence," *Journal of Biomedical Materials Research Part A*, vol. 105, no. 8, pp. 2312–2320, 2017.

[86] K. Kawamoto, H. Miyaji, E. Nishida et al., "Characterization and evaluation of graphene oxide scaffold for periodontal wound healing of class II furcation defects in dog," *International journal of nanomedicine*, vol. 13, p. 2365, 2018.

15

Scaffolds for Pelvic Floor Prolapse: Logical Pathways

Julio Bissoli⊙ and Homero Bruschini

Hospital das Clínicas da Faculdade de Medicina da Universidade de São Paulo (HCFMUSP), 05410-020 São Paulo, SP, Brazil

Correspondence should be addressed to Julio Bissoli; julio_bissoli@yahoo.com.br

Academic Editor: Qiang Wei

Pelvic organ prolapse (POP) has borrowed principles of treatment from hernia repair and in the last two decades we saw reinforcement materials to treat POP with good outcomes in terms of anatomy but with alarming complication rates. Polypropylene meshes to specifically treat POP have been withdrawn from market by manufactures and a blank space was left to be filled with new materials. Macroporous monofilament meshes are ideal candidates and electrospinning emerged as a reliable method capable of delivering production reproducibility and customization. In this review, we point out some pathways that seem logical to be followed but have been only researched in last couple of years.

1. Introduction

Greeks created the term "prosthesis" or "to place before." Hernia repair usually consists of placing meshes to assist the suture control of protrusions [1]. In modern history, Theodore Billroth in 1857 inspired all prosthesis designers with his proposition "If we could artificially produce tissue of the density and toughness of fascia and tendon, the secret of the radical cure of the hernia repair would be discovered" [2], but its concept of repair dates back to Greeks who first described silver strands woven sutured with gold wire to act as a prosthesis [3].

Plastic development revolutionized hernia repair firstly with nylon woven prosthesis, abandoned due to loss of strength caused by hydrolysis, followed by other materials as polypropylene (PP), polytetrafluoroethylene (PTFE), Dacron, and polyethylene (PE) [4].

In anatomical terms, hernias are quite similar to pelvic organ prolapse where one or more vaginal compartments descend downwards vaginal opening causing vaginal bulging [5] and this similarity brought the same treatment concepts from one to another [6–8]. What was ignored was the thin layer of mucosae capable of covering an occasional prosthesis, the elasticity inherent to the organ (specially during sexual activity), and exuberant vaginal local florae [5, 7].

2. Pelvic Organ Prolapse

In recent years, development of a new kind of mesh specifically designed for vaginal surgery earned some interest because of a major withdraw from mesh market by leading industries after FDA safety warnings on complications of polypropylene in prolapse surgery and due to massive losses on court due to litigation [8–10].

POP is a high prevalent disease occurring in up to 37% of asymptomatic women [11], with lifetime risk of intervention up to 80 years estimated to be 11 to 20% [12, 13] and reoperation rates due to symptoms around 30% [14].

With such high prevalence and failure rates with conventional treatments up to 56% [15], it was a natural movement for doctors and researchers to start developing reinforcements to sutures performed during vaginal compartments treatment.

3. Reinforcements Subtypes

The materials used to reinforce POP treatment can be autologous, heterologous, synthetic absorbable, or inabsorbable [7]. The initial tests with fascia lata, acellular dermis, and rectus sheet did not show superior results to conventional techniques, with 38% failure [16], and are limited by donor site, pain, surgery time increase, and quality/quantity variable [17, 18] with the clear advantage to not trigger any immune response. Heterologous grafts keep the same or inferior results potentially acting as a carrier to viral and prion diseases [7, 8].

Among synthetic reinforcement, the polyglactin is shown to be absorbed without matrix remodelling and failure before 2 years of implant [19–21] while inabsorbable polypropylene (most widely used) showed 80% success compared to traditional techniques in short time [22] but with midterm complications such erosions/extrusions around 25% [23] causing FDA to release safety warnings in 2008 and 2011 [9, 10] triggering ethical legal problems and prompting leading companies to withdraw from market [8].

Considering limited results with classical approaches and materials, especially after withdrawing of synthetic meshes manufacturers, a blank space emerged to be filled with new materials.

4. Synthetic Meshes Classification

Meshes are a subtype of synthetic matrices with organized woven or knitted pattern. Amid in 1997 classified meshes accordingly to its porosity and filaments structure [25], predicting complications as bigger pores allow higher vascularization, fibroblasts ingrowth, and immune cells infiltration increasing biocompatibility and infection resilience [26]. On the other way, multifilament fibres (with space between filaments inferior to 10 microns) and microporous meshes are less susceptible to cellular ingrowth and macrophage/lymphocyte action (they measure around 9–20 microns) being prone to bacterial colonization.

Theoretically, an ideal material would be a macroporous monofilament type I mesh of Amid [27–29] and practical applications confirm such statement with types II, III, and IV being used for POP and SUI with short-term complication rates around 20–30% [22, 27, 30].

5. Synthetic Matrices Production

There are several methods for higher porosity matrices production, like self-assembly [31], phase separation, solvent casting and particulate leaching, freeze drying, melt moulding, gas foaming, and solid free-forming [32]. All of them with reproducibility limitations and poor control of characteristics such diameter of fibres and pore size, pore geometry, and fibre orientation with electrospinning as an alternative to surpass all these difficulties [33].

6. Electrospinning

Electrospinning is a physical phenomenon observed when a high viscosity polymer solution is exposed to an intense electric field in a recipient where the liquid is extruded in a slow fashion through one small or multiple small orifices (i.e., spinneret). Usually such solution is made by a high molecular weight polymer in a solvent with high vapour pressure (i.e., capable of easily evaporate in a room temperature) and low conductivity with high dielectric constant (i.e., high resistance to stress before brake and allow passage of electric current) [34].

Besides relatively high viscosity of material used, under high electric field produced by a high tension power source, a superficial tension rupture at the tip of the spinneret occurs (usually a blunt tip needle). This rupture creates an instability region causing stretching of viscous solution towards an earthed collector, parallel to the lines of force of electric field, producing micro/nanometric scale fibres in diameter. Such fibres become dried after evaporation of solvent during its pathway to the collector.

This process is known since the beginning of the last century being observed by Rayleigh in 1897, detailed by Zeleny in 1914, and patented by Formhals in 1934 for textile production. Taylor, in his studies about electrostatics (1969), described the jets produced by the technique (today called by his name, Taylor's cone), but electrospinning as a biomedical application emerged only after the 1990s because of its price and reproducibility, being one of the commonest methods of matrix production in bioengineering.

Parameters controlled during electrospinning process are classically divided into solution properties, controlled variables, and environmental parameters [32]. The solution properties included viscosity, superficial tension, conductivity, and molecular weight, while controlled variables are flow rate, electric field intensity, collector distance to the end of the tip of spinneret, size of the tip, and geometry of collector. Environmental parameters are temperature, pressure, and air speed [34].

From all described variables, the most important is the solution concentration to determine, among other things, the diameter of fibres.

7. Polymers

Polymers are big molecules produced by repetition of numerous subunits called monomers and depending on the number of types of repetitive units can be further classified as homopolymers or copolymers. Its name derives from Greek *polus* (i.e., lots of) and *merossu* (i.e., parts), and its physical characteristics are dependent on the size and length of their chain. Usually as polymer chain increases, its degradation time is prolonged, and its viscosity, strength, and rigidity are higher, being common to allude to it in terms of molecular weight (minimum, maximum, and average value) [35].

They can be further subclassified as naturals or synthetics and absorbable or inabsorbable. For bioengineering applications, there are necessary characteristics: their nontoxicity (direct or indirect), nonimmunogenicity, noncarcinogenicity, and biocompatibility (ability to integrate with living tissues). Desirable characteristics are resilience to infection, low cost, ability to be easily manufactured and stored, and absorption/degradation in a fashion that allows repopulation and integration by living cells from target tissue and mechanical properties compatible with target tissue function from day 0, during repopulation until its complete reabsorption (Figure 1) [24].

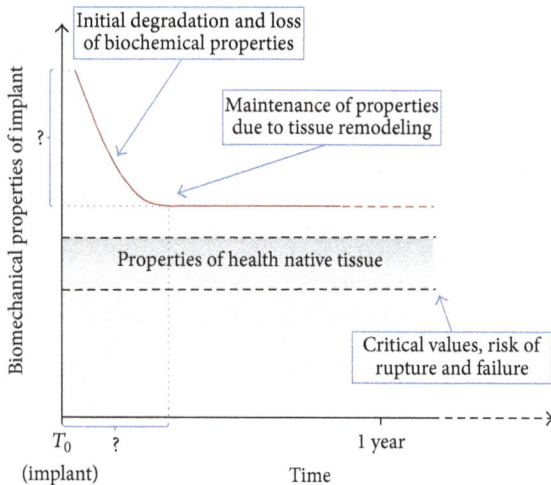

FIGURE 1: Ideal biomechanical properties of biodegradable scaffolds (adapted from Osman et al. [24]).

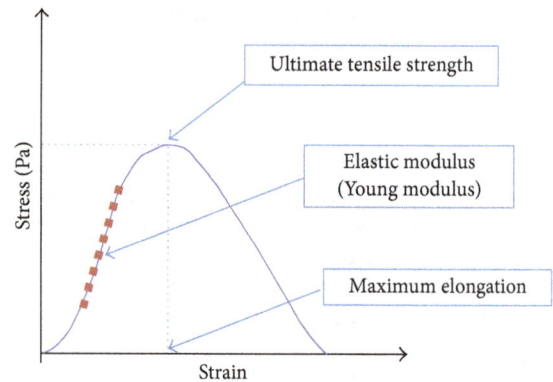

FIGURE 2: Stress-strain example curve with elastic modulus, ultimate tensile strength, and maximum elongation.

For each target tissue the desirable properties are different. For each polymer used in an implant, a drop in its mechanical properties will occur proportionally to its reabsorption and thus depends on its degradation speed and its micro/nanostructure. Ideally, such characteristics to be mimic should be known previously (ultimate tensile strength, elasticity, and absorption time) and adjusted accordingly to each new microenvironment.

Among the most common polymers electrospun are polylactic acid (PLA), polyglycolic acid (PGA), polyhydroxybutyrate covalerate (PHBV), polycaprolactone (PCL), chitosan, collagen, and polyurethanes (PU).

Aliphatic polyesters like polycaprolactone and polyglycolic acid are materials extensively known since 1960 because of its biocompatibility. They are hydrolysed and/or enzymatically digested to nontoxic subproducts. They usually have high elastic modulus and tensile strength with poor elongation (considered rigid polymers), great candidates for tissue engineering [30].

Polyhydroxyalkanoates as poly(3-hydroxybutyrate-co-3-hydroxyvalerate) (PHBV) are recently investigated in tissue engineering because of its resilience to infection and hydrolysis much like polyurethanes (PU) that additionally have calcification resistance and are known in applications as synthetic rubber for more than 30 years. Both present longer absorption times when compared to polyesters being PU known by its great ability to cope with strain.

8. Biomechanical Tests

There are no standard protocols for biomechanical testing of vaginal tissues, being the best methodologies derived from uniaxial tests (i.e., test performed only in one direction), measured in a tensiometer stress-strain curves. Multiaxial tests or biaxial tests potentially would reflect a model closer to its real mechanical properties, but they are more complex and demand specific software and special equipment making its standardization even more complex than uniaxial tests.

Briefly, the uniaxial biomechanical test consists of firmly securing sample between two clamps connected to a tensiometer and a computer and distending the sample in a controlled fashion. Previous known distance of clamps, sample width and thickness, and constant uniaxial distention of clamps (i.e., keeping the same orientation and speed of force) until sample failure or test manually stops will provide data to calculate its biomechanical characteristics.

The ultimate tensile strength is calculated dividing the load applied to the sample (in Newtons, N) by the cross section area of the sample being reported in N/m2 (Pascals). Strain is calculated dividing elongation of the sample (in meters, m) by its initial length between clamps (in meters, m) resulting in a quotient without units or a percentage. Commonly, such data is plotted on a graph and shows a linear portion where tension is direct proportional to strain (respecting Hooke's Law) where strain is reversible or elastic and then a plateau where elongation is irreversible or plastic followed by an inflection (ultimate tensile strength) with correspondence to X-axis defining maximum strain.

The example can be observed at Figure 2, the elastic modulus or Young modulus can be calculated from inclination of linear portion of the stress-strain curve and it is inversely proportional to elasticity of the sample.

To develop an ideal material for a target tissue, knowledge of its properties is crucial and a few studies showed biomechanical properties for human tissues. The vast majority do not normalize its data using cross section area exhibiting values only in Newtons, preventing posterior comparisons [7, 36, 37].

9. Paravaginal Mechanical Properties

Choe et al. compared 2 × 5 cm strips of fascia lata, dermis, rectum sheet, and vaginal mucosae (measure commonly used at sling surgeries [38]) in women operated for various reasons [39] showing that fascia lata had the biggest tensile strength (217 N), followed by human dermis (122 N) and rectum sheet and vaginal mucosae (both with 42 N).

Lei et al. analysed 43 women after hysterectomy and categorized them in groups, premenopause and postmenopause

TABLE 1: Elastic modulus, maximum elongation, and ultimate tensile strength in women with and without prolapse (adapted from Lei et al.).

	Control premenopause	Prolapse premenopause	Control postmenopause	Prolapse postmenopause
Elastic modulus (MPa; mean ± EPM)	6.65 ± 1.48	9.45 ± 0.70	10.26 ± 1.10	12.10 ± 1.10
Maximum elongation (mean ± EPM)	1.68 ± 0.11	1.50 ± 0.02	1.37 ± 0.04	1.14 ± 0.06
Ultimate tensile strength (MPa; mean ± EPM)	0.79 ± 0.05	0.60 ± 0.02	0.42 ± 0.03	0.27 ± 0.03

and with or without prolapse (Table 1), testing with uniaxial tests tissues with 5 mm × 25 mm and plotting their stress-strain curves establishing native values considered maximum (premenopause) and minimum (postmenopause) obtaining maximum elongation and elastic modulus references for most publications in this field by being comparable to other samples (i.e., normalized by cross section area) [40]. As a critic to such conclusions it is important to notice that samples tested were from vaginal mucosae closer to apex and not from suspensory ligaments or mucosae closer to cystocele/rectocele common defects.

10. Discussion

We do not know exactly how much is the demand in Newtons for the pelvic floor, but we estimate forces acting on it to be around 2,2 to 13,4 N/cm from stand still to stand with Valsalva [41]. Meanwhile, we know that the best autologous candidate—rectum fascia—could cope up to 16 N/cm with a 25% vertical strain [42].

It is important to highlight that tensile strength alone is not capable of predicting success in reconstructive urogenital surgery, and fascia lata and acellular dermis (both quite strong) also have high relapse rate before 2 years of surgery [19, 43] showing that, for biocompatible absorbable materials, remodelling in the host is probably of higher importance than the initial tensile strength of the implant [44].

We still do not have a substitute for paravaginal weakened tissues with native characteristics in terms of resistance and flexibility and data have shown that we are not close to find it [9, 10, 23, 45], notably a potential replacement for polypropylene meshes used in POP and SUI surgery, efficient by anatomical point of view but with complications in POP case that precludes its use.

Such complications have multiple factors involved, but polypropylene high resistance, inflexibility, and inelasticity certainly play a role heightened by a contraction tendency after the implant, despite being considered biocompatible [8].

Several good candidates had been prospected to replace it, each scaffold with or without specific cell type being regarded as ideal. Oral fibroblasts [46], adipose derived stem cells [47], vaginal fibroblasts [48], and muscle cells [49] all have its qualities and defects being more similar or not to targeted paravaginal tissue, being more easily or not to harvest and/or to cultivate onto scaffolds produced with variations of PLA, PLGA, PU, and processed small intestine submucosae.

Considering costs, regulations, and facilities needed to widespread cell culture to clinical practice, probably the first material produced to replace polypropylene meshes in POP practice will be an off-the-shelf synthetic scaffold with great cell affinity targeted to a specific biomechanical demand of paravaginal tissue.

Animal experiments of at least 6–12 months in a model physiologically relevant to POP will help to establish how degradation and neotissue formation will affect properties of such material.

11. Conclusion

We have been done a lot to replace polypropylene meshes by matrices not only capable of withstanding tension but also capable of interacting with cells and promote tissue remodelling with fibroblasts ingrowth, extracellular matrix production, and angiogenesis [50] and we already made some progress in this regard with POP.

Respecting Billroth's principle and producing materials strong enough but sufficiently elastic to allow natural distention of vaginal tissue [44] and biocompatible to reflect paravaginal properties [51] are the right pathway and surprisingly only recently started to be followed.

Additional Points

Highlights. (i) Pelvic organ prolapse treatment borrowed anatomical concepts from hernia repair without considering local differences in terms of elasticity, florae, and sexual activity. Major brand meshes for POP have been withdrawn from market due to complications and litigious problems. (ii) Despite the logic to know that the material is to be replaced or reinforced and then use this data to develop a new prosthesis, pelvic prolapse lacks in terms of basic research of biomechanical properties and lacked in safety studies before applying polypropylene meshes for this purpose. (iii) To promote true remodelling and cure we not only should think about reinforcement and biomechanical properties but also must develop a material that interacts with cells and promote fibroblast ingrowth and extracellular matrix production. Lack of these combined qualities invariably will lead to failure.

Conflicts of Interest

The authors declare that there are no conflicts of interest regarding the publication of this manuscript.

Acknowledgments

Thanks are due to Professor Chris Chapple for mentoring during fellowship in Sheffield and whose support was of paramount importance and thanks are due to Professor Miguel Srougi for encouragement and vision during all steps of this work. This work was supported by Conselho Nacional de Pesquisa Tecnologia-Programa Ciência sem Fronteiras.

References

[1] P. L. Carter, N. Lloyd, and S. Rene, "Preperitoneal inguinal pioneers," *American Journal of Surgery*, vol. 211, no. 5, pp. 836–838, 2016.

[2] J. E. Skandalakis, G. L. Colborn, L. J. Skandalakis, D. A. McClusky, R. J. Fitzgibbons, and A. G. Greenburg, "Historic aspects of groin hernia repair," *Nyhus Condons Hernia*, 2002.

[3] Y. Bilsel and I. Abci, "The search for ideal hernia repair; mesh materials and types," *International Journal of Surgery*, vol. 10, no. 6, pp. 317–321, 2012.

[4] M. Kapischke and A. Pries, "Theodor Billroth's vision and Karl Ziegler's action: Commemoration of the 40th day of death and the 50th anniversary of conferment of Nobel Prize for Chemistry of Karl Ziegler," *Surgery (United States)*, vol. 155, no. 2, pp. 347–349, 2014.

[5] J. E. Jelovsek, C. Maher, and M. D. Barber, "Pelvic organ prolapse," *The Lancet*, vol. 369, no. 9566, pp. 1027–1038, 2007.

[6] U. Ulmsten, L. Henriksson, P. Johnson, and G. Varhos, "An ambulatory surgical procedure under local anesthesia for treatment of female urinary incontinence," *International Urogynecology Journal and Pelvic Floor Dysfunction*, vol. 7, no. 2, pp. 81–86, 1996.

[7] P. Dällenbach, "To mesh or not to mesh: A review of pelvic organ reconstructive surgery," *International Journal of Women's Health*, vol. 7, pp. 331–343, 2015.

[8] G. Gigliobianco, S. R. Regueros, N. I. Osman et al., "Biomaterials for pelvic floor reconstructive surgery: How can we do better?" *BioMed Research International*, vol. 2015, Article ID 968087, 20 pages, 2015.

[9] D. G. Schultz, *FDA Public Health Notification: Serious Complications Associated with Transvaginal Placement of Surgical Mesh in Repair of Pelvic Organ Prolapse and Stress Urinary Incontinence*, Food Drug Adm, Silver Spring, Md, USA, 2008.

[10] Administration UF and D, Administration UF and D. FDA safety communication: update on serious complications associated with transvaginal placement of surgical mesh for pelvic organ prolapse. July. 2011;13:2011.

[11] S. Swift, P. Woodman, A. O'Boyle et al., "Pelvic Organ Support Study (POSST): the distribution, clinical definition, and epidemiologic condition of pelvic organ support defects," *American Journal of Obstetrics & Gynecology*, vol. 192, no. 3, pp. 795–806, 2005.

[12] J. M. Wu, C. A. Matthews, M. M. Conover, V. Pate, and M. Jonsson Funk, "Lifetime risk of stress urinary incontinence or pelvic organ prolapse surgery," *Obstetrics and Gynecology*, vol. 123, no. 6, pp. 1201–1206, 2014.

[13] M. F. Fialkow, K. M. Newton, G. M. Lentz, and N. S. Weiss, "Lifetime risk of surgical management for pelvic organ prolapse or urinary incontinence," *International Urogynecology Journal*, vol. 19, no. 3, pp. 437–440, 2008.

[14] A. L. Olsen, V. J. Smith, J. O. Bergstrom, J. C. Colling, and A. L. Clark, "Epidemiology of surgically managed pelvic organ prolapse and urinary incontinence," *Obstetrics and Gynecology*, vol. 89, no. 4, pp. 501–506, 1997.

[15] A. L. Clark, T. Gregory, V. J. Smith, and R. Edwards, "Epidemiologic evaluation of reoperation for surgically treated pelvic organ prolapse and urinary incontinence," *American Journal of Obstetrics and Gynecology*, vol. 189, no. 5, pp. 1261–1267, 2003.

[16] C. Maher, B. Feiner, K. Baessler, C. Christmann-Schmid, N. Haya, and J. Marjoribanks, *Transvaginal mesh or grafts compared with native tissue repair for vaginal prolapse. Transvaginal Mesh Grafts Comp Native Tissue Repair Vaginal Prolapse*, 2016, http://dx.doi.org/10.1002/14651858.CD012079.

[17] M. E. Albo, H. E. Richter, L. Brubaker, P. Norton, S. R. Kraus, P. E. Zimmern et al., "Burch colposuspension versus fascial sling to reduce urinary stress incontinence," *The New England Journal of Medicine*, vol. 356, no. 21, pp. 2143–2155, 2007.

[18] A. J. Walter, J. G. Hentz, J. F. Magrina, and J. L. Cornella, "Harvesting autologous fascia lata for pelvic reconstructive surgery: Techniques and morbidity," *American Journal of Obstetrics and Gynecology*, vol. 185, no. 6, pp. 1354–1359, 2001.

[19] M. P. FitzGerald, J. Mollenhauer, P. Bitterman, and L. Brubaker, "Functional failure of fascia lata allografts," *American Journal of Obstetrics and Gynecology*, vol. 181, no. 6, pp. 1339–1346, 1999.

[20] M. H. Safir, A. E. Gousse, E. S. Rovner, D. A. Ginsberg, and S. Raz, "4-Defect repair of grade 4 cystocele," *Journal of Urology*, vol. 161, no. 2, pp. 587–594, 1999.

[21] A. M. Weber, M. D. Walters, M. R. Piedmonte, and L. A. Ballard, "Anterior colporrhaphy: A randomized trial of three surgical techniques," *American Journal of Obstetrics and Gynecology*, vol. 185, no. 6, pp. 1299–1306, 2001.

[22] T. M. Julian, "The efficacy of Marlex mesh in the repair of severe, recurrent vaginal prolapse of the anterior midvaginal wall," *American Journal of Obstetrics & Gynecology*, vol. 175, no. 6, pp. 1472–1475, 1996.

[23] J. Mahon, D. Varley, and J. Glanville, *Summaries of the safety/adverse effects of vaginal tapes/slings/meshes for stress urinary incontinence and prolapse*, Med Healthc Prod Regul Agency, 2012.

[24] N. Osman, S. Roman, J. Bissoli, F. Sefat, S. MacNeil, and C. Chapple, "Designing a novel tissue inductive bio-absorbable implant for pelvic floor repair: An assessment of tensile and surgical handling properties versus polypropylene mesh and porcine small intestine submucosa," *Neurourol Urodyn*, vol. 32, no. 6, pp. 507–932, 2013.

[25] P. K. Amid, "Classification of biomaterials and their related complications in abdominal wall hernia surgery," *Hernia*, vol. 1, no. 1, pp. 15–21, 1997.

[26] M. Cervigni and F. Natale, "The use of synthetics in the treatment of pelvic organ prolapse," *Current Opinion in Urology*, vol. 11, no. 4, pp. 429–435, 2001.

[27] J. C. Winters, M. P. Fitzgerald, and M. D. Barber, "The use of synthetic mesh in female pelvic reconstructive surgery," *BJU International*, vol. 98, no. 1, pp. 70–77, 2006.

[28] C. Birch and M. M. Fynes, "The role of synthetic and biological prostheses in reconstructive pelvic floor surgery," *Current Opinion in Obstetrics & Gynecology*, vol. 14, no. 5, pp. 527–535, 2002.

[29] M. Slack, J. S. Sandhu, D. R. Staskin, and R. C. Grant, "In vivo comparison of suburethral sling materials," *International Urogynecology Journal and Pelvic Floor Dysfunction*, vol. 17, no. 2, pp. 106–110, 2006.

[30] P. Debodinance, M. Cosson, and G. Burlet, "Tolerance of synthetic tissues in touch with vaginal scars: Review to the point of 287 cases," *European Journal of Obstetrics Gynecology and Reproductive Biology*, vol. 87, no. 1, pp. 23–30, 1999.

[31] Y. Hong, R. L. Legge, S. Zhang, and P. Chen, "Effect of amino acid sequence and pH on nanofiber formation of self-assembling peptides EAK16-II and EAK16-IV," *Biomacromolecules*, vol. 4, no. 5, pp. 1433–1442, 2003.

[32] R. Murugan and S. Ramakrishna, "Nano-featured scaffolds for tissue engineering: A review of spinning methodologies," *Tissue Engineering*, vol. 12, no. 3, pp. 435–447, 2006.

[33] S. Roman, N. Mangir, J. Bissoli, C. R. Chapple, and S. MacNeil, "Biodegradable scaffolds designed to mimic fascia-like properties for the treatment of pelvic organ prolapse and stress urinary incontinence," *Journal of Biomaterials Applications*, vol. 30, no. 10, pp. 1578–1588, 2015.

[34] N. Bhardwaj and S. C. Kundu, "Electrospinning: a fascinating fiber fabrication technique," *Biotechnology Advances*, vol. 28, no. 3, pp. 325–347, 2010.

[35] D. W. Van Krevelen and K. Te Nijenhuis, *Chapter 1 - Polymer Properties. In: Properties of Polymers*, Elsevier, Amsterdam, Netherlands, 4th edition, 2009.

[36] Q. P. Pham, U. Sharma, and A. G. Mikos, "Electrospinning of polymeric nanofibers for tissue engineering applications: a review," *Tissue Engineering*, vol. 12, no. 5, pp. 1197–1211, 2006.

[37] S. Ramakrishna, "An iNtroduction to Electrospinning and Nanofibers," in *World Scientific*, 2005.

[38] M. M. Karram and N. N. Bhatia, "Patch procedure: Modified transvaginal fascia lata sling for recurrent or severe stress urinary incontinence," *Obstetrics and Gynecology*, vol. 75, no. 3, pp. 461–463, 1990.

[39] J. M. Choe, R. Kothandapani, L. James, and D. Bowling, "Autologous, cadaveric, and synthetic materials used in sling surgery: comparative biomechanical analysis," *Urology*, vol. 58, no. 3, pp. 482–486, 2001.

[40] L. Lei, Y. Song, and R. Chen, "Biomechanical properties of prolapsed vaginal tissue in pre- and postmenopausal women," *International Urogynecology Journal*, vol. 18, no. 6, pp. 603–607, 2007.

[41] J. A. Ashton-Miller and J. O. L. DeLancey, "On the biomechanics of vaginal birth and common sequelae," *Annual Review of Biomedical Engineering*, vol. 11, no. 1, pp. 163–176, 2009.

[42] K. Junge, U. Klinge, A. Prescher, P. Giboni, M. Niewiera, and V. Schumpelick, "Elasticity of the anterior abdominal wall and impact for reparation of incisional hernias using mesh implants," *Hernia*, vol. 5, no. 3, pp. 113–118, 2001.

[43] D. C. Owens and J. C. Winters, "Pubovaginal Sling Using Duraderm™ Graft: Intermediate Follow-Up and Patient Satisfaction," *Neurourology and Urodynamics*, vol. 23, no. 2, pp. 115–118, 2004.

[44] A. Mangera, A. J. Bullock, C. R. Chapple, and S. MacNeil, "Are biomechanical properties predictive of the success of prostheses used in stress urinary incontinence and pelvic organ prolapse? A systematic review," *Neurourology and Urodynamics*, vol. 31, no. 1, pp. 13–21, 2012.

[45] K. L. Ward and P. Hilton, "A prospective multicenter randomized trial of tension-free vaginal tape and colposuspension for primary urodynamic stress incontinence: Two-year follow-up," *American Journal of Obstetrics and Gynecology*, vol. 190, no. 2, pp. 324–331, 2004.

[46] A. Mangera, A. J. Bullock, S. Roman, C. R. Chapple, and S. Macneil, "Comparison of candidate scaffolds for tissue engineering for stress urinary incontinence and pelvic organ prolapse repair," *British Journal of Urology*, vol. 112, no. 5, pp. 674–685, 2013.

[47] C. J. Hillary, S. Roman, A. J. Bullock et al., "Developing Repair Materials for Stress Urinary Incontinence to Withstand Dynamic Distension," *PLoS ONE*, vol. 11, no. 3, p. e0149971, 2016.

[48] M. J. Hung, M. C. Wen, C. N. Hung, E. S. Ho, G. D. Chen, and V. C. Yang, "Tissue-engineered fascia from vaginal fibroblasts for patients needing reconstructive pelvic surgery," *International Urogynecology Journal*, vol. 21, no. 9, pp. 1085–1093, 2010.

[49] M. Boennelycke, L. Christensen, L. F. Nielsen, S. Gräs, and G. Lose, "Fresh muscle fiber fragments on a scaffold in ratsa new concept in urogynecology?" *American Journal of Obstetrics and Gynecology*, vol. 205, no. 3, pp. 235–e14, 2011.

[50] S. F. Badylak, J. E. Valentin, A. K. Ravindra, G. P. McCabe, and A. M. Stewart-Akers, "Macrophage phenotype as a determinant of biologic scaffold remodeling," *Tissue Engineering A*, vol. 14, no. 11, pp. 1835–1842, 2008.

[51] T. Aboushwareb, P. Mckenzie, F. Wezel, J. Southgate, and G. Badlani, "Is tissue engineering and biomaterials the future for Lower Urinary Tract Dysfunction (LUTD)/Pelvic Organ Prolapse (POP)?" *Neurourology and Urodynamics*, vol. 30, no. 5, pp. 775–785, 2011.

16

Recent Progress of Fabrication of Cell Scaffold by Electrospinning Technique for Articular Cartilage Tissue Engineering

Yingge Zhou,[1] **Joanna Chyu,**[2] **and Mimi Zumwalt** ⓘ[2]

[1]*Department of Industrial, Manufacturing, and System Engineering, Texas Tech University, Lubbock, TX, USA*
[2]*Department of Orthopedic Surgery and Rehabilitation, Texas Tech University Health Sciences Center, Lubbock, TX, USA*

Correspondence should be addressed to Mimi Zumwalt; mimi.zumwalt@ttuhsc.edu

Academic Editor: Feng-Huei Lin

As a versatile nanofiber manufacturing technique, electrospinning has been widely employed for the fabrication of tissue engineering scaffolds. Since the structure of natural extracellular matrices varies substantially in different tissues, there has been growing awareness of the fact that the hierarchical 3D structure of scaffolds may affect intercellular interactions, material transportation, fluid flow, environmental stimulation, and so forth. Physical blending of the synthetic and natural polymers to form composite materials better mimics the composition and mechanical properties of natural tissues. Scaffolds with element gradient, such as growth factor gradient, have demonstrated good potentials to promote heterogeneous cell growth and differentiation. Compared to 2D scaffolds with limited thicknesses, 3D scaffolds have superior cell differentiation and development rate. The objective of this review paper is to review and discuss the recent trends of electrospinning strategies for cartilage tissue engineering, particularly the biomimetic, gradient, and 3D scaffolds, along with future prospects of potential clinical applications.

1. Introduction

Tissue engineering has emerged as an alternative cell-based approach designed to substitute damaged organs with tissues generated *in vitro*. It overcomes the barriers of conventional allograft transplantation, such as the scarcity of donor organs, the complexity of surgery, and complicated postoperative care [1]. In many cases, it involves the utilization of a scaffold, an engineered porous supporting material for tissue regeneration. Scaffolds mimic the extracellular matrix (ECM) of the original tissues and reproduce or replicate the natural tissue environment. In addition, scaffolds can foster certain mechanical and biological properties to modify the behavior of different cell phases [2–4].

Articular cartilage is tissue located between bones, which can withstand compressive and shear forces several times that of human body weight for years due to its low friction and high load bearing capacity [5, 6], attributed to the unique composition of its ECM [7]. However, it is also the unique ECM that makes the repair of articular cartilage extremely difficult [8]. The micro fissures caused by repetitive cyclic loading might not be visible to the eye but continuously undermine the integrity of collagen fibrils and proteoglycan network [9], causing cartilage destruction over time and leading to degenerative joint disease or osteoarthritis. Osteoarthritis is no longer common only among the elderly but has extended to the younger population who also need a long-term solution for their painful problem [10].

Currently, surgical treatments for knee osteoarthritis include arthroscopic microfracture, Osteoarticular Transfer System (OATS), and autologous chondrocyte implantation (ACI), as well as open procedures such as osteotomy and arthroplasty. Unfortunately, they all fail to provide healthy and vigorous articular cartilage in the long run. For example, osteotomy and arthroplasty can lead to articular exacerbation and impairment of neighboring bones [5]. Cartilage tissue engineering, on the other hand, may provide a potentially better resolution in the treatment of knee osteoarthritis. With scaffolds manufactured especially for articular cartilage, the functional complexity of electrospun scaffolds provides

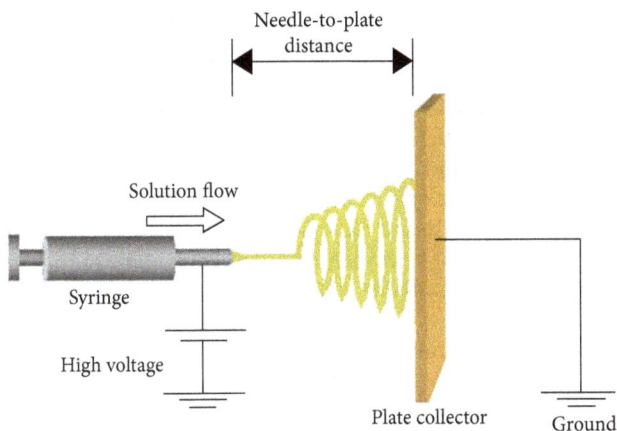

FIGURE 1: Typical electrospinning facility with a plate collector [6].

significant advantages over other techniques such as allowing the mesenchymal stem cells to grow in a way that facilitates the formation of fibrous tissue structure. Electrospinning may be a viable approach to achieving highly aligned and biocompatible scaffolds to meet the demands for cartilage tissue engineering.

1.1. Electrospinning. An electrospinning system is comprised of at least three units: a high voltage power unit, a material delivery unit (a capillary tube with a small spinneret in general), and a fiber collection unit. A typical electrospinning setup is shown in Figure 1 [6]. The randomly distributed nanofibers were collected in Figure 1 on a vertical plate. Usually, an advancement pump is employed to regulate the flow rate of polymeric solution discharged from a syringe. An electrically charged jet of polymer solution or melt is created by a high voltage source. The discharged polymer solution jet goes through an instability process, during which the jet is elongated and the solvent evaporates or solidifies. The jet is finally collected as an interconnected web of ultrathin fibers at the collector that is grounded or connected to an electrode. The diameters of electrospun fibers can range from 10 nm to 100 μm depending on the material, device configuration, and process setup.

Electrospinning is able to generate loosely connected porous mats from submicrometer to nanometer scales in diameter as it is a highly flexible technique to produce continuous fibers. The nanofiber mat yields a high surface area to volume ratio and high porosity [54]. The nanostructure of the mat mimics the structure of the ECM in terms of both morphology and composition [55]. Various patterns of electrospun nanofibers have been fabricated for different biomedical applications ranging from artificial skin to endocrine organs and from the nervous system to cardiovascular tissues [56].

1.2. Electrospun Scaffolds for Cartilage Tissue Engineering. Electrospinning as a feasible and versatile technique can manipulate both natural and artificial polymers in nanoscale and promote the proliferation of mesenchymal stem cells (MSCs). Recent research attention has focused on improving the material design of the composition to produce

biomimetic, gradient, and 3D scaffolds. Biomimetic scaffold is crucial in the reconstruction of connective tissue such as articular cartilage. Electrospinning may achieve the desired biomimetic characteristics with regard to biocompatibility and biomimicry by adding different synthetic and biological materials to the electrospinning nanofibers. Because tissue composition exhibits gradient features from the top layer that withstands high pressure to the bottom layer that connects with the subchondral bone, gradient scaffolds are needed for cartilage tissue engineering. Fibers generated during electrospinning tend to accumulate in a limited thickness and form a denser instead of thicker scaffold; therefore, strategies of fabricating thicker 3D electrospun scaffolds are crucial.

1.3. Objectives. The objective of this paper is to review and discuss the recent trends of electrospinning strategies for cartilage tissue engineering, particularly the biomimetic, gradient, and 3D scaffolds, along with future prospects of potential clinical applications in this field of research.

2. Method of Literature Search

Literature selection for this article was based on PubMed, Scopus, Google Scholar, ScienceDirect, and Web of Science databases from 2011 to the present, as the last review paper on this topic was published in 2012 [57]. The search was conducted using different combinations of the following terms: "tissue engineering," "electrospinning," "cartilage," "scaffold," and "bioscaffold." Articles that were considered to be related to this review and written in English were included.

3. Literature Review

The current literature review reveals that the most prevailing recent trend is the fabrication of biomimetic, gradient, and 3D bioscaffolds, as presented below.

3.1. Biomimetic Composite Scaffolds. The desired characteristics of electrospun scaffold for cartilage regeneration include biocompatibility and biomimicry. Different synthetic and biological materials may be added to electrospinning nanofibers to achieve these characteristics. Cao et al. added graphene oxide with poly(vinyl alcohol) (PVA) and chitosan for better biocompatibility and produced uniform nanofibers with improved cell growth rate compared with nanofibers of chitosan and PVA alone [23]. Bioprinting and electrospinning were combined to fabricate biomimetic scaffolds for cartilage tissue engineering [13]. Electrospun fiber assembled hydrogel was also utilized for supplementation of glycosaminoglycan enriched and mineralized cartilage [18].

3.1.1. Structure Biomimicry. Combination of natural and artificial polymers for scaffolds fabrication is considered promising for repair of cartilage defect or damage. For example, better chondrocytes adhesion and spreading efficiency were achieved with a poly(L-lactic-acid) (PLLA)/silk fibroin (SF) composite scaffold [20]. Poly(L-lactide-co-caprolactone) (P(LL-CL)) and gelatin were used for porous scaffolds fabrication to improve water absorption, cell infiltration, and shape-forming [21]. He et al. investigated the feasibility of collagen/poly(L-lactic acid-co-ε-caprolactone) membranes to facilitate cartilage-like tissue formation [14]. Similarly, Sadeghi et al. blended polyhydroxybutyrate (PHB) with chitosan (CTS) to produce hydrophilic fibrous scaffold for better chondrocytes attachment [22]. The advantages of gelatin/polycaprolactone (PCL) on the formation of 3D cartilage regeneration have also been demonstrated [17, 58].

Biocompatible polymers can be electrospun with wound healing and antimicrobial agents, such as alkanin and shikonin, to achieve high drug capturing efficiencies and multifunctional activities [11]. Yin et al. fabricated core-shell structure nanofibers with embedded kartogenin solution as core fluid to facilitate cartilage regeneration [24]. Mirzaei et al. incorporated glucosamine into PLLA/PEG scaffolds to achieve an enhanced cell proliferation rate [25]. These composite structures can be used to facilitate chondrogenic differentiation and cartilage repair. Oriented electrospinning scaffold thereby exhibits superb biomimicry of articular cartilage by modifying cell adhesion and distribution.

Posttreatment can also benefit biomimetic scaffold fabrication. Liu et al. freeze-dried electrospun fibers with a tricalcium phosphate and produced a trilayered scaffold consisting of oriented fibrous network providing promoted orientation of mesenchymal stem cells [15].

3.1.2. Signal Biomimicry.

3.1.2. Signal Biomimicry. Sustained and stable release of biomolecules such as growth factors, biological signals, and proteins remains an issue that needs to be resolved. Growth factors such as Nel-like molecule-1 have been preloaded in nanofibers via two-phase electrospinning to achieve prolonged release time as well as enhanced MSCs growth rate [26]. A core-sheath structure fibrous electrospun scaffold was fabricated by Man et al. Using a polyvinyl pyrrolidone/bovine serum albumin/rhTGF-β1 composite solution as the core fluid and poly(ε-caprolactone) solution as the sheath fluid, the scaffolds revealed sustained rhTGF-β1 release and promoted chondrogenic differentiation [16]. MSC chondrogenesis was also enhanced on cellulose derived glycosaminoglycan mimetic scaffolds by Huang et al., given that glycosaminoglycans provide signaling cues in native cartilage tissue [19]. Liao et al. coated cartilaginous ECM on electrospun PCL microfiber scaffolds to enhance MSC chondrogenesis [59]. Composite scaffolds such as poly(vinyl alcohol)/polycaprolactone (PVA/PCL) scaffold seeded with MSC also showed better cartilage defects healing effect [12].

Some important electrospinning-based methods to fabricate biomimetic composite scaffolds for cartilage tissue engineering are summarized in Table 1. All of the papers cited in Table 1 are directly or possibly related to articular cartilage tissue engineering applications, such as tracheal cartilage engineering.

3.2. Gradient Scaffolds. Several methods of generating gradient scaffolds have been reported by researchers such as structure and composition gradient. He et al. first deposited a layer of aligned nanofibers of pure poly(L-lactide-co-glycolide) acid (PLGA) on a cylindrical surface and then deposited a layer of aligned nanofibers of PLGA/nanohydroxyapatite

(HA) on top. The microstructural organization and nano-HA content in electrospun scaffolds exhibit gradient change [33] and the anisotropic strain was also eliminated in scaffolds with fiber diameter gradient in Grey et al.'s research [30]. Guided spatial differentiation of MSCs [28] was also observed in composition and biomolecule gradient scaffolds. The scaffolds exhibit superior material properties in burst testing. The biodegradation rate and biological functions were controlled in Angarano et al.'s fiber diameter gradient scaffolds [31].

Fabricating scaffolds with gradients of molecules and element particles for tissue engineering has drawn substantial attention. Chondroitin sulfate (CS) and gentamycin sulfate (GS) gradients have been achieved with mineralized cartilage [18] and spatiotemporal drug release [35]. Ramalingam et al. reported a scaffold with amorphous calcium phosphate nanoparticles (nACP) gradient by coelectrospinning. Adhesion and proliferation of osteogenic cells were enhanced in gradient regions that have higher nACP concentration with a graded osteoblast response [29]. The graded cells' outgrowth rate and osteogenesis were also found to be positively related to the protein concentration gradient [27] and mineral gradient [34] on scaffolds, respectively.

Scaffolds with chemical and mechanical property gradient have also been investigated. Samavedi et al. demonstrated the possibility of fabricating a mechanochemical gradient scaffold with graded meshes by using controlled infusion of encapsulated $CaCO_3$ and TiO_2 nanoparticles [60]. The cell viability has shown preliminary evidence regarding the suitability of these scaffolds for *in vitro* tissue engineering.

Some important electrospinning-based methods to fabricate gradient scaffolds used in cartilage tissue engineering are summarized in Table 2. All of the papers cited in Table 2 are directly or possibly related to articular cartilage tissue engineering applications, such as interfacial tissue engineering and soft tissue regeneration.

3.3. 2D Scaffolds versus 3D Scaffolds. One of the major issues of conventional electrospun nanofibrous scaffolds is the limited thickness (typically < 1 mm) [61]. While 2D scaffolds are widely used in tissue engineering applications, they still fall short of contributing to crucial factors such as cell communication in ECM context, mechanical cues, and nutrient transportation [62]. 3D scaffolds, on the other hand, have superior cell differentiation and development rate, especially for connective tissues such as articular cartilage [53], and can improve cell infiltration rate with larger pore size and higher porosity [50]. The oriented electrospun fibers in 3D scale can guide the MSCs to grow in certain direction and further promote the formation of oriented and connective tissue. Further, in fabricating 3D multiphasic scaffolds where the scaffolds are separated spatially into different sections for guided MSCs differentiation mimicking the structure of connective tissues such as articular cartilage, electrospinning as a less complicated process for producing aligned fibers in 3D scale could demonstrate advantages in fabricating multiphasic scaffolds.

Among a multitude of applications for 3D organization of the nanofibers, processing the nanofibrous mesh into a desirable form after electrospinning has attracted substantial

TABLE 1: Summary of electrospinning-based techniques for fabrication of biomimetic composite scaffolds.

Authors	Year	Technique	Results	Application
Kontogiannopoulos et al. [11]	2011	Incorporated wound healing and antimicrobial agents	High drug entrapment efficiencies and multifunctional activities	Tissue engineering scaffolds
Shafiee et al. [12]	2011	Seed MSC in electrospun scaffold	Improved cartilage defects healing	Bone and cartilage tissue engineering
Xu et al. [13]	2012	Bioprinting and electrospinning	80% survived chondrocytes and cartilage-like tissue formation	Cartilage tissue engineering
He et al. [14]	2013	Electrospinning	Cartilage-like tissue formation	Cartilage tissue engineering
Liu et al. [15]	2014	Freeze-dried electrospinning	Successful regenerated osteochondral defects	Triphasic osteochondral implant
Man et al. [16]	2014	Electrospinning	Sustained rhTGF-β1 release and better chondrogenic differentiation	Cartilage tissue engineering
Zheng et al. [17]	2014	Electrospinning	3D cartilage regeneration	Cartilage tissue engineering
Mohan et al. [18]	2015	Electrospun fiber assembled hydrogel	Sustained release of chondroitin sulfate	Cartilage tissue engineering
Huang et al. [19]	2016	Electrospinning	Enhanced MSC chondrogenesis	Cartilage tissue engineering
Li et al. [20]	2016	Composite electrospinning	Better chondrocytes adhesion and filtration	Cartilage tissue engineering
Kalaithong et al. [21]	2016	Electrospinning and wet spinning	Better water absorption and cell infiltration	Cartilage tissue engineering
Sadeghi et al. [22]	2016	Electrospinning	Better hydrophilic property and cell attachment	Cartilage tissue engineering
Cao et al. [23]	2017	Chitosan/graphene oxide polymer nanofiber	Better biocompatibility and cell growth rate	Cartilage tissue engineering
Yin et al. [24]	2017	Core-shell structure nanofiber with embedded kartogenin solution	Promoted chondrogenic differentiation	Tracheal cartilage regeneration
Mirzaei et al. [25]	2017	Glucosamine incorporated into PLLA/PEG scaffolds	Enhanced cell proliferation rate	Cartilage tissue engineering
Wang et al. [26]	2017	Two-phase electrospinning	Prolonged drug release time and enhanced MSCs growth rate	Cartilage tissue engineering

interest. For example, a 3D scaffold can be created by stacking multiple layers of nanofiber mesh for cell seeding [48]. Garrigues et al. found that multilayer scaffolds enhanced cell infiltration aggrecan (ACAN) gene expression [46]. Yunos et al. designed a bilayered scaffold by combining electrospun fibers with bioglass and obtained viable chondrocyte cells in the scaffold [43]. Xue et al. prepared the electrospun gelatin/PCL membranes into rounded shape with seeded chondrocytes to culture an ear-shaped cartilage, with good elasticity and impressive mechanical strength [45]. Kai et al. incorporated electrospun fibers into hydrogel to improve mechanical properties such as compressive strength [39]. The advantage of this simple method is the high degree of flexibility in creating the ultimate geometry of scaffolds.

Modified electrospinning setups can also be utilized for 3D structure such as electrospinning with external heat source [38] and dual nozzle extrusion system to improve cell infiltration rate. He et al. increased the pore size of scaffolds by near-field electrospinning [63]. Levorson et al. also increased cellular infiltration with multiscale scaffold, which contains microfibers and nanofibers evenly distributed throughout the entire construct [44]. Human mesenchymal cells were found to maintain scaffold cellularity under serum-free conditions with these scaffolds. A fluffy 3D scaffold with randomly and evenly oriented fibers in all directions was also fabricated by utilizing the electrostatic repulsion between fibers; the cell proliferation rate observed was 5 times as fast as the 2D scaffold counterparts after one week [42].

Novel electrospinning solutions have also been investigated for 3D electrospinning. For example, Xu et al. started from a polymer solution with de-cross-linked keratin from chicken feathers to obtain intrinsic water stability for 3D scaffold [47]. Enhancement of cartilage repairing can also be achieved by scaffolds with cross-linked hyaluronic acid [51]

TABLE 2: Summary of electrospinning-based techniques for fabricating gradient scaffolds.

Authors	Year	Gradient parameter(s)	Technique	Results	Application
Zander et al. [27]	2012	Protein concentration gradient	Air-plasma-modified electrospinning	Corresponding cell outgrowth rate	Tissue engineering
Zhang et al. [28]	2012	Composition and biomolecule gradient	Microfluidic assisted electrospinning	Guided spatial cell differentiation	Tissue engineering
Ramalingam et al. [29]	2013	Composition gradient	Coelectrospinning	Enhanced osteogenic cells proliferation and adhesion	Interfacial tissue engineering
Grey et al. [30]	2013	Fiber diameter gradient	Gradient electrospinning	Enhanced mechanical properties	Tissue engineering
Angarano et al. [31]	2013	Fiber diameter gradient	Reactive electrospinning	Gradient biodegradation rate	Soft tissue regeneration
Sundararaghavan et al. [32]	2013	Growth factor gradient	Gradient electrospinning	Directed cell motility in gradient direction	Tissue engineering
He et al. [33]	2014	Structure and composition gradient	Modified coelectrospinning	Gradient cell metabolic activity	Tissue engineering
Liu et al. [34]	2014	Mineral gradient	Graded mineral coating	Graded mesenchymal stem cells osteogenesis	Interfacial tissue engineering
Mohan et al. [18]	2015	CS and BG gradient	Electrospun fiber assembled hydrogel	Glycosaminoglycan enriched and mineralized cartilage formation	Cartilage tissue engineering
Liu et al. [35]	2016	GS and deferoxamine	3D bioprinting and electrospinning	Spatiotemporal drug release	Osteochondral tissue engineering

and biological cue chondroitin sulfate-incorporated electrospinning [37]. Wei et al. combined poly[(butylene succinate)-co-adipate] (PBSA) and PLLA to culture primary human chondrocytes (PHCs) in electrospun 3D fibrous scaffolds for improved cell attachment and proliferation [40].

Another strategy for fabricating 3D scaffold is combining electrospinning with other advanced manufacturing techniques such as 3D printing and freeze-drying. In the research by Xu et al. about hybrid inkjet printing and electrospinning system, the cell viability was 80% after one week, and cartilage-like tissue has formed both in vitro and in vivo [13]. Afonso et al. combined another advanced manufacturing technique, direct writing, with electrospinning to obtain oriented fibers and a highly controlled scaffold structure [52]. Liu et al. enhanced the mechanical strength of electrospinning by freeze-drying, and osteochondral defects were repaired in a rabbit model [49].

Last but not least, 3D structure can be achieved with different postprocess treatments. Holmes et al. improved mechanical properties of the 3D scaffolds with hydrogen treated multiwalled carbon nanotubes (MWCNTs) [41]. Zhu et al.'s scaffolds enhanced chondrogenic differentiation and cell infiltration rate with cold atmospheric plasma treated electrospinning [50]. Chen and Su also demonstrated better viability, proliferation, and differentiation of rabbit articular chondrocytes, with plasma-treatment electrospun scaffold [36].

Some important electrospinning-based methods to construct 3D nanostructures for the scaffolds used in tissue

engineering are summarized in Table 3. All of the papers cited in Table 3 are directly or possibly related to articular cartilage tissue engineering applications, such as osteochondral tissue replacement and MSC chondrogenesis.

4. Discussion

Electrospinning provides an efficient way of fabricating cartilage tissue engineering scaffolds with nanoscale elements and utilization of a wide range of polymers. The process can be easily modified for various biomedical applications. Although there have been progress and improvements in recent years, very limited research has been reported on complete articular cartilage repair or major improvement in tissue engineering for comprehensive treatment of related diseases such as arthritis. There is hardly any parameter optimization for refined fiber strength and geometry. The complexity of electrospinning setups is growing rapidly as it incorporates more and more novel techniques such as biological cue-incorporated electrospinning [37], hydrogel [39], hybrid 3D printing [13], and cold atmospheric plasma treatment [50]. On one hand, it is highly challenging to deal with greater magnitude of parameters and characteristic architectures. On the other hand, these 3D scaffold strategies incorporating different techniques all come with certain weaknesses or vulnerabilities such as small thickness and microscale (instead of nanoscale) fibers (>200 nm). Therefore, the possibility of clinical application of these techniques on a large scale is still fairly slim.

TABLE 3: Summary of electrospinning-based techniques for fabricating 3D nanofibrous structure.

Authors	Year	Technique	Advantage(s)	Disadvantage(s)	Application
Chen and Su [36]	2011	Electrospinning with plasma treatment	Enhanced chondrocytes viability and proliferation	N/A	Cartilage tissue engineering
Coburn et al. [37]	2012	Biological cue chondroitin sulfate incorporated electrospinning	Enhanced cartilaginous formation	Weak mechanical properties	Cartilage tissue engineering
Shabani et al. [38]	2012	Modified setup of electrospinning with heat from halogen light bulbs	Improved cell infiltration rate	Material limitation	Tissue engineering
Kai et al. [39]	2012	Nanofiber with hydrogel	Relatively higher compressive strength	No significant cell proliferation improvement	Tissue regeneration
Xu et al. [13]	2012	Hybrid inkjet printing/electrospinning system	High cell viability, formed cartilage-like structure	Further refinement required	Cartilage tissue engineering
Wei et al. [40]	2012	Electrospinning	Improved cell attachment and proliferation	N/A	Cartilage tissue engineering
Holmes et al. [41]	2013	Hydrogen treated multiwalled carbon nanotubes (MWCNTs)	Higher mechanical strength and cell differentiation	Unclear effect of MWCNTs in vivo	MSC chondrogenesis
Cai et al. [42]	2013	Electrostatic repulsion	Randomly and evenly oriented 3D fibers	Rapid delivery of electrons on fibers required	Cell culture for soft tissues
Yunos et al. [43]	2013	Bilayered scaffold	Chondrocyte cell-supporting ability	Decreased HA formation rate with thicker layer	Osteochondral tissue replacement
Levorson et al. [44]	2013	Dual extrusion electrospinning	Maintained scaffold cellularity	Lack of parameter optimization	Cartilage tissue engineering
Xue et al. [45]	2014	Prepare the electrospun membrane in rounded shape	Formed ear-shaped cartilage tissue	Lack of immunogenicity investigation	Cartilage tissue engineering
Garrigues et al. [46]	2014	Electrospinning	Enhanced cell infiltration	Lower elastic modulus	Cartilage tissue engineering
Xu et al. [47]	2014	Electrospinning solution with de-cross-linked keratin from chicken feathers	Intrinsic water stability	Randomly oriented fibers	Cell penetration and differentiation
Orr et al. [48]	2015	Vertical stacking layers of fiber membrane	Easy to seed cells on surface prior to stacking	Cells unable to penetrate through layers	Compressive loading applications
Liu et al. [49]	2015	Electrospinning and freeze drying	Better mechanical strength	N/A	Cartilage tissue engineering
Zhu et al. [50]	2015	Electrospinning with cold atmospheric plasma treatment	Enhanced chondrogenic differentiation and cell infiltration	Small thickness for 3D scaffold	Cartilage tissue engineering
Chen et al. [51]	2016	Modified scaffold with cross-linked hyaluronic acid	Superabsorbent property and excellent cytocompatibility	Complicated fabrication process	Cartilage tissue engineering
Afonso et al. [52]	2016	Direct writing electrospinning	Directed tissue organization and fibril matrix orientation	Microscale fibers	Tissue engineering
Damaraju et al. [53]	2017	Piezoelectric fibrous scaffolds	Promoted mesenchymal stem cell differentiation	N/A	Cartilage and bone tissue engineering

Several electrospinning strategies have been developed for fabricating biomimetic composite scaffolds. Along with antimicrobial agents and growth factors, both structure and signal biomimetic scaffolds produced exhibit suitable biocompatibility and intermediate level of biomimicry. However, the effects of fiber-guided cell orientation on chondrogenic differentiation, attachment, and mineralization of marrow mesenchymal stem cells need to be further investigated and so should optimization for cell culture and electrospinning process.

Various improvements in mechanical properties of scaffolds, such as microstructural organization [33], tensile strength [60, 64], and anisotropic strain [30], have been achieved with different gradient properties. However, all of the improved mechanical properties, such as tensile strength, are still in the undesired low range (<10 MPa) [13]. And there are no quantitative results for the effect of enhanced mechanical properties on cell proliferation rate. In some cases, the overall mechanical strength was found to be even lower than that of the control scaffolds with no gradient. To address this issue, 3D electrospun fibers were immersed into a medium, such as hydrogel, to improve their mechanical strengths [39]; however, the electrospun fibers dissolve much faster than hydrogel in cell culture. Also, hardly any research has addressed the issue of mechanical properties of 3D scaffolds. These issues need to have experimental justification to verify that the improvement of these mechanical properties is helpful for *in vitro* and *in vivo* cell proliferation.

Using different gradient scaffolds, researchers were able to create graded cell response [29], outgrowth rate [27], and biodegradation rate [31] in a single scaffold. *In vivo* studies are needed to investigate the effects and suitability of these gradient scaffolds employed in cartilage implants, as well as related interfacial tissue engineering.

Although electrospun scaffolds possess higher porosity than other forms of scaffolds, a common drawback is the limited size of pores, which further prevents cells from growing in the scaffold [65]. Several measures have been investigated to address this issue, such as using sacrificial fibers [66], multiscale scaffold [44], hybrid scaffold [67], near-field electrospinning [63], and salt leaching and ice crystals [20]. These techniques could also be incorporated in 3D scaffolds fabrication for better cell infiltration.

The positive effects of 3D scaffolds on cell infiltration rate [44, 63] and cell proliferation rate [42] have been extensively investigated and verified. Other crucial properties in cell culture such as water stability [47] and cartilage-like tissue formation [37, 51] have also been investigated, demonstrating the intrinsic advantages of 3D scaffolds over 2D scaffolds. The biodegradation rate and biomimicry level should be further characterized quantitatively before 3D scaffolds can be implemented as cartilage implants *in vivo*.

Electrospun scaffolds for cartilage tissue engineering still face major challenges in biomechanical and biological properties. More novel synthetic and artificial materials need to be explored and tested with regard to issues such as mechanical strength, biotoxicity, biocompatibility, and biodegradability before research focus can switch to large-scale applications. Fabrication procedures need to be modified and improved in terms of facility and process parameters for better biomimicry and cost effectiveness. After further optimization of biomimetic, gradient, and 3D electrospun scaffolds along with the incorporation of advanced manufacturing techniques and novel materials, electrospun scaffolds exhibit substantial potential for cartilage tissue engineering.

5. Conclusion

This paper presents a variety of electrospinning strategies with novel materials and manufacturing techniques of bioscaffolds for cartilage tissue engineering. The literature review reveals that the most prevailing recent trend is the fabrication of biomimetic, gradient, and 3D bioscaffolds. Natural polymeric and biological materials have been incorporated into electrospinning technique with synthetic materials to improve biomimicry and biocompatibility. Multiple approaches have been taken in electrospinning to form 3D scaffolds with cartilage-like structure and enhanced chondrogenic differentiation. Mechanical properties of scaffolds have been improved by various electrospinning strategies to achieve acceptable levels. Future research should focus on issues such as mechanical strength, biotoxicity, and biodegradability *in vivo* to pave the way for possible human trials.

Conflicts of Interest

The authors declare that there are no conflicts of interest.

References

[1] J. M. Holzwarth and P. X. Ma, "Biomimetic nanofibrous scaffolds for bone tissue engineering," *Biomaterials*, vol. 32, no. 36, pp. 9622–9629, 2011.

[2] B. P. Chan and K. W. Leong, "Scaffolding in tissue engineering: general approaches and tissue-specific considerations," *European Spine Journal*, vol. 17, no. 4, pp. S467–S479, 2008.

[3] F. J. O'Brein, "Biomaterials & scaffolds for tissue engineering," *Materials Today*, vol. 14, no. 3, pp. 88–95, 2011.

[4] E. Carletti, A. Motta, and C. Migliaresi, "Scaffolds for tissue engineering and 3D cell culture.," *Methods in Molecular Biology (Clifton, N.J.)*, vol. 695, pp. 17–39, 2011.

[5] Z. Izadifar, X. Chen, and W. Kulyk, "Strategic design and fabrication of engineered scaffolds for articular cartilage repair," *Journal of Functional Biomaterials*, vol. 3, no. 4, pp. 799–838, 2012.

[6] Q. P. Pham, U. Sharma, and A. G. Mikos, "Electrospinning of polymeric nanofibers for tissue engineering applications: a review," *Tissue Engineering Part A*, vol. 12, no. 5, pp. 1197–1211, 2006.

[7] N. Hamann, F. Zaucke, J. Heilig, K. D. Oberländer, G.-P. Brüggemann, and A. Niehoff, "Effect of different running modes on the morphological, biochemical, and mechanical properties of articular cartilage," *Scandinavian Journal of Medicine & Science in Sports*, vol. 24, no. 1, pp. 179–188, 2014.

[8] A. M. Haleem and C. R. Chu, "Advances in tissue engineering techniques for articular cartilage repair," *Operative Techniques in Orthopaedics*, vol. 20, no. 2, pp. 76–89, 2010.

[9] A. Collins, J. T. Blackburn, C. Olcott, B. Yu, and P. Weinhold, "The impact of stochastic resonance electrical stimulation and knee sleeve on impulsive loading and muscle co-contraction during gait in knee osteoarthritis," *Clinical Biomechanics*, vol. 26, no. 8, pp. 853–858, 2011.

[10] K. Huétink, B. C. Stoel, I. Watt et al., "Identification of factors associated with the development of knee osteoarthritis in a young to middle-aged cohort of patients with knee complaints," *Clinical Rheumatology*, vol. 34, no. 10, pp. 1769–1779, 2015.

[11] K. N. Kontogiannopoulos, A. N. Assimopoulou, I. Tsivintzelis, C. Panayiotou, and V. P. Papageorgiou, "Electrospun fiber mats containing shikonin and derivatives with potential biomedical applications," *International Journal of Pharmaceutics*, vol. 409, no. 1-2, pp. 216–228, 2011.

[12] A. Shafiee, M. Soleimani, G. A. Chamheidari et al., "Electrospun nanofiber-based regeneration of cartilage enhanced by mesenchymal stem cells," *Journal of Biomedical Materials Research Part A*, vol. 99, no. 3, pp. 467–478, 2011.

[13] T. Xu, K. W. Binder, M. Z. Albanna et al., "Hybrid printing of mechanically and biologically improved constructs for cartilage tissue engineering applications," *Biofabrication*, vol. 5, no. 1, Article ID 015001, 2013.

[14] X. He, W. Fu, B. Feng et al., "Electrospun collagen-poly(L-lactic acid-co-ε-caprolactone) membranes for cartilage tissue engineering," *Journal of Regenerative Medicine*, vol. 8, no. 4, pp. 425–436, 2013.

[15] X. Liu, S. Liu, S. Liu, and W. Cui, "Evaluation of oriented electrospun fibers for periosteal flap regeneration in biomimetic triphasic osteochondral implant," *Journal of Biomedical Materials Research Part B: Applied Biomaterials*, vol. 102, no. 7, pp. 1407–1414, 2014.

[16] Z. Man, L. Yin, Z. Shao et al., "The effects of co-delivery of BMSC-affinity peptide and rhTGF-β1 from coaxial electrospun scaffolds on chondrogenic differentiation," *Biomaterials*, vol. 35, no. 19, pp. 5250–5260, 2014.

[17] R. Zheng, H. Duan, J. Xue et al., "The influence of Gelatin/PCL ratio and 3-D construct shape of electrospun membranes on cartilage regeneration," *Biomaterials*, vol. 35, no. 1, pp. 152–164, 2014.

[18] N. Mohan, J. Wilson, D. Joseph, D. Vaikkath, and P. D. Nair, "Biomimetic fiber assembled gradient hydrogel to engineer glycosaminoglycan enriched and mineralized cartilage: An in vitro study," *Journal of Biomedical Materials Research Part A*, vol. 103, no. 12, pp. 3896–3906, 2015.

[19] G. P. Huang, A. Molina, N. Tran, G. Collins, and T. L. Arinzeh, "Investigating cellulose derived glycosaminoglycan mimetic scaffolds for cartilage tissue engineering applications," *Journal of Tissue Engineering and Regenerative Medicine*, 2017.

[20] Z. Li, P. Liu, T. Yang et al., "Composite poly(l-lactic-acid)/silk fibroin scaffold prepared by electrospinning promotes chondro genesis for cartilage tissue engineering," *Journal of Biomaterials Applications*, vol. 30, no. 10, pp. 1552–1565, 2015.

[21] W. Kalaithong, R. Molloy, T. Theerathanagorn, and W. Janvikul, "Novel poly(l-lactide-co-caprolactone)/gelatin porous scaffolds for use in articular cartilage tissue engineering: Comparison of electrospinning and wet spinning processing methods," *Polymer Engineering & Science*, 2016.

[22] D. Sadeghi, S. Karbasi, S. Razavi, S. Mohammadi, M. A. Shokrgozar, and S. Bonakdar, "Electrospun poly(hydroxybutyrate)/chitosan blend fibrous scaffolds for cartilage tissue engineering,"

Journal of Applied Polymer Science, vol. 133, no. 47, Article ID 44171, 2016.

[23] L. Cao, F. Zhang, Q. Wang, and X. Wu, "Fabrication of chitosan/graphene oxide polymer nanofiber and its biocompatibility for cartilage tissue engineering," *Materials Science and Engineering C: Materials for Biological Applications*, vol. 79, pp. 697–701, 2017.

[24] H. Yin, J. Wang, Z. Gu et al., "Evaluation of the potential of kartogenin encapsulated poly(L-lactic acid-co-caprolactone)/collagen nanofibers for tracheal cartilage regeneration," *Journal of Biomaterials Applications*, vol. 32, no. 3, pp. 331–341, 2017.

[25] S. Mirzaei, A. Karkhaneh, M. Soleimani, A. Ardeshirylajimi, H. Seyyed Zonouzi, and H. Hanaee-Ahvaz, "Enhanced chondrogenic differentiation of stem cells using an optimized electrospun nanofibrous PLLA/PEG scaffolds loaded with glucosamine," *Journal of Biomedical Materials Research Part A*, vol. 105, no. 9, pp. 2461–2474, 2017.

[26] C. Wang, W. Hou, X. Guo et al., "Two-phase electrospinning to incorporate growth factors loaded chitosan nanoparticles into electrospun fibrous scaffolds for bioactivity retention and cartilage regeneration," *Materials Science and Engineering C: Materials for Biological Applications*, vol. 79, pp. 507–515, 2017.

[27] N. E. Zander, J. A. Orlicki, A. M. Rawlett, and T. P. Beebe, "Quantification of protein incorporated into electrospun poly-caprolactone tissue engineering scaffolds," *ACS Applied Materials & Interfaces*, vol. 4, no. 4, pp. 2074–2081, 2012.

[28] X. Zhang, X. Gao, L. Jiang, and J. Qin, "Flexible generation of gradient electrospinning nanofibers using a microfluidic assisted approach," *Langmuir*, vol. 28, no. 26, pp. 10026–10032, 2012.

[29] M. Ramalingam, M. F. Young, V. Thomas et al., "Nanofiber scaffold gradients for interfacial tissue engineering," *Journal of Biomaterials Applications*, vol. 27, no. 6, pp. 695–705, 2013.

[30] C. P. Grey, S. T. Newton, G. L. Bowlin, T. W. Haas, and D. G. Simpson, "Gradient fiber electrospinning of layered scaffolds using controlled transitions in fiber diameter," *Biomaterials*, vol. 34, no. 21, pp. 4993–5006, 2013.

[31] M. Angarano, S. Schulz, M. Fabritius et al., "Layered gradient nonwovens of in situ crosslinked electrospun collagenous nanofibers used as modular scaffold systems for soft tissue regeneration," *Advanced Functional Materials*, vol. 23, no. 26, pp. 3277–3285, 2013.

[32] H. G. Sundararaghavan, R. L. Saunders, D. A. Hammer, and J. A. Burdick, "Fiber alignment directs cell motility over chemotactic gradients," *Biotechnology and Bioengineering*, vol. 110, no. 4, pp. 1249–1254, 2013.

[33] J. He, T. Qin, Y. Liu, X. Li, D. Li, and Z. Jin, "Electrospinning of nanofibrous scaffolds with continuous structure and material gradients," *Materials Letters*, vol. 137, pp. 393–397, 2014.

[34] W. Liu, J. Lipner, J. Xie, C. N. Manning, S. Thomopoulos, and Y. Xia, "Nanofiber scaffolds with gradients in mineral content for spatial control of osteogenesis," *ACS Applied Materials & Interfaces*, vol. 6, no. 4, pp. 2842–2849, 2014.

[35] Y.-Y. Liu, H.-C. Yu, Y. Liu, G. Liang, T. Zhang, and Q.-X. Hu, "Dual drug spatiotemporal release from functional gradient scaffolds prepared using 3D bioprinting and electrospinning," *Polymer Engineering & Science*, vol. 56, no. 2, pp. 170–177, 2016.

[36] J.-P. Chen and C.-H. Su, "Surface modification of electrospun PLLA nanofibers by plasma treatment and cationized gelatin immobilization for cartilage tissue engineering," *Acta Biomaterialia*, vol. 7, no. 1, pp. 234–243, 2011.

[37] J. M. Coburn, M. Gibson, S. Monagle, Z. Patterson, and J. H. Elisseeff, "Bioinspired nanofibers support chondrogenesis for articular cartilage repair," *Proceedings of the National Acadamy of Sciences of the United States of America*, vol. 109, no. 25, pp. 10012–10017, 2012.

[38] I. Shabani, V. Haddadi-Asl, E. Seyedjafari, and M. Soleimani, "Cellular infiltration on nanofibrous scaffolds using a modified electrospinning technique," *Biochemical and Biophysical Research Communications*, vol. 423, no. 1, pp. 50–54, 2012.

[39] D. Kai, M. P. Prabhakaran, B. Stahl, M. Eblenkamp, E. Wintermantel, and S. Ramakrishna, "Mechanical properties and *in vitro* behavior of nanofiber–hydrogel composites for tissue engineering applications," *Nanotechnology*, vol. 23, no. 9, Article ID 095705, 2012.

[40] J.-D. Wei, H. Tseng, E. T.-H. Chen et al., "Characterizations of chondrocyte attachment and proliferation on electrospun biodegradable scaffolds of PLLA and PBSA for use in cartilage tissue engineering," *Journal of Biomaterials Applications*, vol. 26, no. 8, pp. 963–985, 2012.

[41] B. Holmes, N. J. Castro, J. Li, M. Keidar, and L. G. Zhang, "Enhanced human bone marrow mesenchymal stem cell functions in novel 3D cartilage scaffolds with hydrogen treated multi-walled carbon nanotubes," *Nanotechnology*, vol. 24, no. 36, Article ID 365102, 2013.

[42] S. Cai, H. Xu, Q. Jiang, and Y. Yang, "Novel 3D electrospun scaffolds with fibers oriented randomly and evenly in three dimensions to closely mimic the unique architectures of extracellular matrices in soft tissues: Fabrication and mechanism study," *Langmuir*, vol. 29, no. 7, pp. 2311–2318, 2013.

[43] D. M. Yunos, Z. Ahmad, V. Salih, and A. R. Boccaccini, "Stratified scaffolds for osteochondral tissue engineering applications: Electrospun PDLLA nanofibre coated Bioglass®-derived foams," *Journal of Biomaterials Applications*, vol. 27, no. 5, pp. 537–551, 2013.

[44] E. J. Levorson, P. R. Sreerekha, K. P. Chennazhi, F. K. Kasper, S. V. Nair, and A. G. Mikos, "Fabrication and characterization of multiscale electrospun scaffolds for cartilage regeneration," *Biomedical Materials*, vol. 8, no. 1, Article ID 014103, 2013.

[45] J. Xue, B. Feng, R. Zheng et al., "Engineering ear-shaped cartilage using electrospun fibrous membranes of gelatin/polycaprolactone," *Biomaterials*, vol. 34, no. 11, pp. 2624–2631, 2013.

[46] N. W. Garrigues, D. Little, J. Sanchez-Adams, D. S. Ruch, and F. Guilak, "Electrospun cartilage-derived matrix scaffolds for cartilage tissue engineering," *Journal of Biomedical Materials Research Part A*, vol. 102, no. 11, pp. 3998–4008, 2014.

[47] H. Xu, S. Cai, L. Xu, and Y. Yang, "Water-stable three-dimensional ultrafine fibrous scaffolds from keratin for cartilage tissue engineering," *Langmuir*, vol. 30, no. 28, pp. 8461–8470, 2014.

[48] S. B. Orr, A. Chainani, K. J. Hippensteel et al., "Aligned multilayered electrospun scaffolds for rotator cuff tendon tissue engineering," *Acta Biomaterialia*, vol. 24, article no. 3738, pp. 117–126, 2015.

[49] S. Liu, J. Wu, X. Liu et al., "Osteochondral regeneration using an oriented nanofiber yarn-collagen type I/hyaluronate hybrid/TCP biphasic scaffold," *Journal of Biomedical Materials Research Part A*, vol. 103, no. 2, pp. 581–592, 2015.

[50] W. Zhu, N. J. Castro, X. Cheng, M. Keidar, and L. G. Zhang, "Cold atmospheric plasma modified electrospun scaffolds with embedded microspheres for improved cartilage regeneration," *PLoS ONE*, vol. 10, no. 7, Article ID e0134729, 2015.

[51] W. Chen, S. Chen, Y. Morsi et al., "Superabsorbent 3D Scaffold Based on Electrospun Nanofibers for Cartilage Tissue Engineering," *ACS Applied Materials & Interfaces*, vol. 8, no. 37, pp. 24415–24425, 2016.

[52] M. Afonso, C. Honglin, W. Paul, M. Carlos, T. Roman, and M. Lorenzo, "Direct writing electrospinning of scaffolds with multi-dimensional fiber architecture for hierarchical tissue engineering," *Frontiers in Bioengineering and Biotechnology*, vol. 4, 2016.

[53] S. M. Damaraju, Y. Shen, E. Elele et al., "Three-dimensional piezoelectric fibrous scaffolds selectively promote mesenchymal stem cell differentiation," *Biomaterials*, vol. 149, pp. 51–62, 2017.

[54] W. Cui, Y. Zhou, and J. Chang, "Electrospun nanofibrous materials for tissue engineering and drug delivery," *Science and Technology of Advanced Materials*, vol. 11, no. 1, Article ID 014108, 2010.

[55] N. G. Rim, C. S. Shin, and H. Shin, "Current approaches to electrospun nanofibers for tissue engineering," *Biomedical Materials*, vol. 8, no. 1, Article ID 014102, 2013.

[56] T. Jiang, E. J. Carbone, K. W.-H. Lo, and C. T. Laurencin, "Electrospinning of polymer nanofibers for tissue regeneration," *Progress in Polymer Science*, vol. 46, pp. 1–24, 2014.

[57] B. Holmes, N. J. Castro, L. G. Zhang, and E. Zussman, "Electrospun fibrous scaffolds for bone and cartilage tissue generation: Recent progress and future developments," *Tissue Engineering - Part B: Reviews*, vol. 18, no. 6, pp. 478–486, 2012.

[58] X. He, B. Feng, C. Huang et al., "Electrospun gelatin/polycaprolactone nanofibrous membranes combined with a coculture of bone marrow stromal cells and chondrocytes for cartilage engineering," *International Journal of Nanomedicine*, vol. 10, pp. 2089–2099, 2015.

[59] J. Liao, X. Guo, K. J. Grande-Allen, F. K. Kasper, and A. G. Mikos, "Bioactive polymer/extracellular matrix scaffolds fabricated with a flow perfusion bioreactor for cartilage tissue engineering," *Biomaterials*, vol. 31, no. 34, pp. 8911–8920, 2010.

[60] S. Samavedi, C. Olsen Horton, S. A. Guelcher, A. S. Goldstein, and A. R. Whittington, "Fabrication of a model continuously graded co-electrospun mesh for regeneration of the ligament-bone interface," *Acta Biomaterialia*, vol. 7, no. 12, pp. 4131–4138, 2011.

[61] M. B. Fisher, E. A. Henning, N. Söegaard, J. L. Esterhai, and R. L. Mauck, "Organized nanofibrous scaffolds that mimic the macroscopic and microscopic architecture of the knee meniscus," *Acta Biomaterialia*, vol. 9, no. 1, pp. 4496–4504, 2013.

[62] T. E. Rinker and J. S. Temenoff, "Micro- and nanotechnology engineering strategies for tissue interface regeneration and repair," *Tissue and Organ Regeneration: Advances in Micro- and Nanotechnology*, pp. 105–155, 2014.

[63] F. He, D. Li, J. He et al., "A novel layer-structured scaffold with large pore sizes suitable for 3D cell culture prepared by near-field electrospinning," *Materials Science and Engineering C: Materials for Biological Applications*, vol. 86, pp. 18–27, 2018.

[64] M. Singh, N. Dormer, J. R. Salash et al., "Three-dimensional macroscopic scaffolds with a gradient in stiffness for functional regeneration of interfacial tissues," *Journal of Biomedical Materials Research Part A*, vol. 94, no. 3, pp. 870–876, 2010.

[65] M. M. Nava, L. Draghi, C. Giordano, and R. Pietrabissa, "The effect of scaffold pore size in cartilage tissue engineering," *Journal of Applied Biomaterials and Functional Materials*, vol. 14, no. 3, pp. e223–e229, 2016.

[66] B. M. Baker, R. P. Shah, A. M. Silverstein, J. L. Esterhai, J. A. Burdick, and R. L. Mauck, "Sacrificial nanofibrous composites provide instruction without impediment and enable functional tissue formation," *Proceedings of the National Acadamy of Sciences of the United States of America*, vol. 109, no. 35, pp. 14176–14181, 2012.

[67] S. Karbasi, F. Fekrat, D. Semnani, S. Razavi, and E. Zargar, "Evaluation of structural and mechanical properties of electrospun nano-micro hybrid of poly hydroxybutyrate-chitosan/silk scaffold for cartilage tissue engineering," *Advanced Biomedical Research*, vol. 5, no. 1, p. 180, 2016.

17

Oxidative Nanopatterning of Titanium Surface Influences mRNA and MicroRNA Expression in Human Alveolar Bone Osteoblastic Cells

Maidy Rehder Wimmers Ferreira,[1] Roger Rodrigo Fernandes,[1] Amanda Freire Assis,[2] Janaína A. Dernowsek,[2] Geraldo A. Passos,[1,2] Fabio Variola,[3] and Karina Fittipaldi Bombonato-Prado[1]

[1]*Cell Culture Laboratory, Department of Morphology, Physiology and Basic Pathology, School of Dentistry of Ribeirão Preto, University of São Paulo, 14040-904 Ribeirão Preto, SP, Brazil*
[2]*Molecular Immunogenetics Group, Department of Genetics, Ribeirão Preto Medical School, University of São Paulo, 14049-900 Ribeirão Preto, SP, Brazil*
[3]*Faculty of Engineering, Department of Mechanical Engineering, University of Ottawa, Ottawa, ON, Canada K1N 6N5*

Correspondence should be addressed to Karina Fittipaldi Bombonato-Prado; karina@forp.usp.br

Academic Editor: Esmaiel Jabbari

Titanium implants have been extensively used in orthopedic and dental applications. It is well known that micro- and nanoscale surface features of biomaterials affect cellular events that control implant-host tissue interactions. To improve our understanding of how multiscale surface features affect cell behavior, we used microarrays to evaluate the transcriptional profile of osteoblastic cells from human alveolar bone cultured on engineered titanium surfaces, exhibiting the following topographies: nanotexture (N), nano+submicrotexture (NS), and rough microtexture (MR), obtained by modulating experimental parameters (temperature and solution composition) of a simple yet efficient chemical treatment with a H_2SO_4/H_2O_2 solution. Biochemical assays showed that cell culture proliferation augmented after 10 days, and cell viability increased gradually over 14 days. Among the treated surfaces, we observed an increase of alkaline phosphatase activity as a function of the surface texture, with higher activity shown by cells adhering onto nanotextured surfaces. Nevertheless, the rough microtexture group showed higher amounts of calcium than nanotextured group. Microarray data showed differential expression of 716 mRNAs and 32 microRNAs with functions associated with osteogenesis. Results suggest that oxidative nanopatterning of titanium surfaces induces changes in the metabolism of osteoblastic cells and contribute to the explanation of the mechanisms that control cell responses to micro- and nanoengineered surfaces.

1. Introduction

Over the last three decades, orthopedics and oral and maxillofacial surgery have used titanium as the metallic material of choice because of its excellent biocompatibility, mainly associated with (1) elastic modulus similar to that of bone, (2) excellent corrosion resistance due to a superficial TiO_2 layer, and (3) biological inertness *in vivo* [1]. These advantages have boosted the application of titanium, ranging from femoral stems to prosthetic devices to replace dental elements [2].

However, specific physiological aspects such as implantation site, blood supply, and quality and quantity of the surrounding bone tissue can interfere with the osseointegration process, ultimately determining the success rate of an implant [3]. In addition to these factors, the metal physicochemical properties (e.g., topography, roughness, chemical composition, and wettability) at various scales will also contribute to the determination of the outcome of the osseointegration process by affecting the cellular and extracellular events that occur during implant-host tissue interactions [4].

The abilities to promote the interactions with adjacent tissues and to elicit the biological response by guiding specific cellular processes along predetermined routes are fundamental characteristics that the next generation of biomaterials should possess [5]. It is now widely accepted that the rational design of surface topography at the micro- and nanoscale is a powerful tool to control and guide cellular response [6]. The topography of a surface can in fact influence cellular response from surrounding tissues by modifying cell adhesion and migration, proliferation, and collagen synthesis at the material-host tissue interface [7]. Similarly, surface chemistry is another key parameter that plays a fundamental role in peri-implant bone apposition [8].

Numerous techniques have been developed to engineer titanium surfaces in ways to promote bone cell growth and ultimately implant fixation. Several studies have shown how different types of titanium surface treatment affect these processes and highlighted how micro- and nanopatterned surfaces exert a differential influence on bone formation and cell behavior obtained from tissues adjacent to the implant surfaces [7]. In this context, cell cultures are a useful tool, because they allow investigation into how cells and matrices interact with the titanium surface [9]. Currently, the investigation of gene expression patterns is increasingly gaining interest, aiming at unveiling the functional roles of genes and enabling new approaches in cell therapies [10]. Tools such as microarrays can now be used to identify gene modulation in cells that are in contact with biomaterials, as reported by Bombonato-Prado et al. [11]. Microarrays can ultimately help to identify differentially regulated genes in osteoblasts exposed to different biomaterials used in bone regeneration/substitution procedures.

The present study relied on biochemical assays and gene expression to evaluate differences in the cellular response of human alveolar bone cells cultured on different titanium surfaces. Our results showed that nanoporous titanium surfaces generated by oxidative nanopatterning influence alveolar bone cells behavior and, distinctively from previous studies, there were investigated differences in the expression of mRNAs and microRNAs of such cells in contact with the distinct topographies.

2. Materials and Methods

2.1. Titanium Surfaces Preparation. Commercially pure grade 2 titanium (Ti) discs, with diameter of 13 mm and thickness of 2 mm, were polished with an Exakt 400 CS machine equipped with 320, 500, 800, 2500, and 4000 grits (Exakt Advanced Technologies, Germany) and successively polished with felt and abrasive particles of alumina paste (Al_2O_3) (0.05 mM). The titanium discs were sonicated in Extran® MA 02 (Merck Millipore, USA) 2% diluted in deionized water, followed by alcohol 70% and deionized water for 30 minutes each. Next, simple yet efficient chemical etching based on a mixture of sulfuric acid (H_2SO_4 at 36 N) and hydrogen peroxide ($H_2O_{2(aq)}$) was applied at varying relative concentrations of the acid and the peroxide, at different temperatures. This procedure afforded three different types of titanium surface,

as described in a previous article [5]. Application of a fully programmable digital hot plate (EchoTherm™ HS40, Torrey Pines Scientific, USA) with automatic feedback ensured temperature control. To obtain the nanotextured surface (N), the titanium discs were submitted to etching with 50 : 50 H_2SO_4 (36 N)/30% aqueous hydrogen peroxide ($H_2O_{2(aq)}$) at 25°C. To achieve the nano+submicrotextured surface (NS), the titanium discs were treated with 50 : 50 H_2SO_4 (36 N)/30% aqueous hydrogen peroxide ($H_2O_{2(aq)}$) at 50°C. Treatment of the titanium discs with 30% aqueous hydrogen peroxide ($H_2O_{2(aq)}$) solution alone at 50°C was used to obtain the rough microtextured (MR) surface. Untreated polished titanium discs served as control (C). Before experiments, treated and untreated (control) titanium discs were rinsed with deionized H_2O, autoclaved, and air-dried. To confirm the presence of different surface topographies, the titanium discs were examined under a field emission scanning electron microscope (Zeiss LEO 440, Cambridge, England) operated at 15 kV.

2.2. Cell Culture. Human alveolar bone fragments were obtained from healthy adult donors with their informed consent, using the research protocols approved by the Committee of Ethics in Research of the School of Dentistry of the University of São Paulo (approval number 2011.1.1015.58.6). The osteoblastic cells were kept in culture flasks until subconfluence and then seeded over titanium discs in 24-well culture plates at a concentration of 2×10^4 cells/well. The growth medium consisted of alpha-minimum essential medium (α-MEM; Invitrogen-Life Technologies, Grand Island, NY) supplemented with 10% fetal calf serum (Gibco-Life Technologies), gentamicin (Gibco) at 50 mg/mL, and fungizone (Gibco) at 0.3 mg/mL, added with ascorbic acid (Gibco-Life Technologies) at 5 mg/mL, β-glycerophosphate (Sigma-Aldrich, St. Louis, MO) at 7 mM, and dexamethasone (Sigma-Aldrich) at 10^{-7} M. The cell cultures were kept at 37°C under humidified atmosphere containing 5% CO_2 and 95% air. The culture medium was changed three times a week.

2.3. Cell Viability. Cell viability was assessed by MTT assay (3-[4,5-dimethylthiazol-2-yl]-2,5-diphenyltetrazolium bromide) 7, 10, and 14 days after the start of the culture. To this end, cells were incubated with 10% MTT (5 mg/mL) in culture medium at 37°C for 4 hours. The medium was then aspirated from the well, and 1 mL of isopropanol (0.04 N HCl in isopropanol) was added to each well. The plates were placed on a shaker for 5 minutes and 200 μL of this solution was transferred to a 96-well plate. The optical density was read at 570 nm (μQuant, BioTek Instruments, Winooski, VT, USA).

2.4. Alkaline Phosphatase Assay. Alkaline phosphatase activity was assayed as the release of thymolphthalein from thymolphthalein monophosphate; a commercial kit (Labtest Diagnóstica, MG, Brazil) was employed for this purpose. Briefly, 50 mL of thymolphthalein monophosphate was mixed with 0.5 mL of diethanolamine buffer (0.3 μmol/mL, pH 10.1), and the resulting solution was kept at 37°C, for 2 min. After that, 50 mL of the lysate was added to each of the wells, which were kept at 37°C, for 10 min. Then, 2 mL of a solution containing Na_2CO_3 (0.09 μmol/mL) and NaOH (0.25 μmol/mL) was added for color development. After 30 min, the absorbance was measured at 590 nm. The alkaline phosphatase activity was calculated from a standard curve with thymolphthalein concentrations ranging from 0.012 to 0.4 μmol of thymolphthalein/h/mL. Data are expressed as the alkaline phosphatase activity normalized by the total protein assay.

2.5. Mineralized Matrix Formation. Mineralized matrix formation was detected at day 21 by means of Alizarin Red S (Sigma-Aldrich) staining for areas rich in calcium. Attached cells were fixed in 10% formalin at 4°C for 2 h followed by one-hour immersion in alcohol for each increasing concentration (30%, 50%, 70%, and 100%). The next step was staining with 2% Alizarin Red S, pH 4.2, for 10 min. The calcium content was evaluated with a colorimetric method formerly described [12]. All biochemical data were compared by the Kruskal-Wallis and Mann-Whitney test. SPSS 17.0 statistical software was used, and differences with $p \leq 0.05$ were considered statistically significant.

2.6. Total RNA Extraction. After 10 days, the total RNA of each culture was extracted with mirVana total RNA isolation kit® (Ambion, NY, USA), according to the manufacturer's instruction. UV spectrophotometry confirmed that the RNA preparations were free of proteins and phenol. RNA degradation was assessed by microfluidic electrophoresis with Agilent 6000 RNA Nano chips, conducted on an Agilent 2100 Bioanalyzer (Agilent Technologies, Santa Clara, CA, USA). Only RNA samples free of proteins and phenol and featuring an RNA Integrity Number (RIN) $\geq$ 9.0 were used.

2.7. Microarray Hybridization

2.7.1. mRNA Expression Profiling with Agilent 4 × 44 K Human Oligoarrays. Changes in gene expression were evaluated with the Agilent one-color (Cy3 fluorochrome) microarray-based gene expression platform according to the manufacturer's instructions. Briefly, 500 ng of the individual total RNA was employed to synthesize double-stranded cDNA and cyanine 3 (Cy3) CTP labeled complementary amplified RNA (cRNA) by means of the Agilent Linear Amplification Kit (Agilent), according to the manufacturer's instructions. By using Agilent human 4 × 44 K oligonucleotide microarrays (Agilent), cyanine-labeled complementary RNA was hybridized to microarrays in SureHyb chambers (Agilent) in a rotator oven at 65°C, for 17 h. Each array contained 44,000 oligonucleotide probes covering the entire human functional genome. The arrays were washed according to the manufacturer's instructions and scanned with an Agilent DNA Microarray scanner.

2.7.2. miRNA Expression Profiling with Agilent 8 × 15 K Mouse Oligoarrays. Briefly, the total RNA samples were labeled with Cy3 by using the Agilent miRNA Complete Labeling and Hybridization Kit (Agilent). To this end, 100 ng of total

RNA was dephosphorylated by incubation with calf intestinal phosphatase at 37°C for 30 min, denatured in 100% DMSO at 100°C for 8 min, and labeled with perCp-Cy3 by using T4 ligase at 16°C for 2 h. Each labeled RNA sample was hybridized to an individual array on 8 × 15 K format Agilent human miRNA array slides. Each array contained probes for 720 human miRNAs. Hybridization was performed in SureHyb chambers (Agilent) at 55°C for 20 h, and the arrays were washed according to the manufacturer's instructions and scanned.

2.8. Microarray Data Analysis. The oligo-mRNA and oligo-miRNA array slides were scanned with a DNA microarray scanner (Agilent), and the hybridization signals were extracted by using the Agilent Feature Extraction software. The microarray numerical quantitative data were normalized to the 75th percentile and were analyzed through the GeneSpring GX bioinformatics platform (http://www.agilent .com/chem/genespring), according to the default instructions. For mRNA analysis, we used ANOVA statistical test ($p \leq 0.05$) with a fold change $\geq$ 2.0 and, for microRNA analysis, we used ANOVA statistical analysis ($p \leq 0.01$) with a fold change $\geq$ 1.5 [13]. A complete file that provides all of the mRNAs and miRNAs present in the arrays used in this study, as well as the experimental conditions, is available online at a public database (http://www.ebi.ac.uk/arrayexpress/), Array Express accession E-MTAB-3091 (for mRNA hybridization) and E-MTAB-3093 (for microRNA hybridization).

2.9. Oligonucleotide Primer Design and Quantitative Real-Time Polymerase Chain Reaction (qRT-PCR). mRNA and microRNA oligoarray data were confirmed by qRT-PCR for five mRNAs (SMURF2, NOTCH1, PHOSPHO1, COL24A1, and FGF1) and six microRNAs (miR-31-3p, miR-134, miR-136-3p, miR-376c-3p, miR-424-5p, and miR-494). The mRNAs or microRNAs were elected on the basis of their expression pattern and biological function associated with the studied model system. The Primer3 web tool (http://biotools.umassmed.edu/bioapps/primer3_www.cgi) was used to select pairs of oligonucleotide primers with an optimal melting temperature of 60°C. Table 1 lists the oligonucleotide primers used in qRT-PCRs primers for mRNAs and Table 2 lists the accession number and mature sequences for microRNAs. All the qRT-PCRs experiments were conducted in triplicate. One-way ANOVA statistical test was performed with the statistical software GraphPad Prism 5.0 (http://www.graphpad.com/prism/Prism.htm).

3. Results

3.1. Scanning Electron Microscopy (SEM). SEM helped to characterize the surface morphology of the titanium discs. The polished control titanium disc displayed smooth surface. The treated titanium discs presented topographic surfaces bearing nanocavities with distinct distribution along the disc surface. The nano+submicrotextured surface group contained the greatest number of nanocavities. Except for surface roughness, the rough microtextured surface group and the

TABLE 1: Primers used in qRT-PCR reactions and their respective sense and antisense sequences.

mRNA symbol	Primer	Sequences	T_m
SMURF2	Forward	5′-CATGTCTAACCCCGGAGGC-3′	60°C
	Reverse	5′-TCCATCAACCACCACCTTAGC-3′	
NOTCH1	Forward	5′-TACAAGTGCAACTGCCTGCT-3′	60°C
	Reverse	5′-ATAGTCCTCGGATTGCCTGC-3′	
PHOSPHO1	Forward	5′-ATACCTCAGCTAGCCCCCTT-3′	60°C
	Reverse	5′-TGTAGGGACTCTGTTGGCCT-3′	
COL24A1	Forward	5′-CCCCACGGCAAAAACGAAAT-3′	60°C
	Reverse	5′-GCCTCCAAGGCCTAGTTGAT-3′	
FGF1	Reverse	5′-ACGGGCTTTTATACGGCTCA-3′	60°C
	Forward	5′-ATGGTTCTCCTCCAGCCTTTC-3′	
GAPDH	Forward	5′-GGGTGTGAACCACGAGAAAT-3′	60°C
	Reverse	5′-CCTTCCACAATGCCAAAGTT-3′	

nanotextured surface group had quite similar surface images (Figure 1).

3.2. Cell Viability.
Cell viability was similar for all the groups at the various time points. The exception was the rough microtextured surface group, which exhibited significantly higher viability at day 7 after the start of the culture ($p \leq 0.05$) as compared with the other groups (Figure 2).

3.3. Alkaline Phosphatase Activity.
In all the experimental groups, the activity of the enzyme alkaline phosphatase, related to the mineralization process, was higher at day 14 after the start of the culture (Figure 3). Among the treated groups, the nanotextured surface group showed the highest alkaline phosphatase production, which was statistically significant as compared with the nano+submicrotextured group at day 14 after the start of the culture ($p \leq 0.05$).

3.4. Mineralized Nodules.
Based on Alizarin Red S quantification, the treated titanium surfaces contained different amounts of calcium deposits. The rough microtextured surface group had the greatest amount of calcium as compared with the control group and the nanotextured surface group ($p \leq 0.05$) (Figure 4).

3.5. Analysis of Differentially Expressed mRNAs.
A total of 716 genes showed fold change ≥ 2.0 and $p \leq 0.05$ (Figure 5). Table 3 displays differentially expressed genes associated with osteogenesis, cell adhesion, apoptosis, cell growth, and cell differentiation. Table 4 depicts the differences in the expression of these genes as a function of the titanium surface topography.

3.6. Analysis of Differentially Expressed MicroRNAs.
According to the results, 32 miRNAs showed fold change ≥ 1.5 and $p \leq 0.01$ (Figure 6). Table 5 presents differentially expressed miRNAs associated with osteogenesis, apoptosis, and cell growth. Table 6 summarizes the different expression of these miRNAs in the studied titanium surfaces.

3.7. mRNA Data Confirmation by Real-Time Quantitative PCR.
qRT-PCR confirmed five differentially expressed mRNAs in the experimental groups; these mRNAs had been previously detected by oligomicroarray analysis (Table 5). The confirmed mRNAs are associated with apoptosis (NOTCH1), bone tissue and bone tissue mineralization (SMURF2, NOTCH1, and PHOSPHO1), cell adhesion (COL24A1 and FGF1), and cell proliferation (FGF1) (http://geneontology.org/, accessed 20/02/2014) (Figure 7).

3.8. MicroRNA Data Confirmation by Real-Time Quantitative PCR.
Real-time quantitative PCR allowed analysis of six miRNAs that were differentially expressed in the experimental groups: miR-31-3p, miR-134, miR-136-3p, miR-376C-3p, miR-494, and miR-424-5p (Table 6). These miRNAs are associated with several functions like apoptosis (miR-134 and miR-494), bone mineralization (miR-31-3p, miR-136-3p, miR-376C-3p, and miR-424-5p), and cell growth and proliferation (miR-134) (http://geneontology.org/, accessed 02/20/2014) (Figure 8).

4. Discussion

Several studies aimed at analyzing and comparing the response of different cell types in contact with titanium surfaces modified by numerous methods [14]. In the present study, it was observed that oxidative nanopatterning of titanium engenders micro- and nanotextured surfaces that influence the metabolism of human alveolar bone cells. It has been shown that [15] nanotextured titanium surfaces generated by this chemical method promote osteoblast proliferation, making this a promising technique for the regulation of cellular activities in biological environments. A previous investigation revealed that chemical oxidation with $H_2SO_4^{conc}/H_2O_2^{aq}$ solution is an efficient tool to achieve various physical and chemical configurations on titanium surface, demonstrating that, by varying etching parameters such as solution composition, temperature, and exposure time, it is possible to modify the topography, oxide thickness, and wettability of commercially pure titanium [5]. In the

TABLE 2: Accession numbers and mature sequences of the microRNAs used in qRT-PCR reactions.

MicroRNA	ID-miRBASE	Sequences
hsa-miR-31-3p	MIMAT0004504	UGCUAUGCCAACAUAUUGCCAU
hsa-miR-134	MI0000474	CAGGGUGUGUGACUGGUUGACCAGAGGGGCAUGCACUGUGUUCACCCUGUGGGCCACCUAGUCACCAACCCUC
hsa-miR-136-3p	MIMAT0004606	CAUCAUCGUCUCAAAUGAGUCU
hsa-miR-376c-3p	MIMAT0000720	AACAUAGAGGAAAUUCCACGU
hsa-miR-424-5p	MIMAT0001341	CAGCAGCAAUUCAUGUUUUGAA
hsa-miR-494	MI0003134	GAUACUCGAAGGAGGAGGUUGUCCGUGUUGUCUUCUCUUUAUUUAUGAUGAAACAUACACGGGAAACCUCUUUUUAGUAUC

TABLE 3: Genes differentially expressed among the following groups: control, nanotexture, nano+submicrotexture, and rough microtexture after 10 days of culture, with associated functions according to the Gene Ontology database.

Osteogenesis	Cell adhesion	Apoptosis	Cell growth and proliferation	Cell differentiation
ATP6VOA4	AMICA1	AIPL1	E2F5	C10orf27
CDX1	SORBS1	LUC7L3	KIAA1109	SOX21
CYP27B1	NRXN1	ERBB3	LGI1	CYP24A1
SMURF2	PCDHA11	NOTCH1	CYP27B1	NANOG
NOTCH1	NRXN1	PSMD3	GRIN2A	HSF4
PHOSPHO1	FGF1	BTK	GPAM	LAMC3
	ERBB3	PTPRC	KDR	
	PCDHB10	ABCB9	MMP7	
	EPB41L5	KDR	PDE3A	
	UTRN	MAPT	CENPF	
	NOTCH1	GPAM	FGF1	
	NCKAP1L	RNF7	MECOM	
	COL24A1	CD28	PTPRK	

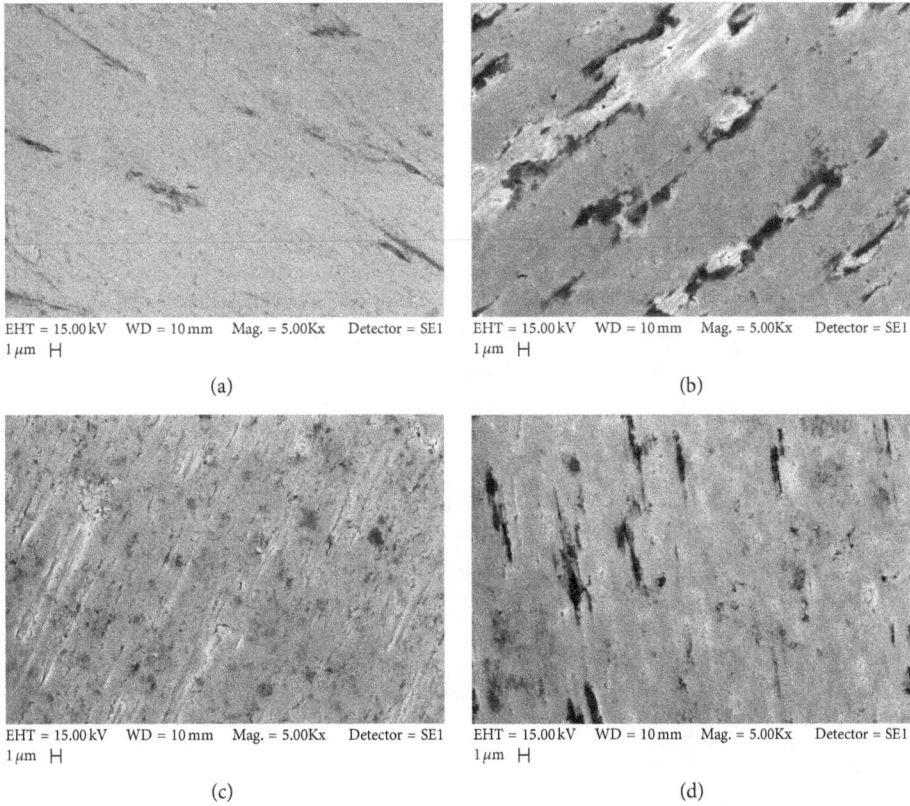

EHT = 15.00 kV   WD = 10 mm   Mag. = 5.00Kx   Detector = SE1
1 μm H

(a)

EHT = 15.00 kV   WD = 10 mm   Mag. = 5.00Kx   Detector = SE1
1 μm H

(b)

EHT = 15.00 kV   WD = 10 mm   Mag. = 5.00Kx   Detector = SE1
1 μm H

(c)

EHT = 15.00 kV   WD = 10 mm   Mag. = 5.00Kx   Detector = SE1
1 μm H

(d)

FIGURE 1: Scanning electron microscopy of different titanium surfaces: (a) control surface, (b) nanotextured surface, (c) nano+submicrotextured surface, and (d) rough microtextured surface.

present study, by using these parameters, we showed that oxidative nanopatterning promotes similar responses for cell viability for all groups at all time points tested, except for the rough microtexture, which showed significantly higher viability compared to the other groups after 7 days. Besides viability, biochemical assays like alkaline phosphatase activity are important to associate biomaterials topography with cell differentiation and osseointegration [8]. The ALP is among the first functional genes expressed in the calcification process, so it is possible that one of their roles in the mineralization process occurs at an early stage [16]. Nevertheless, the increased production of alkaline phosphatase was observed just after 14 days of culture in all experimental groups. Among the etched surfaces, cells cultured on nanotextured titanium

TABLE 4: Relative expression levels of mRNAs differentially expressed among control (C), nanotexture (N), nano+submicrotexture (NS), and rough microtexture (MR) groups after 10 days of culture. FC: fold change.

Gene	C versus MR FC	C versus S FC	C versus N FC	MR versus NS FC	MR versus N FC	NS versus N FC
ABCB9	2,423	1,045	1,106	−2,319	−2,191	1,059
AIPL1	1,055	−2,369	1,164	−2,500	1,103	2,758
AMICA1	2,308	1,693	2,524	−1,364	1,094	1,491
ATP6V0A4	−5,137	−1,532	−3,833	3,353	1,340	−2,502
BTK	−4,130	−1,516	−6,530	2,725	−1,581	−4,308
C10orf27	−2,049	−1,277	−1,827	1,605	1,122	−1,431
CD28	−3,104	−1,478	−2,068	2,101	1,501	−1,400
CDX1	−2,178	−1,136	−1,572	1,918	1,386	−1,384
CENPF	2,296	1,639	1,574	−1,401	−1,459	−1,041
COL24A1	2,308	1,076	−1,584	−2,144	−3,655	−1,705
CYP27B1	2,325	−1,042	4,278	−2,422	1,840	4,457
E2F5	2,572	1,012	3,240	−2,542	1,260	3,203
EPB41L5	2,725	1,796	2,440	−1,518	−1,117	1,359
ERBB3	−2,036	1,391	−1,934	2,832	1,052	−2,691
FGF1	1,929	1,309	2,792	−1,473	1,447	2,132
GPAM	3,438	−1,467	2,420	−5,045	−1,421	3,551
GRIN2A	2,065	−1,066	2,005	−2,202	−1,030	2,138
HSF4	−2,358	−2,213	−1,331	1,066	1,772	1,663
KDR	1,563	−1,134	2,118	−1,772	1,355	2,401
KIAA1109	1,948	1,340	2,146	−1,453	1,102	1,602
LAMC3	2,138	2,921	2,086	1,366	−1,025	−1,400
LGI1	3,533	1,452	4,124	−2,433	1,167	2,839
LUC7L3	2,064	1,673	1,507	−1,234	−1,370	−1,110
MAPT	6,567	1,653	4,711	−3,973	−1,394	2,849
MECOM	−2,952	−3,421	−8,666	−1,159	−2,936	−2,533
MMP7	2,731	1,120	2,208	−2,438	−1,237	1,972
NANOG	3,365	2,137	3,791	−1,575	1,127	1,774
NCKAP1L	−1,722	−2,257	−1,736	−1,311	−1,008	1,300
NOTCH1	1,306	−1,812	1,416	−2,367	1,084	2,565
NRXN1	2,462	1,377	−1,203	−1,788	−2,963	−1,657
PCDHA11	2,074	−1,266	2,259	−2,626	1,089	2,859
PCDHB10	3,328	−1,025	3,581	−3,411	1,076	3,669
PDE3A	−3,260	1,121	−3,908	3,656	−1,199	−4,382
PHOSPHO1	−2,024	−1,214	−1,757	1,667	1,152	−1,448
PSMD3	−2,075	−2,131	−1,353	−1,027	1,533	1,574
PTPRC	1,453	−1,392	1,475	−2,022	1,016	2,054
PTPRK	−3,909	−1,952	1,034	2,002	4,040	2,018
RNF7	2,803	1,199	1,387	−2,339	−2,021	1,157
SMURF2	1,193	−2,344	−1,142	−2,797	−1,362	2,053
SORBS1	−3,880	−3,008	−3,342	1,290	1,161	−1,111
SOX21	2,006	1,583	2,829	−1,267	1,411	1,787
UTRN	−1,602	−1,884	−2,062	−1,176	−1,287	−1,094

discs showed the highest ALP activity, which was statistically significant compared to the NS group after 14 days. By exploiting microarray methodology, we found 716 differentially expressed genes of alveolar bone osteoblastic cells in contact with different titanium surfaces, most of them associated with the process of osteogenesis (e.g., mineralization, adhesion, apoptosis, proliferation, and differentiation). Among them, it was observed that NOTCH1 gene increased its expression in nano+submicrotexture surfaces when compared to the other experimental conditions. NOTCH is a key target in the osteoblastic cells and in osteoclastogenesis, as well as skeletal development and bone remodeling [17]. NOTCH is

TABLE 5: MicroRNAs differentially expressed among control, nanotexture, nano+submicrotexture, and rough microtexture groups after 10 days of culture, with associated functions according to Gene Ontology database.

Osteogenesis	Apoptosis	Cell growth and proliferation
hsa-miR-424-5p	hsa-miR-494	hsa-miR-134
hsa-miR-136-3p	hsa-miR-134	
hsa-miR-136-5p		
hsa-miR-31-3p		
hsa-miR-376c-3p		
hsa-miR-19b-3p		
hsa-miR-21-3p		
hsa-miR-21-5p		
hsa-miR-218-5p		
hsa-miR-29b-3p		

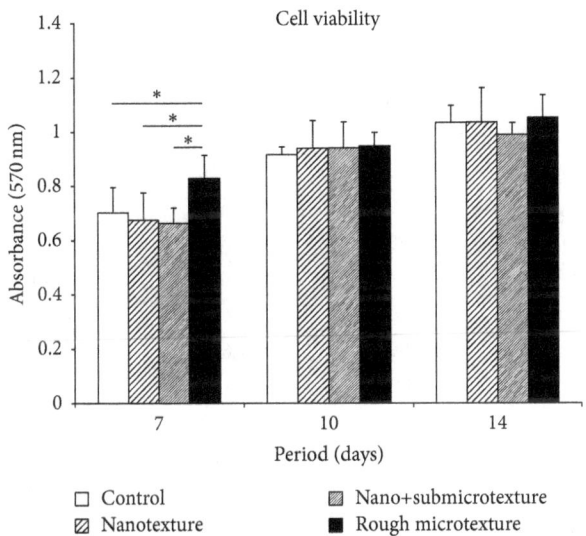

FIGURE 2: Cell viability of human alveolar osteoblastic cells at days 7, 10, and 14 of the start of the culture on polished titanium discs and on titanium surfaces etched by oxidative nanopatterning. Mann-Whitney test for $^*p \leq 0.05$.

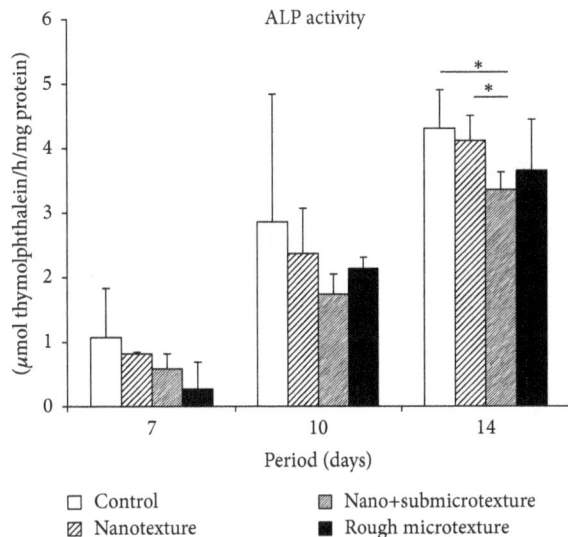

FIGURE 3: Alkaline phosphatase (ALP) activity of human alveolar osteoblastic cells after 7, 10, and 14 days of the start of the culture on polished titanium discs and on titanium surfaces etched by oxidative nanopatterning. Mann-Whitney test for $^*p \leq 0.05$.

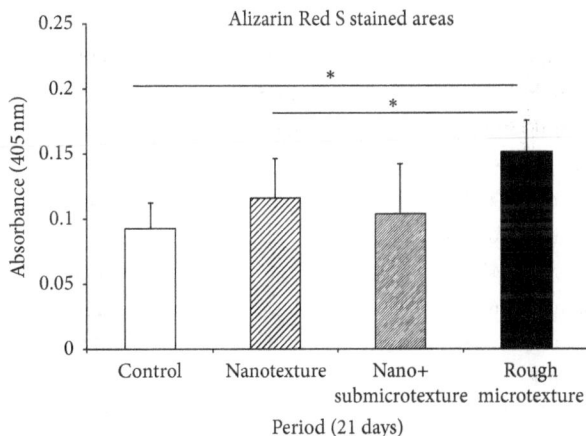

FIGURE 4: Quantitative analysis of Alizarin Red S stained areas of calcified nodules 21 days after the start of the culture on polished titanium discs and on titanium surfaces etched by oxidative nanopatterning. Mann-Whitney test for $^*p \leq 0.05$.

an additional pathway that is triggered early in the response to modified titanium surfaces, as it would be involved in the process of osteogenesis and in vivo bone formation and healing [18]. These findings are in agreement with our microarray results, despite the fact that qRT-PCR validation method showed similar responses only in the increased expression between C/MR and C/N groups. Another differentially expressed gene was PHOSPHO1, which showed higher expression in cells adhering onto rough microtexture. This gene codifies a protein involved in the initial deposition of hydroxyapatite crystals in early events of matrix mineralization [19]. Interestingly, the microarray data were in good agreement with our biochemical results, where the cultures in contact with rough microtexture showed higher calcium deposition, despite not being similar to qRT-PCR validation method. SMURF2 interacts with SMADS and induces ubiquitin-mediated degradation, preventing the

signaling of TGF-β (transforming growth factor) and BMP (bone morphogenetic protein). Both the TGF-β and BMP are multifunctional proteins important for regulation of proliferation, differentiation, migration, and apoptosis [20]. Microarray data showed a higher expression of SMURF2 in cells adhering onto nano+submicrotexture surfaces when comparing to the other etched surfaces, whereas qRT-PCR revealed a higher expression of cells seeded on nanotextured titanium. These results suggest that the consequently lower expression of SMURF2 in cells seeded on MR surfaces might influence positively the osteoblast differentiation and Runx2 stability, promoting cell-biomaterial interaction, as seen after 7 and 21 days in MTT and Alizarin Red S staining assays, respectively. Another gene important to cell adhesion and

TABLE 6: Relative expression levels of microRNAs differentially expressed among control (C), nanotexture (N), nano+submicrotexture (NS), and rough microtexture (MR) groups after 10 days of culture. FC: fold change.

miRNA	C versus MR FC	C versus NS FC	C versus N FC	MR versus NS FC	MR versus N FC	NS versus N FC
hsa-miR-101-3p	2,635	1,033	1,036	−2,551	−2,542	1,003
hsa-miR-106b-5p	1,511	−1,018	1,000	−1,539	−1,510	1,019
hsa-miR-1246	1,173	1,963	1,133	1,674	−1,036	−1,733
hsa-miR-1290	1,057	2,161	1,113	2,045	1,053	−1,941
hsa-miR-134	−1,478	−1,119	1,025	1,321	1,515	1,147
hsa-miR-136-3p	182,144	−1,044	4,031	−190,073	−45,188	4,206
hsa-miR-136-5p	67,552	1,001	−1,132	−67,496	−76,442	−1,133
hsa-miR-15a-5p	1,633	1,009	−1,027	−1,619	−1,678	−1,036
hsa-miR-1826_v15.0	−1,174	1,173	1,299	1,377	1,525	1,107
hsa-miR-1914-3p	−1,401	−1,001	1,186	1,399	1,662	1,188
hsa-miR-193a-3p	2,493	−1,059	−1,021	−2,640	−2,546	1,037
hsa-miR-19a-3p	2,033	−1,060	−1,069	−2,154	−2,173	−1,009
hsa-miR-19b-3p	1,544	−1,035	−1,053	−1,598	−1,627	−1,018
hsa-miR-21-3p	1,402	−1,205	−1,097	−1,688	−1,538	1,098
hsa-miR-21-5p	1,782	−1,000	−1,000	−1,782	−1,782	−1,000
hsa-miR-218-5p	1,931	1,054	−1,005	−1,832	−1,940	−1,059
hsa-miR-26b-5p	1,635	1,036	1,024	−1,579	−1,598	−1,012
hsa-miR-27a-3p	1,588	−1,000	−1,000	−1,588	−1,588	−1,000
hsa-miR-29b-3p	2,128	1,083	1,014	−1,964	−2,098	−1,068
hsa-miR-301a-3p	1,685	−1,041	1,034	−1,754	−1,630	1,076
hsa-miR-31-3p	1,464	−1,092	−1,120	−1,599	−1,640	−1,026
hsa-miR-374a-5p	1,868	−1,083	−1,093	−2,023	−2,041	−1,009
hsa-miR-376a-3p	1,411	−1,081	−1,046	−1,526	−1,477	1,033
hsa-miR-376b-3p	2,361	−1,105	1,085	−2,608	−2,176	1,199
hsa-miR-376c-3p	1,424	−1,141	−1,077	−1,625	−1,533	1,060
hsa-miR-377-3p	1,754	−1,019	1,005	−1,788	−1,745	1,025
hsa-miR-424-5p	2,125	1,051	1,043	−2,021	−2,038	−1,008
hsa-miR-450a-5p	202,710	1,121	3,964	−180,812	−51,138	3,536
hsa-miR-494	−1,416	1,006	1,120	1,424	1,585	1,113
hur_5	−1,357	1,120	1,132	1,519	1,537	1,011
miRNABrightCorner30	1,657	2,170	1,197	1,309	−1,385	−1,813
mr_1	−1,519	−1,000	−1,000	1,519	1,519	−1,000

cell proliferation is FGF1, which plays a crucial role in the proliferation and differentiation of osteoblasts [21]. FGF1 expression in microarray was increased in cells cultivated on control and NS surfaces, whereas qRT-PCR showed higher expression in cells seeded on nanotextured discs. Our biochemical assays on proliferation and viability did not reveal differences on cell adhesion and proliferation. COL24A1 gene proved to have some control on osteoblast differentiation and mineralization, through interaction with integrin $\beta3$ and the transforming growth factor beta (TGF-β)/SMADS signaling pathway [22]. It was found that COL24A1 gene is activated at the same time as the gene encoding osteocalcin, and its expression increases gradually as osteoblasts begin to deposit

mineralized matrix [23]. Our microarray analysis showed increased expression in cells seeded on nanotexture titanium discs when compared to the other etched surfaces, which is in agreement with other investigations [24].

It has been observed that during bone formation several miRNAs participate both in early and in late stage of osteoblast differentiation regulating pathways *in vivo* [25]. The present study identified 32 miRNAs differentially expressed among the experimental groups. In particular, these miRNAs are involved in mineralization, apoptosis, and regulation of cell growth and proliferation. We showed that miR-136-3p exhibits a lower expression in cells seeded on rough microtexture surfaces when compared to N and MR

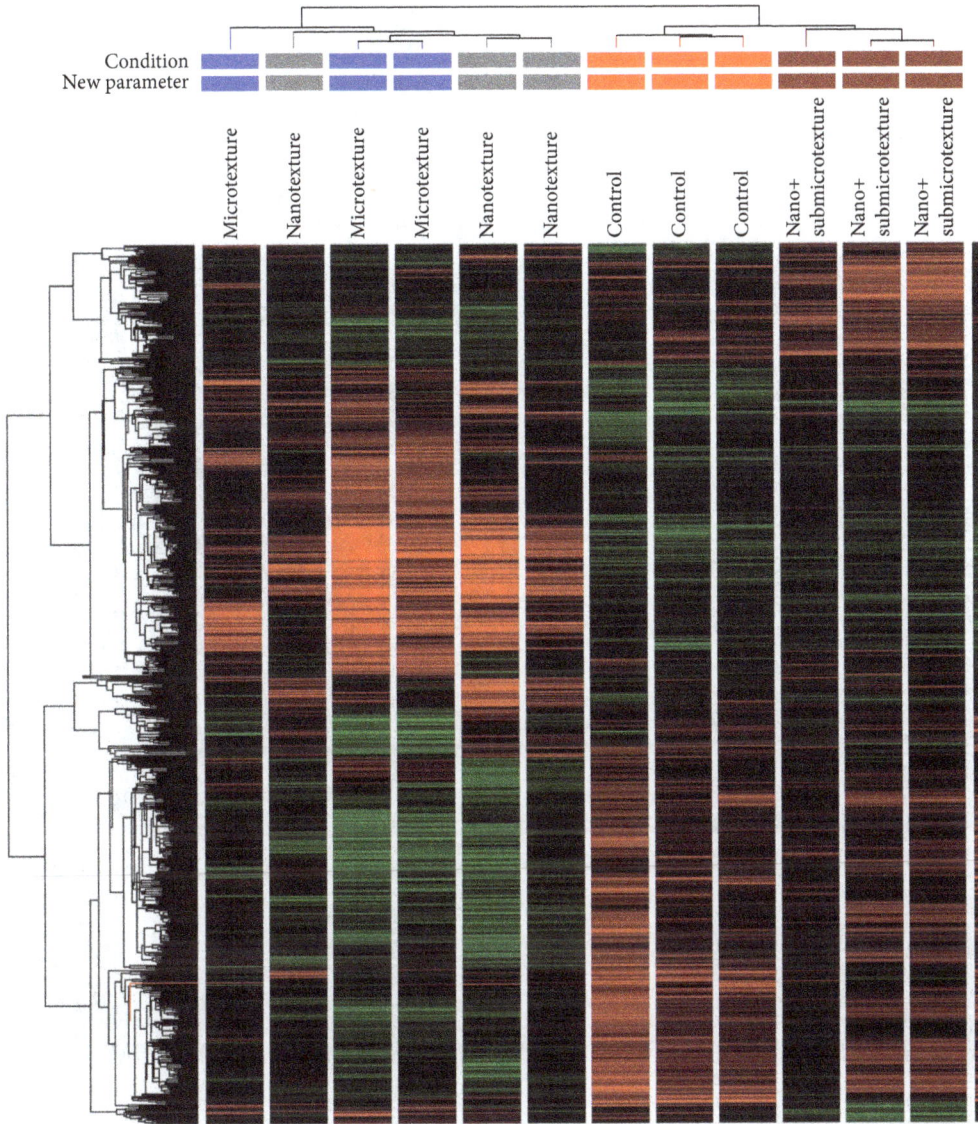

FIGURE 5: Hierarchical clustering of 716 differentially expressed genes of human alveolar osteoblastic cells on polished and etched titanium surfaces (nanotextured, nano+submicrotextured, and rough microtextured) 10 days after the start of the culture. Red: upregulation; green: downregulation; black: unmodulated.

surfaces in microarray and qRT-PCR data. The literature shows that miR-136 promotes downregulation in osteoblast differentiation [26], suggesting that in the present work this microRNA benefited the deposition of extracellular matrix in MR group as seen in our biochemical results. On the other hand, the higher expression of miR-134 in MR group revealed by microarray data may have resulted in diminished cell adhesion, as some microRNAs promote regulation of integrins, which are transmembrane cell adhesion receptors and have great importance for the unity of the cell to extracellular matrix, as well as participating in interactions between cells [26]. Poitz et al. [27] demonstrated that miR-134 promoted $\beta1$ integrin negative regulation, resulting in reduction of adhesion to fibronectin of mesenchymal stem cells (MSCs). Despite that, our MTT assay does not show any decrease in the viability of cells seeded on MR surface. Microarray

and qRT-PCR methods also showed an increased expression of miR-494 in rough microtexture when compared to nanotexture. This microRNA mediates apoptosis and necrosis in different cell types [28], but this effect was not seen in the present work, with all cells seeded on etched surfaces showing similar or higher viability than control cells. We observed involvement of miR-31 in the regulation of transcription factor osterix in human bone marrow cells (MSCs) differentiated into osteoblasts [29]. Osterix is a key regulator of bone cell differentiation and plays an essential role in bone homeostasis [30]. Deng et al. [31] also observed that overexpression of miR-31 significantly reduced the expression of osteogenic transcription factors like OPN, BSP, OCN, and OSX, but not Runx2. Our data from microarray analysis as well as from qRT-PCR revealed that miR-31-3p had an increase in its expression in cells seeded on nanotexturized surface and

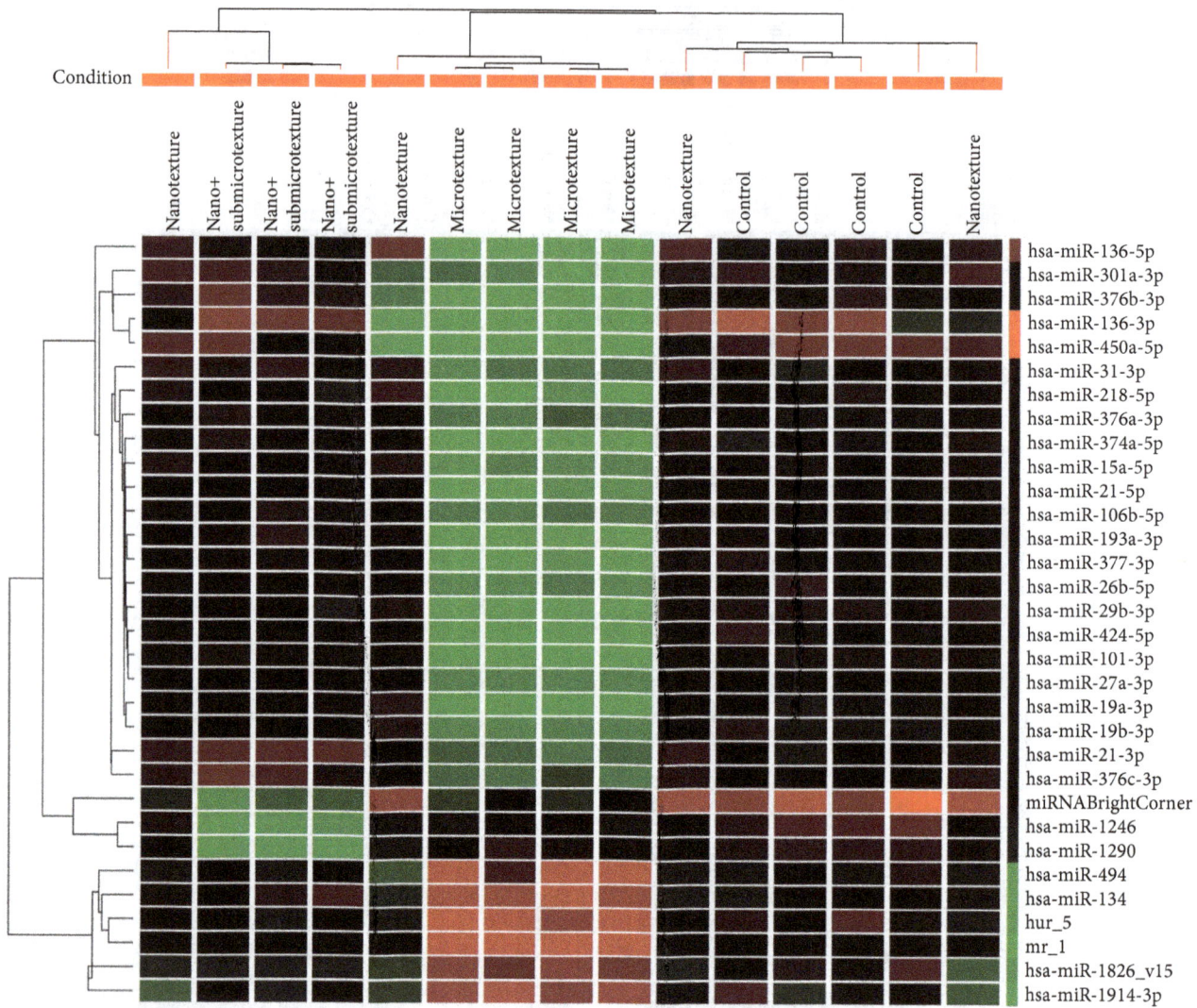

FIGURE 6: Hierarchical clustering of 32 differentially expressed microRNAs of human alveolar osteoblastic cells on polished and etched titanium surfaces (nanotextured, nano+submicrotextured, and rough microtextured) 10 days after the start of the culture. Red: upregulation; green: downregulation; black: unmodulated.

a decrease on MR titanium discs. In the same way, cells on rough microtextured surface showed lower expression of miR-424-5p when compared to other etched surfaces whereas nanotextured surface revealed higher expression, confirmed by microarray and qRT-PCR. miR-424 also has regulatory roles in the differentiation of human bone marrow cells into osteoblasts [32] and these results may have contributed to the enhanced cell differentiation seen in the first group by means of ALP activity after 14 days. Other miRNAs, which influence bone metabolism, were differentially expressed in our study, such as miR-19b, miR-21, miR-218, and miR-29b.

The qRT-PCR method helps to validate the microarray data. In this study, we validated microarray methodology by conducting qRT-PCR of five genes and six microRNAs elected on the basis of their expression pattern and of their association with osteogenesis. However, qRT-PCR did not confirm all the results obtained by microarray analysis, which is shown also by other studies [33, 34]. According

to the literature, differences between data obtained by the two methods may occur because microarray hybridization protocol might avoid the detection of subtle differences in gene expression, which could be detected by qRT-PCR technique [35]. Another cited reason is that these differences could occur as a consequence of increased separation between the locations of the PCR primers and the microarray probes [36]. Besides, microarray and qRT-PCR protocols perform normalization by different software (i.e., global gene expression and endogenous control, resp.) [37].

The results shown in this investigation reveal that several other mRNAs and miRNAs can be modulated as a consequence of surface modification, and more studies should be addressed to elucidate their role in osteoblast metabolism. In conclusion, it has been shown that oxidative nanopatterning of titanium surface influences alveolar bone osteoblastic cell metabolism and modulates the expression of genes encoding proteins that are important for osteogenesis.

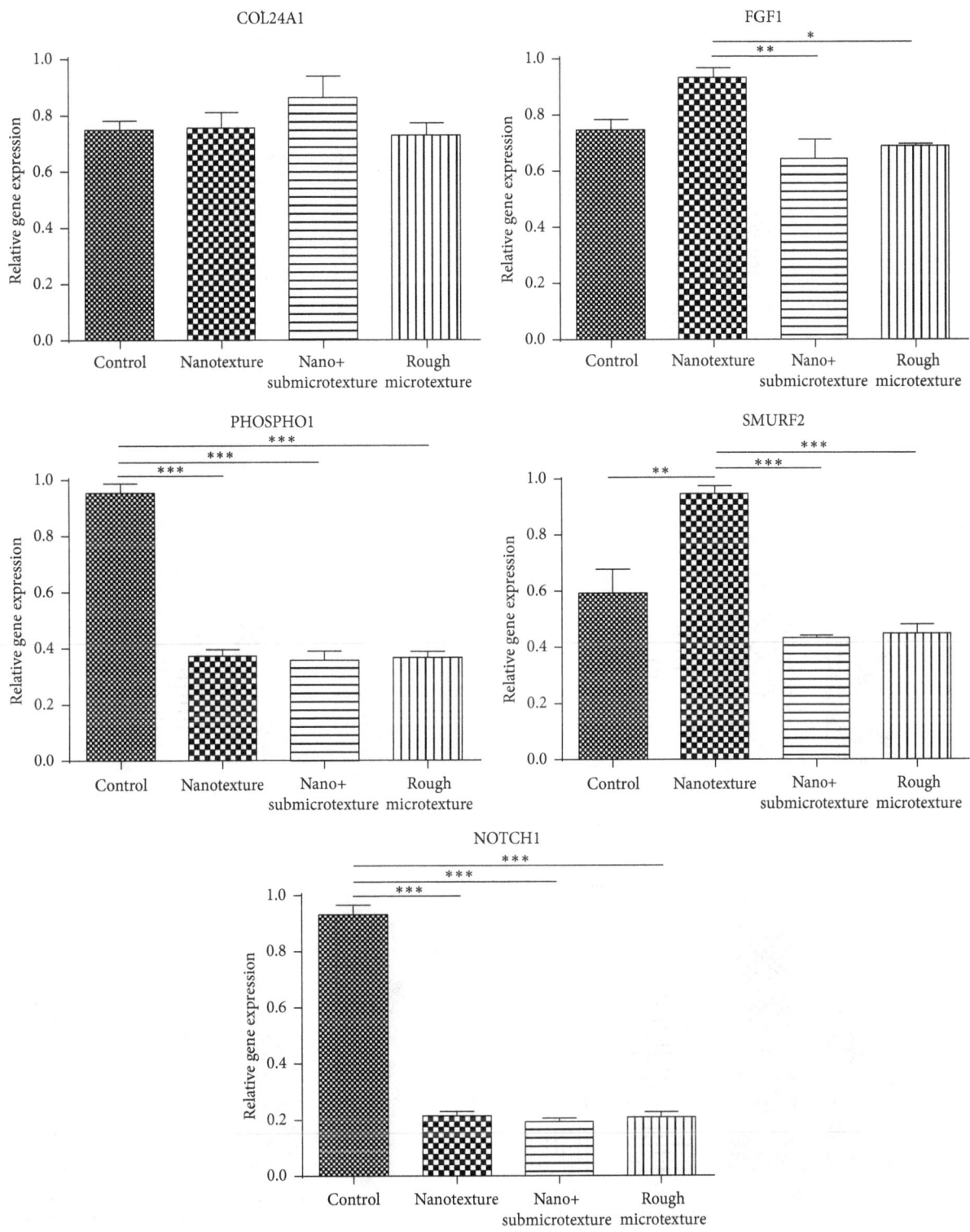

FIGURE 7: Quantitative expression of mRNAs of human alveolar osteoblastic cells on polished and etched titanium surfaces (nanotextured, nano+submicrotextured, and rough microtextured). Statistical analysis by Tukey Multiple Comparison Test; $^{*}p < 0.05$, $^{**}p < 0.01$, and $^{***}p < 0.001$.

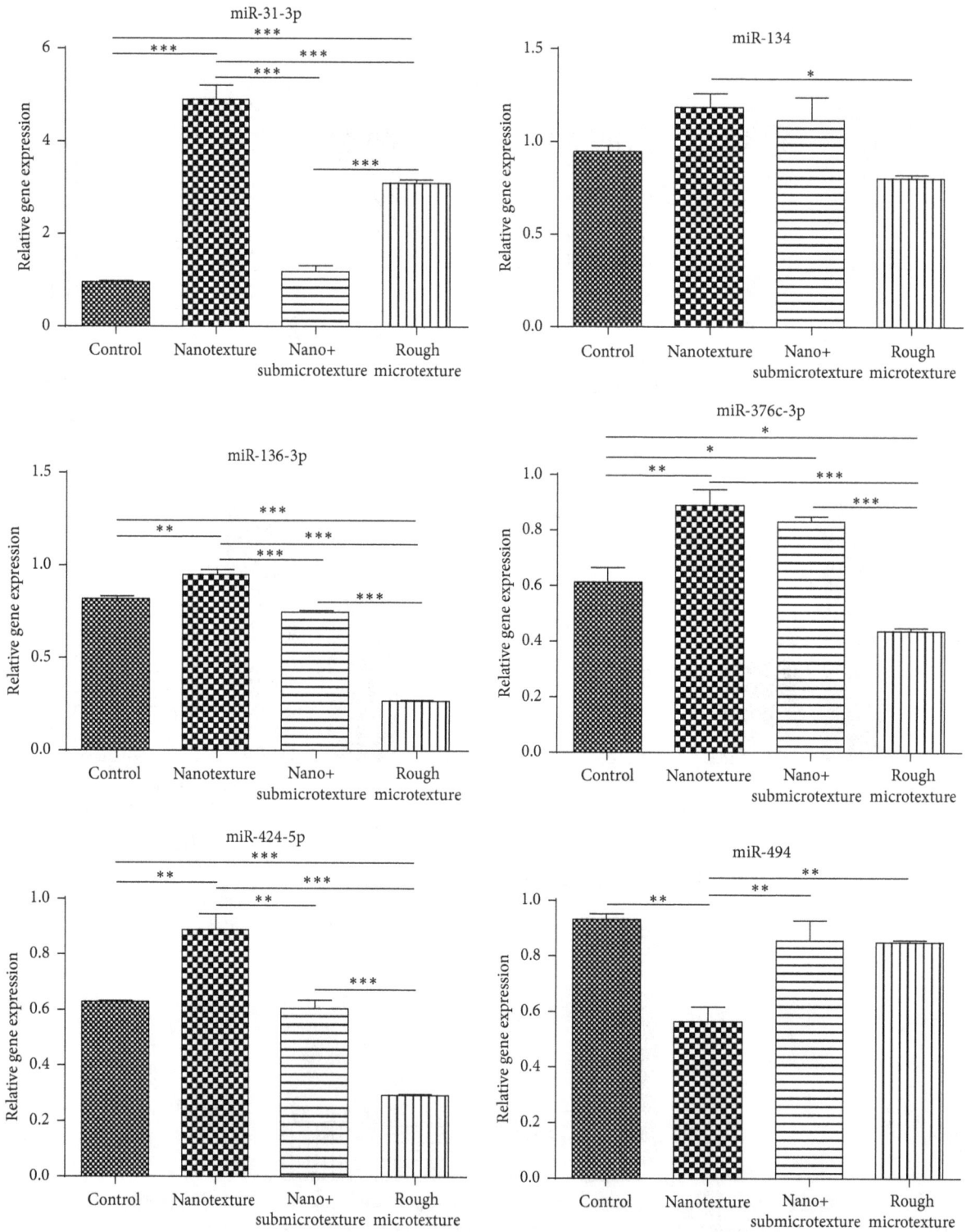

FIGURE 8: Quantitative expression of miRNAs of human alveolar osteoblastic cells on polished and etched titanium surfaces (nanotextured, nano+submicrotextured, and rough microtextured). Statistical analysis by Tukey Multiple Comparison Test; $^{*}p < 0.05$, $^{**}p < 0.01$, and $^{***}p < 0.001$.

Competing Interests

The authors declare that they have no competing interests.

Acknowledgments

The authors would like to thank São Paulo Research Foundation (FAPESP) for financial support (Processes 2011/00702-0 and 2011/21982-0).

References

[1] S. T. Stern and S. E. McNeil, "Nanotechnology safety concerns revisited," *Toxicological Sciences*, vol. 101, no. 1, pp. 4–21, 2008.

[2] R. Bhola, F. Su, and C. E. Krull, "Functionalization of titanium based metallic biomaterials for implant applications," *Journal of Materials Science: Materials in Medicine*, vol. 22, no. 5, pp. 1147–1159, 2011.

[3] L. Tolstunov, "Implant zones of the jaws: implant location and related success rate," *The Journal of Oral Implantology*, vol. 33, no. 4, pp. 211–220, 2007.

[4] K. Nishio, M. Neo, H. Akiyama et al., "The effect of alkali- and heat-treated titanium and apatite-formed titanium on osteoblastic differentiation of bone marrow cells," *Journal of Biomedical Materials Research*, vol. 52, no. 4, pp. 652–661, 2000.

[5] F. Variola, A. Lauria, A. Nanci, and F. Rosei, "Influence of treatment conditions on the chemical oxidative activity of H_2SO_4/H_2O_2 mixtures for modulating the topography of titanium," *Advanced Engineering Materials*, vol. 11, no. 12, pp. B227–B234, 2009.

[6] R. Kriparamanan, P. Aswath, A. Zhou, L. Tang, and K. T. Nguyen, "Nanotopography: cellular responses to nanostructured materials," *Journal of Nanoscience and Nanotechnology*, vol. 6, no. 7, pp. 1905–1919, 2006.

[7] J. S. Colombo, A. Carley, G. J. P. Fleming, S. J. Crean, A. J. Sloan, and R. J. Waddington, "Osteogenic potential of bone marrow stromal cells on smooth, roughened, and tricalcium phosphate-modified titanium alloy surfaces," *The International Journal of Oral & Maxillofacial Implants*, vol. 27, no. 5, pp. 1029–1042, 2012.

[8] B.-A. Lee, C.-H. Kang, M.-S. Vang et al., "Surface characteristics and osteoblastic cell response of alkali-and heat-treated titanium-8tantalum-3niobium alloy," *Journal of Periodontal and Implant Science*, vol. 42, no. 6, pp. 248–255, 2012.

[9] X. Rausch-Fan, Z. Qu, M. Wieland, M. Matejka, and A. Schedle, "Differentiation and cytokine synthesis of human alveolar osteoblasts compared to osteoblast-like cells (MG63) in response to titanium surfaces," *Dental Materials*, vol. 24, no. 1, pp. 102–110, 2008.

[10] M. R. W. Ferreira, J. Dernowsek, G. A. Passos, and K. F. Bombonato-Prado, "Undifferentiated pulp cells and odontoblast-like cells share genes involved in the process of odontogenesis," *Archives of Oral Biology*, vol. 60, no. 4, pp. 593–599, 2015.

[11] K. F. Bombonato-Prado, L. S. Bellesini, C. M. Junta, M. M. Marques, G. A. Passos, and A. L. Rosa, "Microarray-based gene expression analysis of human osteoblasts in response to different biomaterials," *Journal of Biomedical Materials Research Part A*, vol. 88, no. 2, pp. 401–408, 2009.

[12] C. A. Gregory, W. G. Gunn, A. Peister, and D. J. Prockop, "An Alizarin red-based assay of mineralization by adherent cells in culture: comparison with cetylpyridinium chloride extraction," *Analytical Biochemistry*, vol. 329, no. 1, pp. 77–84, 2004.

[13] M. B. Eisen, P. T. Spellman, P. O. Brown, and D. Botstein, "Cluster analysis and display of genome-wide expression patterns," *Proceedings of the National Academy of Sciences of the United States of America*, vol. 95, no. 25, pp. 14863–14868, 1998.

[14] F. Vetrone, F. Variola, P. T. De Oliveira et al., "Nanoscale oxidative patterning of metallic surfaces to modulate cell activity and fate," *Nano Letters*, vol. 9, no. 2, pp. 659–665, 2009.

[15] F. Variola, J.-H. Yi, L. Richert, J. D. Wuest, F. Rosei, and A. Nanci, "Tailoring the surface properties of Ti6Al4V by controlled chemical oxidation," *Biomaterials*, vol. 29, no. 10, pp. 1285–1298, 2008.

[16] P. Santiago-Medina, P. A. Sundaram, and N. Diffoot-Carlo, "The effects of micro arc oxidation of gamma titanium aluminide surfaces on osteoblast adhesion and differentiation," *Journal of Materials Science: Materials in Medicine*, vol. 25, no. 6, pp. 1577–1587, 2014.

[17] S. Zanotti and E. Canalis, "Notch and the skeleton," *Molecular and Cellular Biology*, vol. 30, no. 4, pp. 886–896, 2010.

[18] N. Chakravorty, S. Hamlet, A. Jaiprakash et al., "Pro-osteogenic topographical cues promote early activation of osteoprogenitor differentiation via enhanced TGFβ, Wnt, and Notch signaling," *Clinical Oral Implants Research*, vol. 25, no. 4, pp. 475–486, 2014.

[19] S. Roberts, S. Narisawa, D. Harmey, J. L. Millán, and C. Farquharson, "Functional involvement of PHOSPHO1 in matrix vesicle-mediated skeletal mineralization," *Journal of Bone and Mineral Research*, vol. 22, no. 4, pp. 617–627, 2007.

[20] A. Nakano, D. Koinuma, K. Miyazawa et al., "Pin1 down-regulates transforming growth factor-β (TGF-β) signaling by inducing degradation of Smad proteins," *The Journal of Biological Chemistry*, vol. 284, no. 10, pp. 6109–6115, 2009.

[21] M. J. Feito, R. M. Lozano, M. Alcaide et al., "Immobilization and bioactivity evaluation of FGF-1 and FGF-2 on powdered silicon-doped hydroxyapatite and their scaffolds for bone tissue engineering," *Journal of Materials Science: Materials in Medicine*, vol. 22, no. 2, pp. 405–416, 2011.

[22] W. Wang, D. Olson, G. Liang et al., "Collagen XXIV (Col24α1) promotes osteoblastic differentiation and mineralization through TGF-β/smads signaling pathway," *International Journal of Biological Sciences*, vol. 8, no. 10, pp. 1310–1322, 2012.

[23] N. Matsuo, S. Tanaka, H. Yoshioka, M. Koch, M. K. Gordon, and F. Ramirez, "Collagen XXIV (Col24a1) gene expression is a specific marker of osteoblast differentiation and bone formation," *Connective Tissue Research*, vol. 49, no. 2, pp. 68–75, 2008.

[24] L. Prodanov, C. M. Semeins, J. J. W. A. van Loon et al., "Influence of nanostructural environment and fluid flow on osteoblast-like cell behavior: a model for cell-mechanics studies," *Acta Biomaterialia*, vol. 9, no. 5, pp. 6653–6662, 2013.

[25] X. Zhao, D. Xu, Y. Li et al., "MicroRNAs regulate bone metabolism," *Journal of Bone and Mineral Metabolism*, vol. 32, no. 3, pp. 221–231, 2014.

[26] J. H. An, J. H. Ohn, J. A. Song et al., "Changes of microRNA profile and microRNA-mRNA regulatory network in bones of ovariectomized mice," *Journal of Bone and Mineral Research*, vol. 29, no. 3, pp. 644–656, 2014.

[27] D. M. Poitz, F. Stölzel, L. Arabanian et al., "MiR-134-mediated β1 integrin expression and function in mesenchymal stem cells," *Biochimica et Biophysica Acta (BBA)—Molecular Cell Research*, vol. 1833, no. 12, pp. 3396–3404, 2013.

[28] Y.-F. Lan, H.-H. Chen, P.-F. Lai et al., "MicroRNA-494 reduces ATF3 expression and promotes AKI," *Journal of the American Society of Nephrology*, vol. 23, no. 12, pp. 2012–2023, 2012.

[29] S. R. Baglìo, V. Devescovi, D. Granchi, and N. Baldini, "MicroRNA expression profiling of human bone marrow mesenchymal stem cells during osteogenic differentiation reveals osterix regulation by miR-31," *Gene*, vol. 527, no. 1, pp. 321–331, 2013.

[30] Y. Gao, A. Jheon, H. Nourkeyhani, H. Kobayashi, and B. Ganss, "Molecular cloning, structure, expression, and chromosomal localization of the human osterix (SP7) gene," *Gene*, vol. 341, no. 1-2, pp. 101–110, 2004.

[31] Y. Deng, S. Wu, X. Bi et al., "Effects of a miR-31, Runx2, and Satb2 regulatory loop on the osteogenic differentiation of bone mesenchymal stem cells," *Stem Cells and Development*, vol. 22, no. 16, pp. 2278–2286, 2013.

[32] J. Gao, T. Yang, J. Han et al., "MicroRNA expression during osteogenic differentiation of human multipotent mesenchymal stromal cells from Bone Marrow," *Journal of Cellular Biochemistry*, vol. 112, no. 7, pp. 1844–1856, 2011.

[33] M. S. Rajeevan, S. D. Vernon, N. Taysavang, and E. R. Unger, "Validation of array-based gene expression profiles by real-time (kinetic) RT-PCR," *Journal of Molecular Diagnostics*, vol. 3, no. 1, pp. 26–31, 2001.

[34] P. B. Dallas, N. G. Gottardo, M. J. Firth et al., "Gene expression levels assessed by oligonucleotide microarray analysis and quantitative real-time RT-PCR—how well do they correlate?" *BMC Genomics*, vol. 6, article 59, 2005.

[35] K. Allanach, M. Mengel, G. Einecke et al., "Comparing microarray versus RT-PCR assessment of renal allograft biopsies: similar performance despite different dynamic ranges," *American Journal of Transplantation*, vol. 8, no. 5, pp. 1006–1015, 2008.

[36] W. Etienne, M. H. Meyer, J. Peppers, and R. A. Meyer Jr., "Comparison of mRNA gene expression by RT-PCR and DNA microarray," *BioTechniques*, vol. 36, no. 4, pp. 618–626, 2004.

[37] J. S. Morey, J. C. Ryan, and F. M. Van Dolah, "Microarray validation: factors influencing correlation between oligonucleotide microarrays and real-time PCR," *Biological Procedures Online*, vol. 8, no. 1, pp. 175–193, 2006.

18

Preparation and Evaluation of Gelatin-Chitosan-Nanobioglass 3D Porous Scaffold for Bone Tissue Engineering

Kanchan Maji,[1] Sudip Dasgupta,[1] Krishna Pramanik,[2] and Akalabya Bissoyi[2]

[1]Department of Ceramic Engineering, National Institute of Technology, Rourkela 769008, India
[2]Department of Biotechnology and Medical Engineering, National Institute of Technology, Rourkela 769008, India

Correspondence should be addressed to Sudip Dasgupta; dasguptas@nitrkl.ac.in

Academic Editor: Esmaiel Jabbari

The aim of the present study was to prepare and characterize bioglass-natural biopolymer based composite scaffold and evaluate its bone regeneration ability. Bioactive glass nanoparticles (58S) in the size range of 20–30 nm were synthesized using sol-gel method. Porous scaffolds with varying bioglass composition from 10 to 30 wt% in chitosan, gelatin matrix were fabricated using the method of freeze drying of its slurry at 40 wt% solids loading. Samples were cross-linked with glutaraldehyde to obtain interconnected porous 3D microstructure with improved mechanical strength. The prepared scaffolds exhibited >80% porosity with a mean pore size range between 100 and 300 microns. Scaffold containing 30 wt% bioglass (GCB 30) showed a maximum compressive strength of 2.2 ± 0.1 MPa. Swelling and degradation studies showed that the scaffold had excellent properties of hydrophilicity and biodegradability. GCB 30 scaffold was shown to be noncytotoxic and supported mesenchymal stem cell attachment, proliferation, and differentiation as indicated by MTT assay and RUNX-2 expression. Higher cellular activity was observed in GCB 30 scaffold as compared to GCB 0 scaffold suggesting the fact that 58S bioglass nanoparticles addition into the scaffold promoted better cell adhesion, proliferation, and differentiation. Thus, the study showed that the developed composite scaffolds are potential candidates for regenerating damaged bone tissue.

1. Introduction

Designing new biomaterials that can mimic the micro/nanostructure and chemical composition of the native extracellular matrix (ECM) is one of the most attractive and important research areas in the field of tissue engineering. ECM governs cell attachment, growth, migration, and differentiation, as well as the formation of new tissues [1]. Scaffolds that mimic ECM of bone should ideally be processed from biomaterials with adequate properties such as biocompatibility, osteoconduction, bioactivity, osteoinduction, and biodegradation [2]. Three-dimensional (3D) porous structure of scaffolds provides the necessary support for cells to proliferate and maintain their differential function, and its architecture ultimately governs the geometry of a new bone [3]. Moreover, scaffolds for bone regeneration should mimic bone morphology, structure, and function. Bone is composed of calcium phosphate (69–80 wt%, mainly hydroxyapatite), collagen (17–20 wt%), and other components (water, proteins, etc.) [4]. For

this reason, composites based on apatite crystals and natural biopolymers have received increasing attention in bone tissue engineering applications [5, 6].

Biocomposites based on biodegradable polymers and bioactive ceramics have been developed for applications in bone repair and reconstruction. Several polymers such as polylactic acid (PLA), polyglycolic acid (PGA), polylacticcoglycolic acid (PLGA), gelatin, alginate, and chitosan (CH) are widely used for this purpose because of their proven biocompatibility and complete bioresorbability [7–13].

Chitosan (CS) is a natural biopolymer extracted from crustacean. CS is a polysaccharide-type biological polymer possessing reactive amine and hydroxyl groups that promote osteoblast growth and in vivo bone formation [14–17]. Chitosan plays an important role in the attachment, differentiation, and morphogenesis of osteoblasts, the bone forming cells, because of its structural similarities with glycosaminoglycans, a major component of bone and cartilage [18–21]. Despite its tremendous promise in bone tissue engineering application, the poor mechanical properties of chitosan limits its clinical application in weight bearing bones, which has been addressed by the addition of bioceramics in chitosan scaffolds [22].

Gelatin, a product from partial hydrolysis of collagen, has gained interest in biomedical engineering, mainly because of its biocompatibility and biodegradability. Since it contains Arg-Gly-Asp (RGD)-like sequences that promote bone cell adhesion and migration, it has been blended with chitosan to improve the biological activity of composite scaffold [23, 24]. Gelatin-chitosan scaffold has been tested for the regeneration of various tissues including skin [25], cartilage [26], and bone [27].

Bioactive glasses are a group of bioactive ceramic materials with good bioresorbability. When immersed in a physiological solution, these bioactive materials can form hydroxyl carbonate apatite (HCA) layer that is chemically and compositionally similar to the mineral phase of human bone. The formation of apatite layer triggers chemical bonding between implant biomaterials and bone tissues [28–31]. Shalumon et al. [32] investigated the effect of nanoscale bioactive glass and hydroxyapatite incorporation in PCL/chitosan nanofiber for bone and periodontal tissue engineering. Gentile et al. [33] in their study found that SiO_2-P_2O_5-CaO-MgO-Na_2O-K_2O containing bioglass in 70 wt% amount in chitosan-gelatin composite exhibit excellent bioactivity. Bioactivity and mechanical properties of composite scaffold comprising chitosan (CS) and 55S bioactive glass ceramic nanoparticles were reported by Peter et al. [34]. In another study Peter et al. [35] investigated the effect of 55S bioglass nanoparticle addition on physiochemical properties of chitosan-gelatin scaffold. The effect of 45S5 BG in Chi-Gel based scaffold on mesenchymal stem cell activity was reported elsewhere [36]. Previous reports demonstrated that 58S-BG containing 60 mol% SiO_2– 33 mol% CaO– 7 mol% P_2O_5 is suitable for bone repair due to its excellent biocompatibility, bioactivity, and biodegradability [37, 38]. None of the studies was focused on investigating the effect of addition of 58S bioglass nanoparticles into chitosan-gelatin based scaffold on mechanical properties and mesenchymal stem cell activity onto the scaffold.

Here we report the effect of compositional variation amongst chitosan, gelatin, and synthesized 58S bioglass nanoparticles on physicochemical properties, microstructure, mechanical properties, and bioactivity of the prepared scaffold. Optimization of the composition of the prepared scaffold was performed evaluating the pore size and its distribution, biodegradation kinetics, and mechanical properties. In detail scaffold-MSCs *in vitro* interaction was investigated using SEM study on cell cultured specimens, MTT assay, and immunocytochemistry after optimization of its physicochemical and mechanical properties.

2. Materials and Method

2.1. Materials. Chitosan (Mw = 2.46×10^5, degree of deacetylation = 85%) and gelatin were purchased from Sigma Aldrich (USA). Tetraethyl orthosilicate (TEOS: $C_8H_{20}O_4Si$), triethyl phosphate (TEP: $C_6H_{15}O_4P$), and calcium nitrate

tetra hydrate ($Ca(NO_3)_2 \cdot 4H_2O$) were purchased from Merck (India). Glacial acetic acid (96%) and ammonia solution (NH_4OH) were procured from LOBA chemical (India). Glutaraldehyde (GA) ($C_5H_8O_2$) and 0.1 M nitric acid (HNO_3) were purchased from Merck Inc. (India).

2.2. Sol-Gel Preparation of 58S Bioactive Glass. The composition of the studied bioactive glass was 57.44% SiO_2, 35.42% CaO, and 7.15% P_2O_5 in molar percentages and composition was chosen based on the ternary phase diagram of CaO-SiO_2-P_2O_5 [39, 40]. In brief, 14.8 g of tetraethoxysilane (TEOS) was added to 30 mL of 0.1 M nitric acid, and the mixture was allowed to react for 30 min for maximum completion of acid hydrolysis of TEOS. Distilled water was added to the solution and allowed to mix until the solution became clear. The H_2O : (TEOS) molar ratio was 12 : 1. After 30 minutes, 0.85 gm of triethyl phosphate (TEP) was added to the stirring solution followed by the addition of 7.75 gm of calcium nitrate after another 20 minutes. The solution was then stirred for an hour. The mixture was cast in a cylindrical Teflon container and kept sealed for a week at room temperature to allow hydrolysis and a polycondensation reaction to occur in course of gelation of the sol. The gel was dried at 120°C for two days in an oven. The dried powder was heated for 24 h at 700°C for nitrate elimination and stabilization of gel. Subsequently, the powders were ball milled by planetary milling (Fritsch Company, Germany) at 400 rpm for six h.

2.3. Preparation of Porous Gelatin-Chitosan/Bioglass (GCB) Composite Scaffold. The GCB scaffolds were prepared with varying amount of synthesized 58S-BG nanopowder keeping the amount of gelatin constant at 30 wt% as listed in Table 2. Our previous work [41] indicated that increase in nanoceramic phase content beyond 30 wt% in gelatin-chitosan matrix resulted in <50 μm pore size in the prepared scaffold and rendered it not ideal for cell ingrowth, hence exhibiting osteoinduction. Similarly, scaffold's average pore size decreased below 40 μm on increasing gelatin content higher than 30 wt% with nanoceramic phase content varying between 10 and 30 wt% in the scaffold. As we found that 30 wt% gelatin content scaffold exhibited the highest compressive strength, our objective in this study is to evaluate the physicochemical and mechanical properties of the scaffold on varying bioglass content from 10 wt% to 30 wt% keeping gelatin content fixed at 30 wt%.

Figure 1 illustrates the preparation processes of GCB scaffolds. First, chitosan (CS) solution was prepared by dissolving required amount of medium molecular weight chitosan in a solution containing 4 mL of acetic acid and 96 mL of deionized water with stirring for 5 h to get a perfectly transparent solution, as we have prepared in our previous work [42]. Separately, glutaraldehyde (GA) solution was prepared by dissolving 0.5 mL (50%) glutaraldehyde in 100 mL deionized water. Next required amount of gelatin was dissolved in deionized water with continuous stirring at 40°C for four h. This solution was then added to the 2 wt% chitosan solution. Different percentage (10%, 20%, and 30%) of 58S-BG was then added into the gelatin-chitosan solution and stirred for 4 h. The resulting slurry (gelatin-chitosan-bioglass) was put

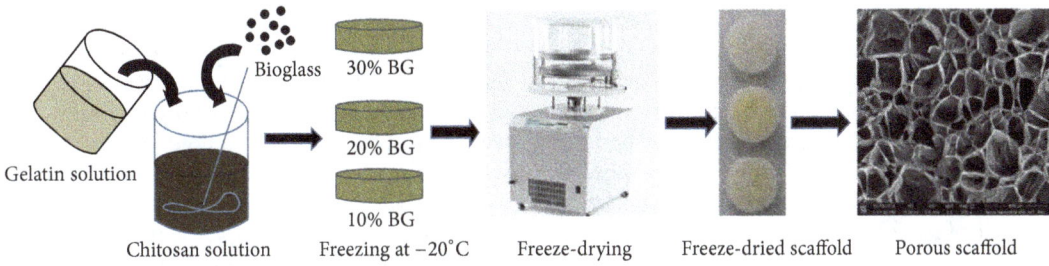

FIGURE 1: Fabrication procedure for the GCB scaffolds.

in PTFE cylindrical mould and then rapidly prefreezed at −20°C to solidify the water and kept overnight. Next, the frozen samples were lyophilized using a freeze-dryer (Labconco, USA) at −52°C for 24 h. Finally, samples were soaked in 1 wt% GA solution for 24 h and then carefully washed with deionized water to remove the remaining amount of GA. The cross-linked scaffold was further treated with sodium borohydride aqueous solution to block residual aldehyde groups and then washed with ethanol-NaOH followed by deionized water. The washed porous scaffold was again freeze-dried to obtain the hybrid network of gelatin-chitosan-bioglass.

2.4. Characterization

2.4.1. Particle Size Analysis Using Dynamic Light Scattering (DLS).
Particle size of synthesized bioglass nanopowder was measured using (DLS) technique by Malvern Zetasizer (ZEN3690, USA). Two mg of dried powder was suspended in 50 mL water after ultrasonication, and 2 mL of the suspension was taken in a cuvette and used for the measurement of particle size.

2.4.2. X-Ray Diffraction (XRD) Analysis.
The phases of composite scaffolds were characterized by X-ray diffraction (XRD) using fully automated X-ray diffractometer (Panalytical, USA) fitted with Ni filter. The diffraction patterns were recorded with a XRD analyser using Cu-Kα radiation (λ = 1.542 Å) at 35 kV and 40 mA. The samples were scanned in the interval of $10° < 2\theta < 70°$ at a scan speed of 2°/min with step size .02° in a continuous mode.

2.4.3. Fourier Transform Infrared Spectroscopy (FTIR) Analysis.
The characteristic peaks of the composite scaffold were analysed using Fourier transform infrared (FTIR) (Perkin Elmer, USA) spectroscopy. The scaffolds were put in a vacuum oven at 50°C for 48 h before they were ground to a suitable size for IR analysis with spectrometer. The pellet for the FTIR measurement was prepared by mixing the sample (2 mg) with 200 mg of IR-grade KBr. The absorption spectra were measured using IR spectrometer at a wavenumber of $4000–400\,cm^{-1}$ with a resolution of $1\,cm^{-1}$.

2.4.4. Scanning Electron Microscope (SEM) Analysis.
The composition of prepared bioglass powder was studied using energy dispersion X-ray (EDX) analysis. The morphology and microstructure of the scaffolds have been observed using field emission scanning electron microscopy (FESEM) (Novanano 450, FEI, Netherlands) operated at 78 μA, 15 KV. The surface of the scaffolds was coated with thin layer of gold (Au) and then placed inside the FESEM chamber.

2.4.5. Mechanical Behaviour (Compression Test).
The mechanical properties of the composite scaffolds were determined using a Universal Testing Machine (Tinius Olsen, UK) with a crosshead velocity of $1\,mm\,min^{-1}$ and a 1000 N load cell. For compressive testing, the samples (n = 5) were cylinders of approximately 15 mm in diameter and 6 mm in height in accordance with the compression mechanical test guidelines set in American Standard Test and Measurement (ASTM F 451-95). Specimens were compressed to ~40% of their original thickness and the values were expressed as the means ± standard error.

2.4.6. Porosity of Scaffold.
The porosity of composite scaffold ($1 \times 1 \times 1\,cm^3$) was measured using Archimedes principle with xylene as liquid medium using the following equation:

$$\text{Porosity } (\%) = \frac{\text{Soaked weight of the sample} - \text{Dry weight of the sample}}{\text{Soaked weight of the sample} - \text{Suspended Weight of the sample in liquid}} \times 100\%, \tag{1}$$

where W_1 is the weight of the sample in air, W_2 is the soaked weight of the sample in xylene, and W_3 is the weight of the sample suspended in xylene. Pore size distribution in the scaffold was measured using Hg porosimeter (Quanta Chrome, USA).

2.4.7. Swelling Index.
The swelling properties of the chitosan-gelatin scaffolds were investigated according to a method described earlier [43]. Briefly, the scaffolds were fully immersed in the phosphate buffer solution (pH 7.4) at room temperature for 60 min. Samples were recovered after

every 10 minutes of soaking in PBS and excess water onto scaffold surface was extracted out using distilled paper. The wet weight of the scaffold (W_s) was determined using an electronic balance and recorded. Next the swollen scaffold was dried in an oven at 50°C overnight and weighed and the dry weight was recorded as W_d. Three scaffolds were selected for each period and weighed under the same conditions. The swelling percentage of scaffolds was calculated as follows:

$$S = \left[\frac{(W_s - W_d)}{W_d} \right] \times 100\%. \tag{2}$$

2.4.8. In Vitro Biodegradability. The scaffolds were immersed into the phosphate buffer fluid PBS solutions for degradation assessment by monitoring the weight loss. The scaffolds were immersed in PBS (pH 7.4) at 37°C. At a predetermined day interval up to 14 days, after being incubated for various time durations the scaffolds were taken out from the medium, washed with distilled water, and freeze-dried. The degradability ratio D was calculated as follows:

$$D = \left[\frac{(W_o - W_t)}{W_o} \right] \times 100\%, \tag{3}$$

where W_o denotes the original weight and W_t is the weight of the sample after immersing in PBS up to day t. Each biodegradation experiment was conducted on three samples, and the average value was taken as the percentage of biodegradation.

2.5. Cell Culture Study

2.5.1. Attachment and Morphology of Cell on GCB Composite Scaffolds. Human umbilical cord mesenchymal stem cells (MSCs) were grown in a Dulbecco's modified Eagle's medium (DMEM, Sigma Aldrich, UK) supplemented with 10% fetal bovine serum (FBS, Sigma Aldrich, UK) and 100 U/mL penicillin-streptomycin. Cells were propagated in a 75-cm^2 T-flask (Corning, NY, USA) in a humidified atmosphere containing 5% CO_2 at 37°C. Ethylene oxide sterilized GCB 30 scaffold disks were placed inside a 24-well plate and MSCs were seeded at 1×10^6 cells per well in the scaffolds. Cell culture medium was replaced after every 3 days. At 3, 5, and 7 days after cell seeding, GCB 30 disks were taken out from the cell culture medium. GCB 30 disks were first briefly rinsed in cold 0.1 M phosphate buffer solution (pH 7.4) twice. The cells were then fixed in 2.5% glutaraldehyde in 0.1 M phosphate buffer for 3 h. The fixative was removed, and cells were rinsed in 0.1 M phosphate buffer and in distilled water. Afterwards, the MSC cultured GCB 30 disks were dehydrated in a graded series of ethanol solution (50%, 70%, 80%, and 90% absolute ethanol), 100% pure acetone, and hexamethyldisilazane (HMDS). Next the samples were desiccated under vacuum and were coated with gold before the cell attachment; spreading and morphology were investigated using FESEM at 10 KV.

2.5.2. Analysis of Viability Differentiation of MSc Cultured onto the Scaffolds. Cell viability of MSCs cultured onto

the scaffolds was determined after performing [3-(4,5-dimethylthiazol-2-yl)-2,5-diphenyltetrazolium bromide] (MTT) assay. We note that, for the MTT assay, the amount of produced formazan crystal is proportional to the number of viable cells present in the scaffold [44]. In brief, the scaffolds were submerged in DMEM cell culture media and MSCs were seeded on each scaffold at 1×10^6 cells per well and incubated for 3, 5, and 7 days with replacement of DMEM after every 3 days of culture. After each time interval of cell culture, 20 μL (0.5 mg/mL) MTT solution was added onto each scaffold placed in 24-well plate. The scaffolds containing MSCs and MTT solution were then incubated for another 3.5 h at 37°C. The precipitated purple coloured formazan crystals were dissolved in 150 μL of MTT solvent by shaking the plate for 15 min. Then the solutions were taken out in ependrofs and centrifuged and the supernatant was transferred to another 96-well plate to record the absorbance at 595 nm using a microplate reader (Bio-Tek ELx800).

2.5.3. Inmunofluorescent Imaging of Cell Markers: Osteocalcin and RUN-X2. The expression of RunX2, an osteogenic marker, was examined using immunocytochemistry. MSCs were seeded on scaffolds and incubated as mentioned earlier. Cells on scaffolds were fixed and permeabilized using 100% methanol (precooled to −20°C) for 10 min. Blocking was done using FBS. Staining of RUN-X2 was achieved using rabbit RUN-X2 antibody (1 : 200, Abcam) and using Alexa 647 anti-rabbit (1 : 400, Invitrogen) as secondary antibody. Staining for osteocalcin was performed using rabbit anti-mouse osteocalcin (1 : 100, Abcam) and Alexa 488 anti-mouse (1 : 400, Invitrogen) as a secondary antibody. Counter staining with diammonium propidium iodide (DAPI) was performed for 2 min after washing with PBS. To trace actin filament in the cell, green phalloidin Alexa Fluor 488 counter stained with DAPI was used. Finally the samples were mounted with Fluorsave Vectashield mounting medium with DAPI. Samples were mounted on glass slides and confocal images of the samples were acquired using Leica TCS SP5 X Supercontinuum Confocal Microscope.

2.5.4. Statistical Studies. All data are expressed as mean ± standard deviation. The data were compared using Student's t-test and differences were considered significant when $^*p < 0.05$. A p value more than 0.05 ($p > 0.05$) was taken, indicating no significant difference.

3. Result and Discussion

3.1. XRD and FTIR Analysis. Figure 2(a) shows the FT-IR spectrum of synthesized bioglass powder. The characteristic bands for different functional groups present in the powder are shown in Table 1. The band at 1638 cm^{-1} corresponds to carbonate ($CO_3{}^{2-}$) coming from atmospheric CO_2 and attached to Ca^{2+}. The band at 1022 cm^{-1} arises from ν^3 PO_4, whereas the band at 653 cm^{-1} suggests the presence of ν^4 PO_4. The band at 836 cm^{-1} was ascribed to the stretching vibration of Si-O− groups in bioglass. The absorption bands observed at 1076 cm^{-1} and 540 cm^{-1} were assigned

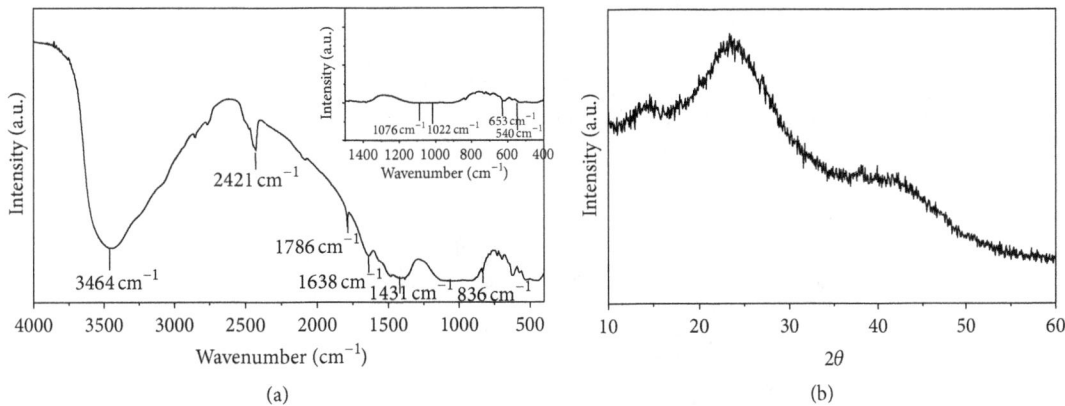

(a)   (b)

FIGURE 2: FTIR and XRD pattern of 58S bioglass powder prepared via sol-gel process.

TABLE 1: Component of prepared scaffolds.

Specimen name	Gel concentration (w/w) %	Chitosan concentration (w/w %)	CS-Gel/BG ratio (w/w) %	Solid loading (%)
GCB-0	30	70	100/0	40
GCB-10	30	60	90/10	40
GCB-20	30	50	80/20	40
GCB-30	30	40	70/30	40

TABLE 2: Peaks of infrared spectra assigned to synthesized bioglass powder.

Bond	Infrared frequency (cm^{-1})
Si–O–Si bending	540, 1076 cm^{-1}
P–O bending vibration	1022, 653 cm^{-1}
–OH	3456 cm^{-1}
CO_3^{2-}	1638 cm^{-1}
Adsorbed molecular water	1786, 1431 cm^{-1}

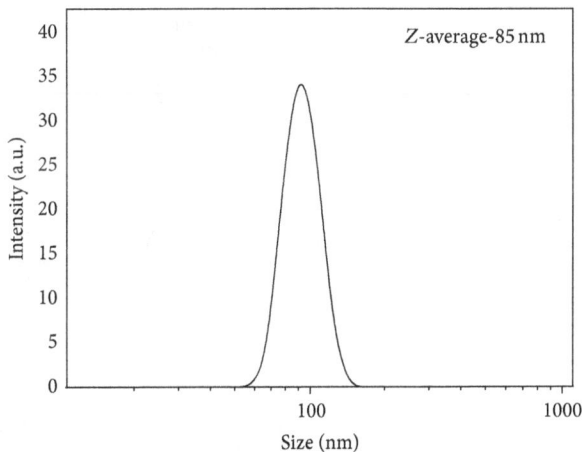

FIGURE 3: The particle size distribution of NBG measured by DLS.

to stretching vibration of Si–O–Si, respectively, and bending of Si–O–Si [45]. Typical absorption band at 1786 cm^{-1}, 1431 cm^{-1} is attributed to the deformation mode of adsorbed molecular water in the pores. Phase analysis of the composite nanopowder was performed using XRD. Figure 2(b) clearly indicates the amorphous nature of the synthesized bioglass powder.

3.2. Particle Size, Morphology, and Composition of Nanobioglass Powder. Figure 3 represents DLS particle size measurement data of synthesized 58S bioglass nanopowder. An average particle size of 85 nm as obtained from the DLS measurement was little bit higher as compared to the particle size shown in the FESEM image in Figure 4(a). Particle size measurement by DLS technique gives the hydrodynamic diameter of the particle and also bioglass particles were to some extent in agglomerated state in the suspension that suggest higher average particle size value in the DLS measurement. From FESEM micrograph in Figure 4(b) it is evident that the bioglass particle was spherical in morphology and was agglomerated in course of drying on a carbon film (Figure 4(a)). EDS analysis in Figure 4(c) and Table 3

suggests that the synthesized powder with a composition of Si : Ca : P = 36 : 21 : 7 closely resembled the theoretical composition.

3.3. XRD Analysis of Composite Scaffold. Phase analysis of the composite scaffolds shown in Figure 5 was performed using XRD. The characteristic diffraction peaks for both chitosan and gelatin were suppressed by the huge amorphous peak of bioglass observed in the range between 2θ equal to $20°$–$40°$. The broad amorphous peaks of bioglass confirmed that the synthesized scaffolds were predominantly amorphous.

3.4. Chemical Structure Study of Composite Scaffold. The chemical structure of composite scaffolds was investigated

(a)   (b)

(c)

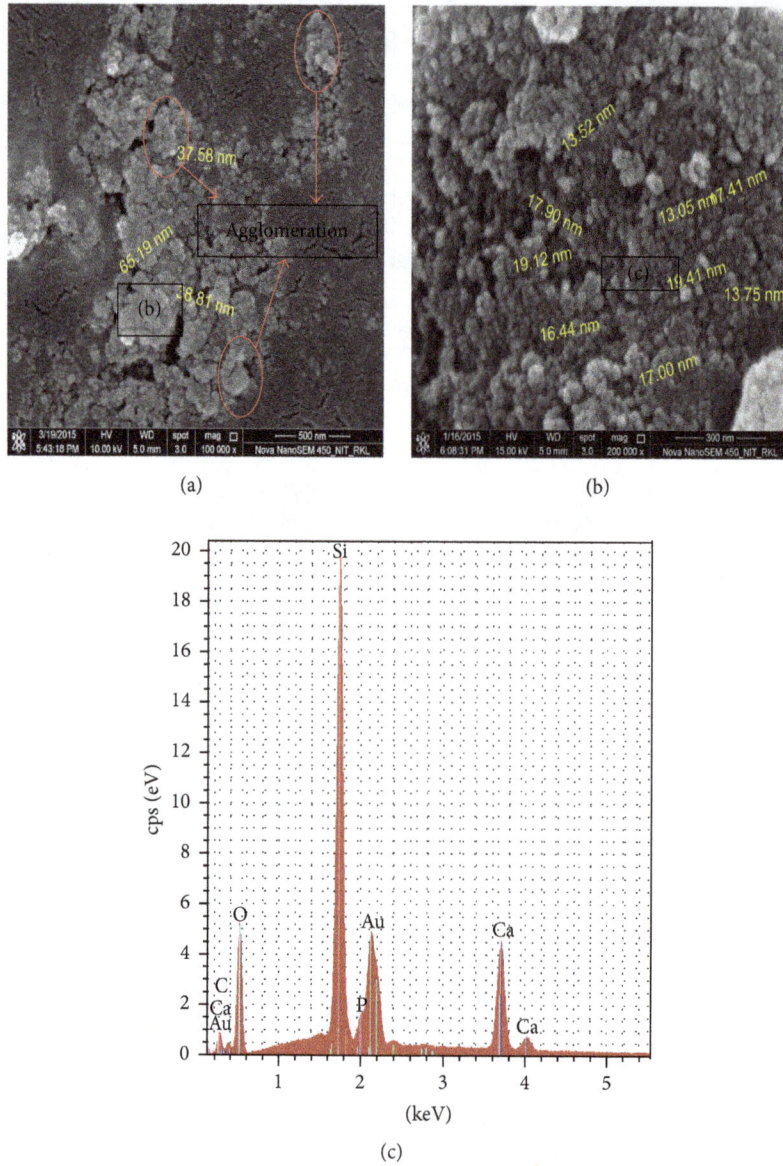

FIGURE 4: (a) FESSEM micrograph of synthesized bioglass nanopowders, (b) magnified image of particle agglomerate, and (c) elemental composition analysis of the synthesized bioglass nanoparticles.

TABLE 3: EDS analysis of bioglass powder.

Elements	Bioglass atomic percent (%)
Si	36.14
Ca	20.88
P	7.02
O	35.89

by FTIR, in order to examine the chemical interactions between the gelatin, chitosan, and bioglass phases, as shown in Figure 6. Apart from corresponding peaks of PO_4^{3-} at 462 and 660 cm^{-1}, Si–O–Si in bioglass at 1076 cm^{-1} and CO_3^{2-} bands at 828 cm^{-1}, bands for amide I at 1662 cm^{-1}, amide III at 1232 cm^{-1}, and carboxylate at 1446 cm^{-1} in

gelatin, a distinct band at 1552 cm^{-1} was detected. This absorption band signifies the formation of –C=N– bond due to intergelatin and chitosan-gelatin cross-linking to develop 3D interconnected network in the scaffold [46]. The appearance of an amide I mode at 1662 cm^{-1} indicates that bioglass-gelatin composites adopted a predominantly α-helical protein configuration, and this was further confirmed by the appearance of amide II mode at 1552 cm^{-1}. The band at 1330 cm^{-1} is attributed to the chemical bond formation between carboxylate group in gelatin and Ca^{+2} ion in BG to bind the particulate reinforced composite scaffold together [47, 48]. These interactions between the gelatin, chitosan, and silica phases would enhance the stability of the composite gelatin-chitosan-bioglass hybrid scaffolds in a water-based environment.

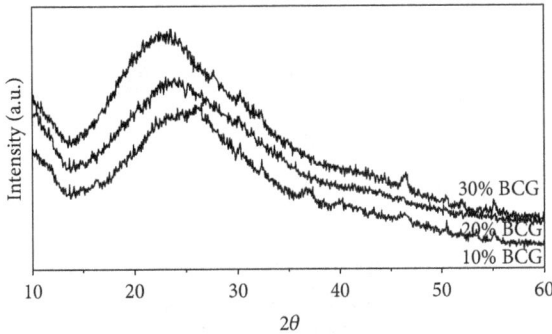

FIGURE 5: XRD analysis of composite scaffold with varying amount of bioglass powder.

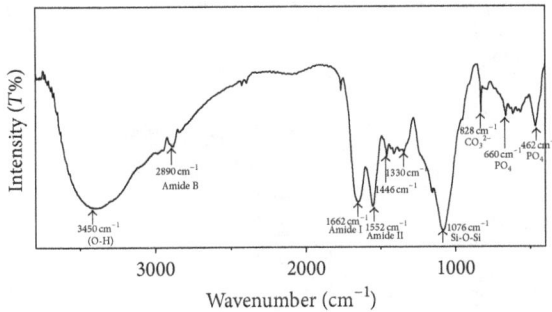

FIGURE 6: FTIR spectra of gelatin/chitosan/bioglass (GCB 10) composite scaffold.

TABLE 4: Pore diameter and porosity of composite scaffold.

Sample name	Average pore diameter (μm)	Porosity
BCG 0	400 ± 13.6	89.3 ± 7.8
BCG 10	250 ± 26.3	82.4 ± 5.0
BCG 20	160 ± 17.8	80.8 ± 3.3
BCG 30	100 ± 25.9	81.3 ± 6.1

confirmed subsequently by FESEM examination. All results demonstrate that the prepared scaffolds exhibited a pore size distribution in the range of 100–250 microns possessing rough pore wall, crucial for protein and cell adhesion.

3.6. Analysis of Pore size and Its Distribution in the Scaffold. Table 4 indicates the total porosity and means pore diameter values of the different scaffold with varying composition. The mean porosity value which varied between 81% and 89% suggests that compositional variation did not affect the total porosity in a significant manner (all p values > 0.05). Pore size distribution data in Figure 8 also shows the unimodal pore size distribution for all the prepared scaffold. The most frequent pore diameter in the scaffold was decreased from $253\,\mu m$ to nearly $100\,\mu m$ as we increase bioglass content from 10% to 30% in GCB slurry. All the scaffold showed the ideal pore size distribution for exhibiting osteoinductivity and osteoblast or stem cell ingrowth into the scaffold as the pore size distribution in the scaffold falls within $100–400\,\mu m$.

3.5. Microstructure Study. The physical characteristic of a scaffold can be described by its average pore size, pore interconnectivity, and pore shape. Pores are necessary for the bone tissue formation because they allow migration and proliferation of osteoblast and mesenchymal stem cells, as well as the proper vascularization of the implant [49]. The morphology of BCG scaffolds before and after n-bioglass incorporation is shown in Figure 7. FESEM micrograph of composite scaffolds (Figure 7) showed that scaffolds were macroporous in nature. The porosity was found to vary from 81 to 89% depending on the percentage addition of bioglass powder. The pore size of GCB scaffolds with 10 wt% bioglass was around 200–250 μm. As the bioglass content increased to 30 wt%, the pore size was reduced to 100–120 μm and pore shape became irregular. With the augmentation of bioglass content in GCB slurry, the interactions between bioglass particles and gelatin increased with consequent increase in solution viscosity. Thus as we go from 10 to 30 wt% bioglass content in scaffold, a higher and higher force was necessary for GCB slurry to be expelled by water molecules, so the ice crystals growth was hindered during freezing of slurry resulting in reduced size of intertrapped ice blocks and scaffolds with smaller pores were formed during subsequent sublimation [50]. Top views (Figures 7(a)–7(d)) and side views (Figures 7(a_1)–7(d_1)) representations of the scaffold are shown in Figure 7 that were used to calculate the average pore size. In particular, with an increase in bioglass content, increasing amount of bioglass particles was deposited onto the chitosan-gelatin walls (Figures 7(a_{11})–7(d_{11})) as

3.7. Mechanical Properties. Figure 9(a) describes the stress-strain behaviour of GCB composite scaffolds under compression. The stress-strain curve can be decomposed in two stages. The first is the elastic region before yield point and second is the postyield stage, that is, deformation region and then densifying region where the pore wall collapses. In this case, compressive strength of the scaffold is denoted by the onset of deformation region in load elongation curve. With increase of bioglass concentration from 10% to 30%, both compressive strength and elastic modulus continued to enhance significantly as in Table 5. The compressive strength and elastic modulus of pure gelatin scaffold are only 0.8 ± 0.16 MPa and 50 ± 5.23 MPa, respectively (Table 5). After incorporating 30 wt% bioactive glass particles in gelatin-chitosan matrix, the compressive strength and modulus, respectively, increased to 2.2 ± 0.02 MPa and 111 ± 12.09 MPa, respectively. The composite scaffold was elastic to yield point but after yield point some microcrack occurred. The interaction between bioglass particles and polymer network bridges the cracks in the scaffold which result in postponed final fracture. It is noteworthy that the slope of the deformation region in 30 wt% bioglass loaded scaffold was the maximum that was due to enhanced resistance from the ceramic particle to crack propagation.

3.8. Biodegradation Study. Scaffold materials are expected to be biodegradable and bioresorbable with a proper rate to match the speed of new bone tissues formation. The controlled and steady degradation behaviour of bone scaffold

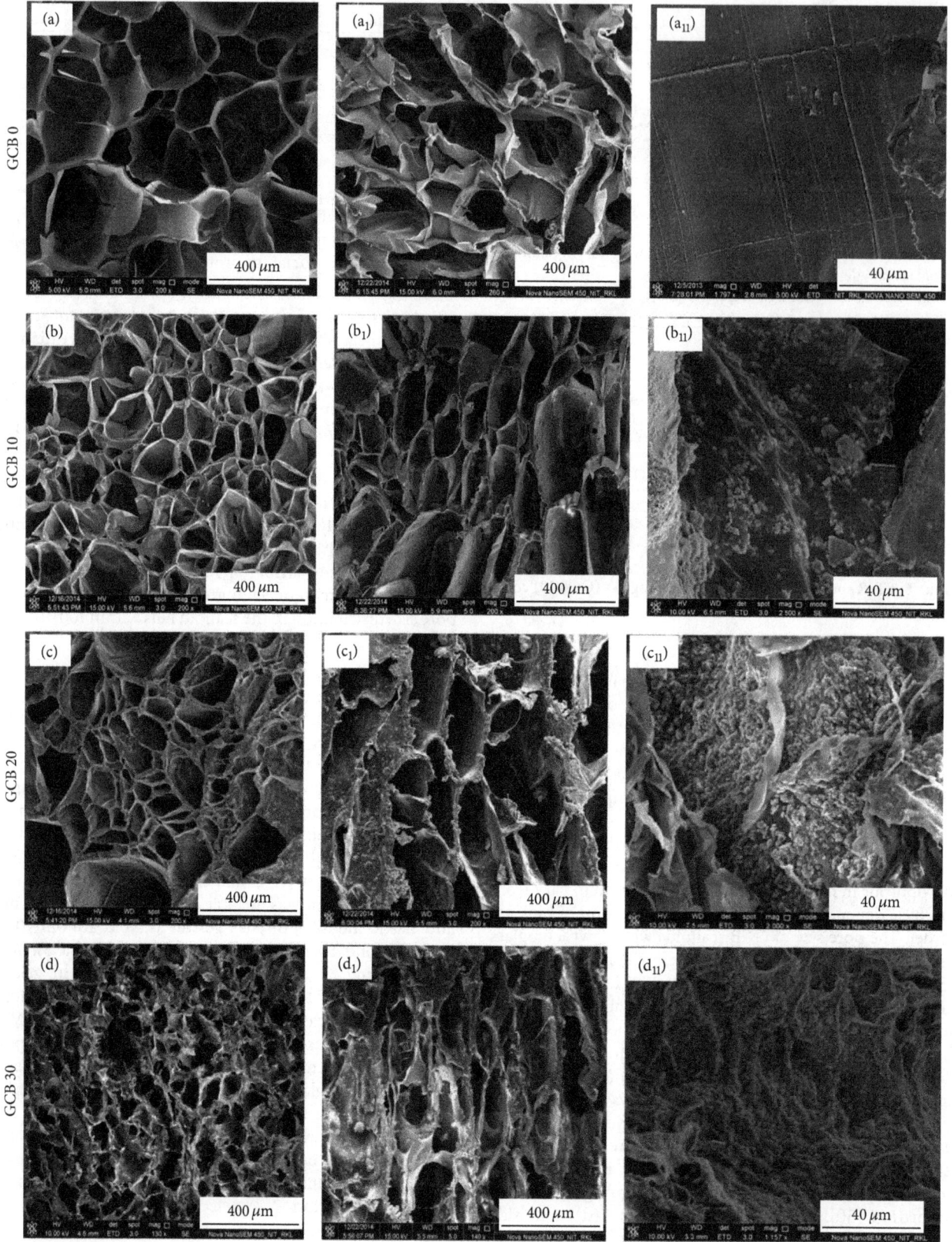

FIGURE 7: FESEM microstructure of composite scaffolds. Cross-sectional Figures 7((a)–(d)), longitudinal Figures 7((a$_1$)–(d$_1$)), and pore wall distribution Figures 7((a$_{11}$)–(d$_{11}$)).

TABLE 5: Summery of mechanical properties of GCB scaffolds prepared from solutions of different bioglass concentrations (wt%). Data refer to mean value ± standard deviation.

Bioglass scaffolds specimens	Porosity (%)	Mechanical properties	
		Elastic modulus* (MPa)	Compressive strength* (MPa) (at 40% strain)
BG 0%	89.3 ± 7.8	50 ± 5.23	0.8 ± 0.16
BG 10%	82.4 ± 5.0	55 ± 7.12	1.2 ± 0.01
BG 20%	80.8 ± 3.3	68 ± 10.10	1.6 ± 0.01
BG 30%	81.3 ± 6.1	111 ± 12.09	2.2 ± 0.02

*Except between sample BG 0% and BG 10% ($p > 0.05$), all values in each mechanical property category were found to be significantly different from each other ($p < 0.05$, by Student's t-test, $n = 5$).

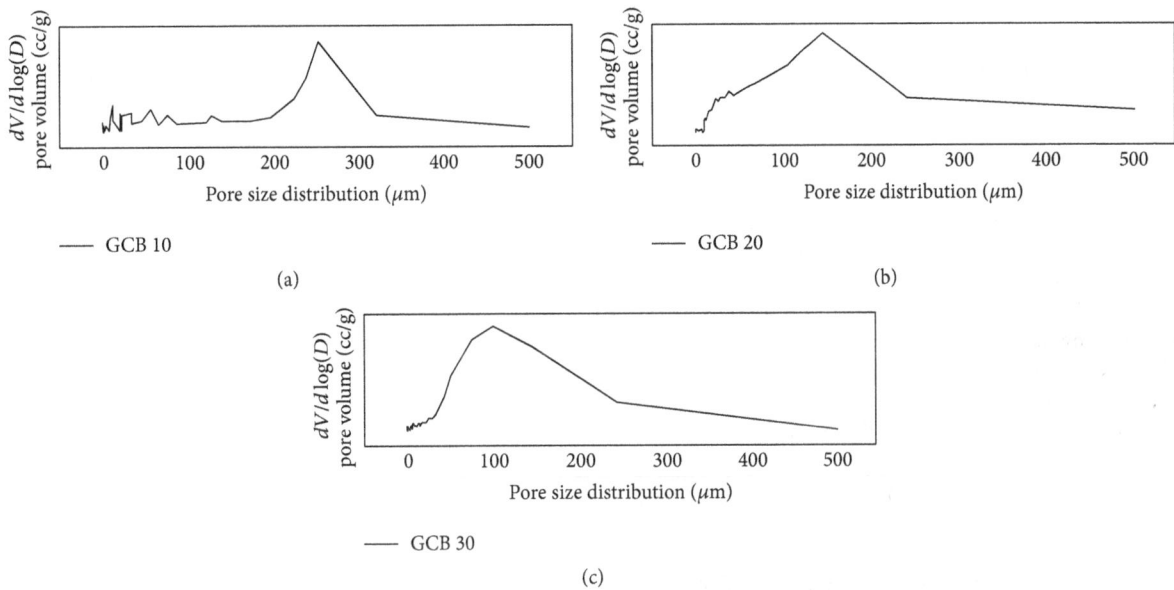

— GCB 10

(a)

— GCB 20

(b)

— GCB 30

(c)

FIGURE 8: Pore size distribution data of gelatin-chitosan-bioglass scaffold: (a) GCB 10, (b) GCB 20, and (c) GCB 30.

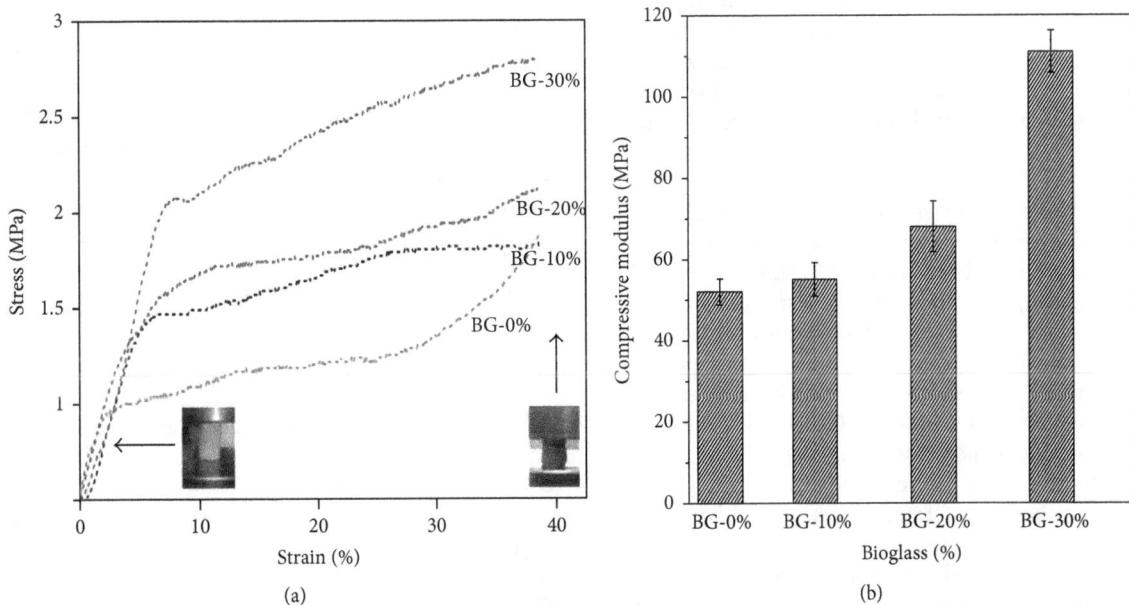

(a)

(b)

FIGURE 9: Mechanical properties of GCB scaffolds prepared at different bioglass concentrations and acquired in (a) compression test and (b) Young's modulus. The corresponding porosity and moduli are shown in Table 5.

in physiological environments plays an important role in the regeneration of new bone tissues. The *in vitro* degradation of GCB scaffold in phosphate buffer solution (PBS) at 37°C is an important study to assess its resorbability under physiologically relevant conditions. For long-term mechanical stability and reliability, scaffold material should exhibit slow and controlled degradation behaviour. Figure 10 demonstrates the effect of bioglass contents on the degradation rate of the scaffold. With an increase in the bioglass content, the degradation rate in PBS solution at 37°C decreased significantly. The bioglass particulate interacted through hydrogen bonded and ionic interaction with the water molecules and weakened the interaction between the gelatin macromolecule and water. The presence of bioglass also acted as a physical crosslinking sites to enhance the stability of the chitosan-gelatin network. Thus, the degradation rate and as a result strength retention in the physiological environment may be controlled by adjusting the bioglass contents in the scaffold.

3.9. Swelling Studies. The ability of the scaffold to swell plays an important role during the *in vitro* cell culture studies. Swelling of scaffold allows absorption of body fluid and transfer of cell nutrients and metabolites inside the scaffold. Swelling also increases the pore size and total porosity, thus maximizing the internal surface area of the scaffolds for cell infusion and attachment. However, swelling under physiological condition must be controlled; otherwise it may cause weakening and rapid degradation of the bone scaffold.

Figure 11 shows that swelling in the GCB scaffold increased initially and gradually attained saturation after 10 hours of soaking in PBS at 37°C. The degree of swelling was found to decrease with increasing bioglass amount in the scaffold that may be attributed to the stronger bonding interactions between cationic sites of inorganic phase and carboxylate group of gelatin in the polymeric imparting higher elasticity to the composite scaffold. Thus the increase in bioglass and decrease in hydrophilic chitosan content in the scaffold resulted in a decreased water sorption. Moreover, higher elasticity in composite scaffolds due to enhanced polymer-bioactive glass interactions resulted in slower relaxation of polymer chains that also accounted for decreasing the swelling ratio.

3.10. Cell Attachment Study on Scaffolds. Figure 12 shows cell density and morphology after culturing MSCs on GCB 30 for different days of culture time. Cell densities, as well as number of lamellipodia and filopodia extensions from the cytoskeleton of MSCs, were increased with progress in cell culture time on the composite scaffold. The cell presented a round shape initially (Figure 12(a)) and became elongated with increasing time of culture. After 14 days, MSCs cells adopted a polygonal morphology and spread well on the scaffolds (Figure 12(c)). These results clearly exhibit that MSCs were attached, proliferated, and spread with increasing cell culture time.

Figures 12(a_1)–12(c_1) show confocal micrographs of MSc cells on the GCB 30 scaffolds after 1, 3, and 14 days of culture. Confocal images revealed that a higher number of cells were attached to the scaffold on increasing days of cell culture time.

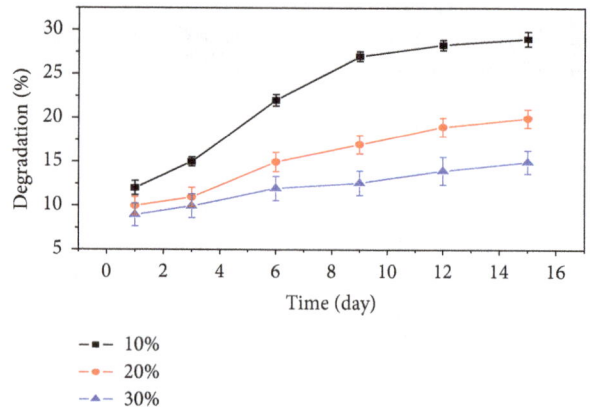

FIGURE 10: *In vitro* degradation of the GCB with different bioglass contents in PBS.

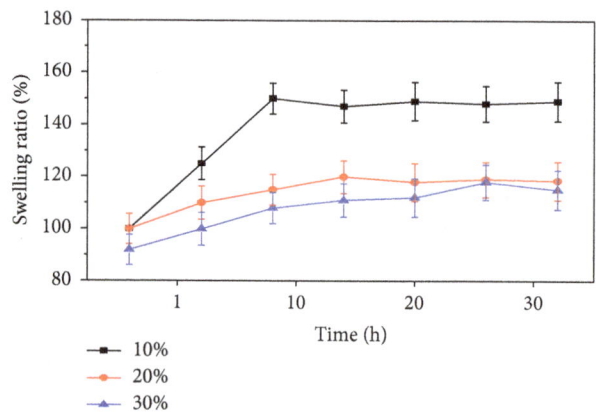

FIGURE 11: Effect of bioglass on swelling behaviour of GCB composite scaffolds as a function of time.

Figures 12(b_1) and 12(c_1) show a uniform interconnecting cytoskeleton of MSCs network on the GCB 30 scaffolds after culturing cells for 3 and 14 days. The increasing numbers of lamellipodia and filopodia extensions from the cytoskeleton of MSCs were evident with progress in cell culture time as in Figures 12(a_1)–12(c_1). Extensive networks of polymerized β-tubulin and F-actin filaments as well as multiple cell-cell contacts indicate a higher degree of active cell spreading, movement, and signalling events with progress in cell culture. All the results revealed higher proliferation of MSCs on GCB 30 scaffolds after 14 days of cell culture.

3.11. MTT Assay Study. After culturing it on the scaffold, MSCs viability study for 3–7 days was performed using MTT assay with cell culture media as negative control and cultured MSCs on sole tissue culture plate as a positive control. Keeping up with the same trend as exhibited by the positive control the scaffold material showed higher and higher number of the viable cell with progress in cell culture time as in Figure 13. The cell density on five-day cell cultured GCB 30 sample was significantly higher (p value = 0.018) from that on three-day sample. Again seven-day cell cultured GCB 30 sample showed significantly higher ($p < 0.001$)

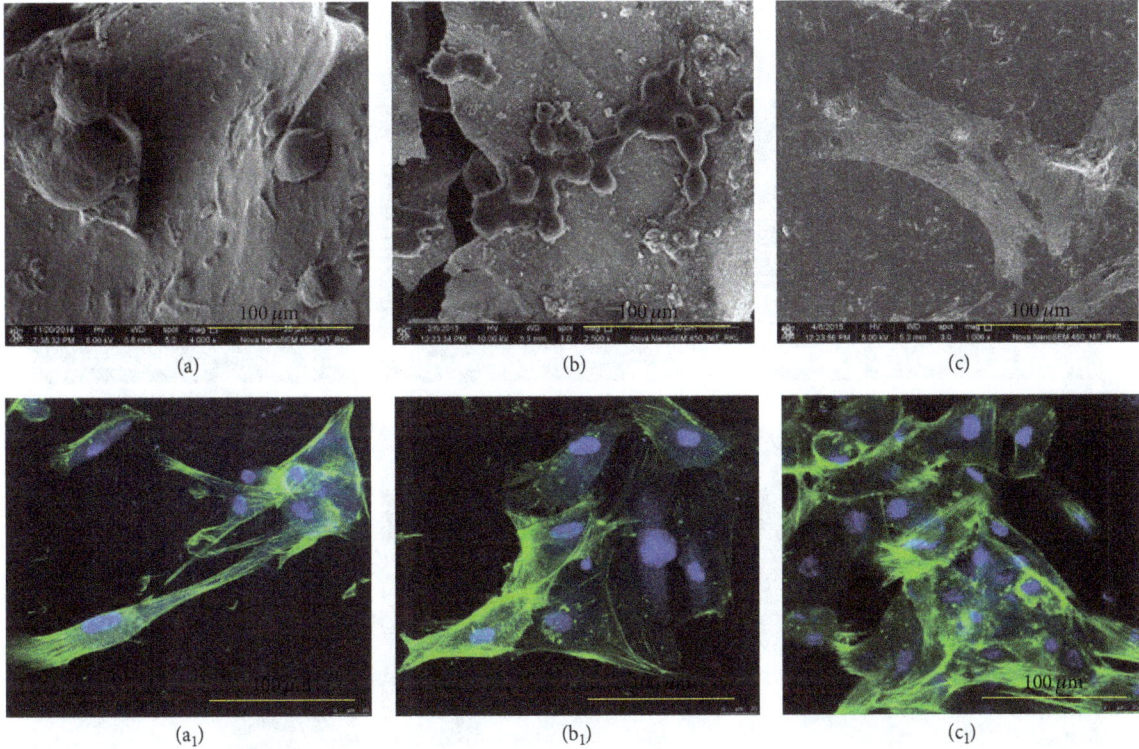

FIGURE 12: FESEM image of MSC on gelatin-chitosan-bioglass (GCB 30) for (a) 1 d, (b) 3 d, and (c) 14 d of cell culture. Confocal images of cytoskeleton of MSC on GCB 30 ((a$_1$), (b$_1$), and (c$_1$)).

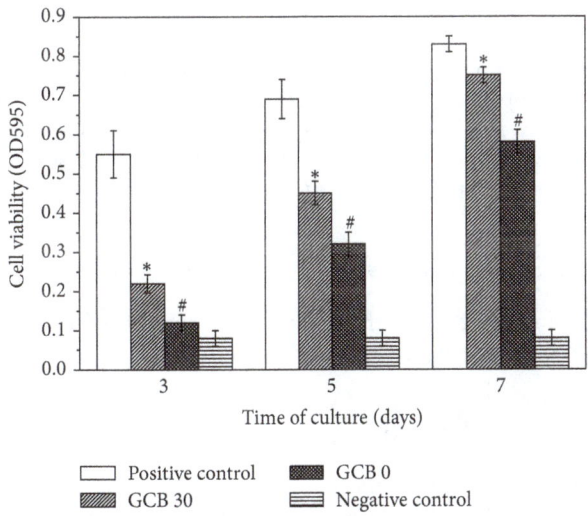

FIGURE 13: Cell viability after 3, 5, and 7 days of culture on composite scaffold of GCB 30 at 30/40/30 weight ratio as determined by MTT assay compared with glass substrate (positive control; paired with each material) on days 3 and 7. The control values were normalized to 100%. $^*p < 0.05$, $^\#p < 0.05$ compared with the corresponding control group (mean ± SE).

viable cell density as compared to that on day 5 sample. For all incubation periods, GCB 30 presented significantly higher ($p = 0.026$) cell viability than that on GCB 0 scaffold which suggests that addition of 58S bioglass nanoparticles in the scaffold promoted better cell adhesion and proliferation. 58S nanoparticles helped in apatite mineralization onto the scaffold in the presence of cell culture media and acted as sites for cell adhesion through integrin mediated interactions from the MSCs. This showed that the selected scaffold was conducive to cell attachment and proliferation.

3.12. RUNX2 and Osteocalcin Expression. In order to assess the effect of material composition on cell differentiation and osteogenesis, expression of osteoblastic transcription factor RUN-X2 and bone noncollagenous protein osteocalcin (OC) were studied using immunofluorescent markers. As can be seen in Figures 14(a)-14(a$_1$) no RUNX2 marker was visible from 1-day cell cultured sample indicating nascent and premature stage of osteoprecursor cells. RUNX2 expression appeared strongly positive in the GCB 30 scaffold cultured for 7 days, supporting the differentiation of MSCs into preosteoblast and osteoblast. There was severalfold decrease in the level of expression of most of the genes on GCB 0 scaffold (Figures 14(b)-14(c)) as compared to the GCB 30 (Figures 14(b$_1$)-14(c$_1$)) at different stages of differentiation, which supports the fact that bioglass addition actually increased the bioactivity of the overall scaffold. RUNX2 is a marker for osteoblast differentiation, and an increase in the specific activity of RUNX2 with progress in cell culture time in a population of mesenchymal stem cell indicates a corresponding shift to a more differentiated state (Figures 14(b$_1$)-14(c$_1$)).

Advanced Principles and Techniques in Tissue Engineering

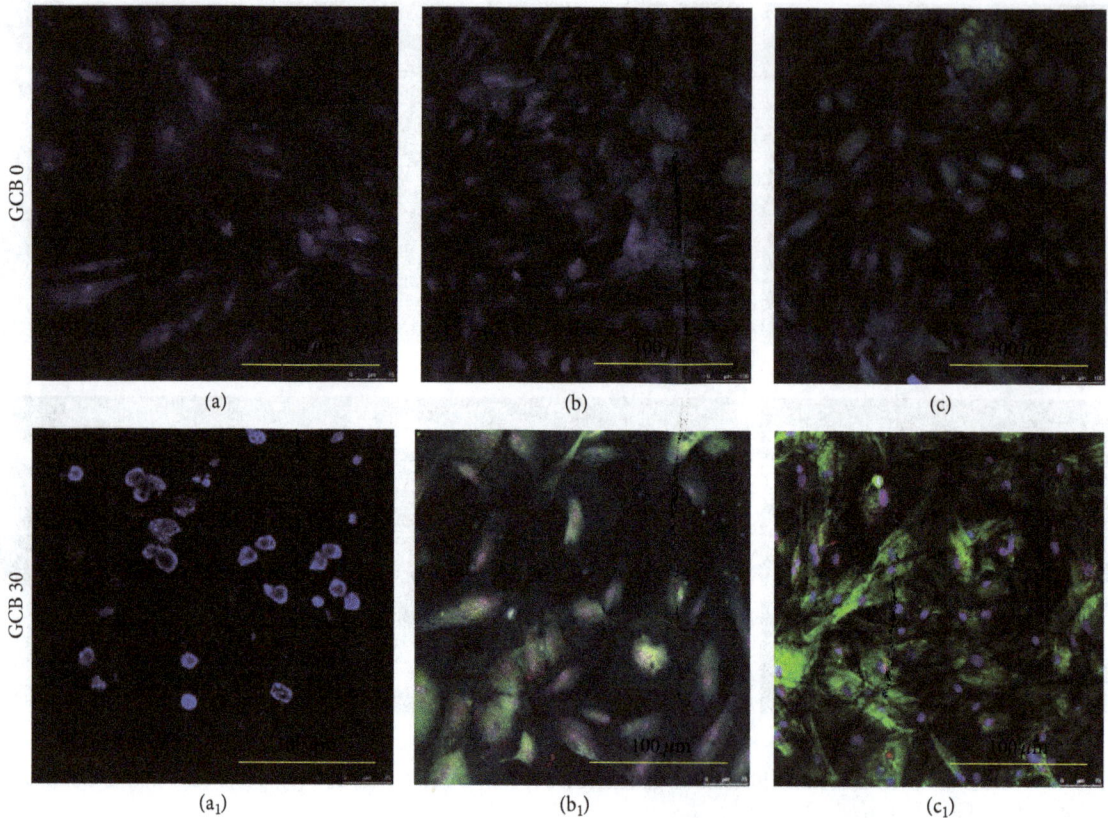

FIGURE 14: Fluorescence image of MSC's cells cultured on GCB 30 scaffolds prepared from solutions with bioglass concentration of 30% on (a) 1 day, (b) 7 days, and (c) 14 days. In fluorescence images, osteocalcin is stained green, RUNX2 is stained red, and nuclei are stained blue.

Osteocalcin is a primary noncollagenous protein produced by osteoblasts, which signals termination of osteogenic differentiation and is commonly used to measure bone cell lineage and new bone formation [51]. Greater osteocalcin (green) deposits were seen in scaffolds of 14-day cell culture indicating the presence of higher amount of osteoprogenitor cells with progress in cell culture time (Figure 14(c_1)). Thus, our result suggests that MSCs in GCB30 scaffold were well committed to osteogenic lineage.

4. Conclusions

In summary, we have successfully fabricated GCB nanocomposite scaffolds using freeze-drying method. The scaffolds were highly porous with total porosity of about 80% and average pore size in the scaffold fell to nearly 100 μm from 250 μm with increase in bioglass content from 10 wt% to 30 wt% in the gelatin-chitosan matrix. The bioglass particles (BG) were well distributed in gelatin-chitosan matrix, significantly improving the compressive strength and elastic modulus of the scaffolds. Thus, GCB 30 scaffold showed a compressive strength and elastic modulus value comparative to that of natural cancellous bone. It was found that the swelling behaviour of the scaffolds was reduced on the increase in 58S-BG nanopowder content in the scaffold. Biodegradation test in PBS showed that the increase in 58S-BG content resisted the biodegradability of the scaffold. Preliminary results on

cell culture using MSCs suggested that cells could adhere, spread, proliferate, and differentiate very well onto GCB 30 scaffolds. MSCs were also found to transform into the new bone within 14 days of cell culture on the GCB 30 scaffold making them promising artificial bone grafts.

Conflict of Interests

The authors declare that there is no conflict of interests regarding the publication of this paper.

Acknowledgment

The authors acknowledge the Department of Ceramic Engineering at NIT, Rourkela, for providing the necessary supports to carry out above research work.

References

[1] M. M. Stevens and J. H. George, "Exploring and engineering the cell surface interface," Science, vol. 310, no. 5751, pp. 1135–1138, 2005.

[2] S. Gronthos, P. J. Simmons, S. E. Graves, and P. G. Robey, "Integrin-mediated interactions between human bone marrow stromal precursor cells and the extracellular matrix," Bone, vol. 28, no. 2, pp. 178–181, 2001.

[3] J. D. Chen, Y. J. Wang, K. Wei, S. H. Zhang, and X. T. Shi, "Self-organization of hydroxyapatite nanorods through oriented attachment," *Biomaterials*, vol. 28, no. 14, pp. 2275–2280, 2007.

[4] Y. Wang, C. Yang, X. Chen, and N. Zhao, "Development and characterization of novel biomimetic composite scaffolds based on bioglass-collagen-hyaluronic acid-phosphatidylserine for tissue engineering applications," *Macromolecular Materials and Engineering*, vol. 291, no. 3, pp. 254–262, 2006.

[5] C. V. M. Rodrigues, P. Serricella, A. B. R. Linhares et al., "Characterization of a bovine collagen-hydroxyapatite composite scaffold for bone tissue engineering," *Biomaterials*, vol. 24, no. 27, pp. 4987–4997, 2003.

[6] S. Mann, "Molecular tectonics in biomineralization and biomimetic materials chemistry," *Nature*, vol. 365, no. 6446, pp. 499–505, 1993.

[7] J. Rich, T. Jaakkola, T. Tirri, T. Närhi, A. Yli-Urpo, and J. Seppälä, "In vitro evaluation of poly(epsilon-caprolactone-co-DL-lactide)/bioactive glass composites," *Biomaterials*, vol. 23, no. 10, pp. 2143–2150, 2002.

[8] D. Walsh, T. Furuzono, and J. Tanaka, "Preparation of porous composite implant materials by in situ polymerization of porous apatite containing ε-caprolactone or methyl methacrylate," *Biomaterials*, vol. 22, no. 11, pp. 1205–1212, 2001.

[9] Q.-Q. Qiu, P. Ducheyne, and P. S. Ayyaswamy, "New bioactive, degradable composite microspheres as tissue engineering substrates," *Journal of Biomedical Materials Research*, vol. 52, no. 1, pp. 66–76, 2000.

[10] Y. Shikinami and M. Okuno, "Bioresorbable devices made of forged composites of hydroxyapatite (HA) particles and poly-L-lactide (PLLA). Part II: practical properties of miniscrews and miniplates," *Biomaterials*, vol. 22, pp. 3197–3211, 2001.

[11] J. E. Devin, M. A. Attawia, and C. T. Laurencin, "Three-dimensional degradable porous polymer-ceramic matrices for use in bone repair," *Journal of Biomaterials Science, Polymer Edition*, vol. 7, no. 8, pp. 661–669, 1996.

[12] X. Deng, J. Hao, and C. Wang, "Preparation and mechanical properties of nanocomposites of poly(D,L-lactide) with Ca-deficient hydroxyapatite nanocrystals," *Biomaterials*, vol. 22, no. 21, pp. 2867–2873, 2001.

[13] S. C. Rizzi, D. J. Heath, A. G. A. Coombes, N. Bock, M. Textor, and S. Downes, "Biodegradable polymer/hydroxyapatite composites: surface analysis and initial attachment of human osteoblasts," *Journal of Biomedical Materials Research*, vol. 55, no. 4, pp. 475–486, 2001.

[14] S. K. L. Levengood and M. Q. Zhang, "Chitosan-based scaffolds for bone tissue engineering," *Journal of Materials Chemistry B*, vol. 2, no. 21, pp. 3161–3184, 2014.

[15] F. Croisier and C. Jérôme, "Chitosan-based biomaterials for tissue engineering," *European Polymer Journal*, vol. 49, no. 4, pp. 780–792, 2013.

[16] S. K. Nandi, B. Kundu, and D. Basu, "Protein growth factors loaded highly porous chitosan scaffold: a comparison of bone healing properties," *Materials Science and Engineering C*, vol. 33, no. 3, pp. 1267–1275, 2013.

[17] P. Ghosh, A. P. Rameshbabu, N. Dogra, and S. Dhara, "2,5-dimethoxy 2,5-dihydrofuran crosslinked chitosan fibers enhance bone regeneration in rabbit femur defects," *RSC Advances*, vol. 4, no. 37, pp. 19516–19524, 2014.

[18] Y. Shen, Y. Zhan, J. Tang et al., "Multifunctioning pH-responsive nanoparticle from hierarchical self-assembly of polymer brush for cancer drug delivery," *AIChE Journal*, vol. 54, pp. 2979–2989, 2008.

[19] J. P. Vacanti and R. Langer, "Tissue engineering: the design and fabrication of living replacement devices for surgical reconstruction and transplantation," *The Lancet*, vol. 354, no. 1, pp. 32–34, 1999.

[20] H. Nagahama, H. Maeda, T. Kashiki, R. Jayakumar, T. Furuike, and H. Tamura, "Preparation and characterization of novel chitosan/gelatin membranes using chitosan hydrogel," *Carbohydrate Polymers*, vol. 76, no. 2, pp. 255–260, 2009.

[21] P. J. Vandevord, H. W. T. Matthew, S. P. Desilva, L. Mayton, B. Wu, and P. H. Wooley, "Evaluation of the biocompatibility of a chitosan scaffold in mice," *Journal of Biomedical Materials Research*, vol. 59, no. 3, pp. 585–590, 2002.

[22] M. Gravel, T. Gross, R. Vago, and M. Tabrizian, "Responses of mesenchymal stem cell to chitosan-coralline composites microstructured using coralline as gas forming agent," *Biomaterials*, vol. 27, no. 9, pp. 1899–1906, 2006.

[23] J. S. Mao, L. G. Zhao, Y. J. Yin, and K. D. Yao, "Structure and properties of bilayer chitosan-gelatin scaffolds," *Biomaterials*, vol. 24, no. 6, pp. 1067–1074, 2003.

[24] T. Chen, H. D. Embree, E. M. Brown, M. M. Taylor, and G. F. Payne, "Enzyme-catalyzed gel formation of gelatin and chitosan: potential for in situ applications," *Biomaterials*, vol. 24, no. 17, pp. 2831–2841, 2003.

[25] J. Mao, L. Zhao, K. de Yao, Q. Shang, G. Yang, and Y. Cao, "Study of novel chitosan-gelatin artificial skin in vitro," *Journal of Biomedical Materials Research A*, vol. 64, no. 2, pp. 301–308, 2003.

[26] W. Xia, W. Liu, L. Cui et al., "Tissue engineering of cartilage with the use of chitosan-gelatin complex scaffolds," *Journal of Biomedical Materials Research B*, vol. 71, no. 2, pp. 373–380, 2004.

[27] Y. Yin, F. Ye, J. Cui, F. Zhang, X. Li, and K. Yao, "Preparation and characterization of macroporous chitosan-gelatin/beta-tricalcium phosphate composite scaffolds for bone tissue engineering," *Journal of Biomedical Materials Research Part A*, vol. 67, no. 3, pp. 844–855, 2003.

[28] L. L. Hench, R. J. Splinter, W. C. Allen, and T. K. Greenlee, "Bonding mechanisms at the interface of ceramic prosthetic materials," *Journal of Biomedical Materials Research*, vol. 5, no. 6, pp. 117–141, 1971.

[29] H. Oudadesse, M. Mami, R. Doebez-Sridi et al., "Study of various mineral compositions and their bioactivity of bioactive glasses," *Bioceramics*, vol. 22, pp. 379–382, 2009.

[30] L. L. Hench, "The story of bioglass," *Journal of Materials Science: Materials in Medicine*, vol. 17, no. 11, pp. 967–978, 2006.

[31] L. L. Hench and J. K. West, "Biological applications of bioactive glasses," *Life Chemistry Reports*, vol. 13, pp. 187–241, 1996.

[32] K. T. Shalumon, S. Sowmya, D. Sathish, K. P. Chennazhi, S. V. Nair, and R. Jayakumar, "Effect of incorporation of nanoscale bioactive glass and hydroxyapatite in PCL/chitosan nanofibers for bone and periodontal tissue engineering," *Journal of Biomedical Nanotechnology*, vol. 9, no. 3, pp. 430–440, 2013.

[33] P. Gentile, M. Mattioli-Belmonte, V. Chiono et al., "Bioactive glass/polymer composite scaffolds mimicking bone tissue," *Journal of Biomedical Materials Research Part A*, vol. 100, no. 10, pp. 2654–2667, 2012.

[34] M. Peter, N. S. Binulal, S. Soumya et al., "Nanocomposite scaffolds of bioactive glass ceramic nanoparticles disseminated

chitosan matrix for tissue engineering applications," *Carbohydrate Polymers*, vol. 79, no. 2, pp. 284–289, 2010.

[35] M. Peter, N. S. Binulal, S. V. Nair, N. Selvamurugan, H. Tamura, and R. Jayakumar, "Novel biodegradable chitosan-gelatin/nano-bioactive glass ceramic composite scaffolds for alveolar bone tissue engineering," *Carbohydrate Polymers*, vol. 80, pp. 687–694, 2010.

[36] I. D. Xynos, M. V. J. Hukkanen, J. J. Batten, L. D. Buttery, L. L. Hench, and J. M. Polak, "Bioglass 45S5 stimulates osteoblast turnover and enhances bone formation *in vitro*: implications and applications for bone tissue engineering," *Calcified Tissue International*, vol. 67, no. 4, pp. 321–329, 2000.

[37] R. C. Bielby, R. S. Pryce, L. L. Hench, and J. M. Polak, "Enhanced derivation of osteogenic cells from murine embryonic stem cells after treatment with ionic dissolution products of 58S bioactive sol-gel glass," *Tissue Engineering*, vol. 11, no. 3-4, pp. 479–488, 2005.

[38] H.-Y. Li, R.-L. Du, and J. Chang, "Fabrication, characterization, and *in vitro* degradation of composite scaffolds based on PHBV and bioactive glass," *Journal of Biomaterials Applications*, vol. 20, no. 2, pp. 137–155, 2005.

[39] C. García, S. Ceré, and A. Durán, "Bioactive coatings prepared by sol–gel on stainless steel 316L," *Journal of Non-Crystalline Solids*, vol. 348, pp. 218–224, 2004.

[40] L. L. Hench and J. Wilson, *An Introduction to Bioceramic*, World Scientific, Singapore, 1993.

[41] K. Maji, S. Dasgupta, B. Kundu, and A. Bissoyi, "Development of gelatin-chitosan-hydroxyapatite based bioactive bone scaffold with controlled pore size and mechanical strength," *Journal of Biomaterials Science, Polymer Edition*, vol. 26, no. 16, pp. 1190–1209, 2015.

[42] K. Maji and S. Dasgupta, "Hydroxyapatite-chitosan and gelatin based scaffold for bone tissue engineering," *Transactions of the Indian Ceramic Society*, vol. 73, no. 2, pp. 110–114, 2014.

[43] F.-L. Mi, "Synthesis and characterization of a novel chitosan-gelatin bioconjugate with fluorescence emission," *Biomacromolecules*, vol. 6, no. 2, pp. 975–987, 2005.

[44] Q.-X. Niu, C.-Y. Zhao, and Z.-A. Jing, "An evaluation of the colorimetric assays based on enzymatic reactions used in the measurement of human natural cytotoxicity," *Journal of Immunological Methods*, vol. 251, no. 1-2, pp. 11–19, 2001.

[45] S. R. Federman, V. C. Costa, D. C. L. Vasconcelos, and W. L. Vasconcelos, "Sol-gel SiO_2-CaO-P_2O_5 biofilm with surface engineered for medical application," *Materials Research*, vol. 10, no. 2, pp. 177–181, 2007.

[46] Y. J. Yin, F. Zhao, X. F. Song, K. D. Yao, W. W. Lu, and J. G. Leong, "Preparation and characterization of hydroxyapatite/chitosan-gelatin network composite," *Journal of Applied Polymer Science*, vol. 77, no. 13, pp. 2929–2938, 2000.

[47] S. Itoh, M. Kikuchi, Y. Koyama et al., "Development of a novel biomaterial, hydroxyapatite/collagen (HAp/Col) composite for medical use," *Bio-Medical Materials and Engineering*, vol. 15, no. 1-2, pp. 29–41, 2005.

[48] M. Kikuchi, H. N. Matsumoto, T. Yamada, Y. Koyama, K. Takakuda, and J. Tanaka, "Glutaraldehyde cross-linked hydroxyapatite/collagen self-organized nanocomposites," *Biomaterials*, vol. 25, no. 1, pp. 63–69, 2004.

[49] Y. Kuboki, H. Takita, D. Kobayashi et al., "BMP-induced osteogenesis on the surface of hydroxyapatite with geometrically feasible and nonfeasible structures: topology of osteogenesis," *Journal of Biomedical Materials Research*, vol. 39, no. 2, pp. 190–199, 1998.

[50] N. Arabi and A. Zamanian, "Effect of cooling rate and gelatin concentration on the microstructural and mechanical properties of ice template gelatin scaffolds," *Biotechnology and Applied Biochemistry*, vol. 60, no. 6, pp. 573–579, 2013.

[51] J. M. Oliveira, M. T. Rodrigues, S. S. Silva et al., "Novel hydroxyapatite/chitosan bilayered scaffold for osteochondral tissue-engineering applications: scaffold design and its performance when seeded with goat bone marrow stromal cells," *Biomaterials*, vol. 27, no. 36, pp. 6123–6137, 2006.

19

In Vitro and *In Vivo* Characterization of N-Acetyl-L-Cysteine Loaded Beta-Tricalcium Phosphate Scaffolds

Yong-Seok Jang(ID),[1] **Phonelavanh Manivong,**[1] **Yu-Kyoung Kim**(ID),[1] **Kyung-Seon Kim,**[2] **Sook-Jeong Lee,**[3] **Tae-Sung Bae,**[1] and **Min-Ho Lee**(ID)[1]

[1]*Department of Dental Biomaterials and Institute of Biodegradable Materials, Institute of Oral Bioscience and School of Dentistry (Plus BK21 Program), Chonbuk National University, Jeonju, Republic of Korea*
[2]*Department of Dental Hygiene, Jeonju Kijeon College, Jeonju, Republic of Korea*
[3]*Department of Bioactive Material Science, Chonbuk National University, Jeonju, Republic of Korea*

Correspondence should be addressed to Min-Ho Lee; mh@jbnu.ac.kr

Academic Editor: Traian V. Chirila

Beta-tricalcium phosphate bioceramics are widely used as bone replacement scaffolds in bone tissue engineering. The purpose of this study is to develop beta-tricalcium phosphate scaffold with the optimum mechanical properties and porosity and to identify the effect of N-acetyl-L-cysteine loaded to beta-tricalcium phosphate scaffold on the enhancement of biocompatibility. The various interconnected porous scaffolds were fabricated using slurries containing various concentrations of beta-tricalcium phosphate and different coating times by replica method using polyurethane foam as a passing material. It was confirmed that the scaffold of 40 w/v% beta-tricalcium phosphate with three coating times had optimum microstructure and mechanical properties for bone tissue engineering application. The various concentration of N-acetyl-L-cysteine was loaded on 40 w/v% beta-tricalcium phosphate scaffold. Scaffold group loaded 5 mM N-acetyl-L-cysteine showed the best viability of MC3T3-E1 preosteoblastic cells in the water-soluble tetrazolium salt assay test.

1. Introduction

Bone tissue engineering is emerging as a significant potential alternative or complementary solution to repairing diseased or damaged tissue, enabling full recovery of the original state and function [1]. Calcium phosphate ceramics (CPCs) have been widely used as biomaterials for the regeneration of bone tissue because of their biocompatibility and ability to induce osteoblastic differentiation in progenitor cells [2, 3]. Among the CPCs, beta-tricalcium phosphate (β-TCP) bioceramics are widely used for hard tissue regeneration due to their excellent biocompatibility and their close similarity to biological apatite present in human bones [4, 5]. β-TCP is known to be highly resorbable *in vivo* with new bone ingrowths replacing the implanted β-TCP, which contributes significant advantage to β-TCP compared to unresorbable biomedical materials [4, 6]. In addition to the bioactivity and biocompatibility of the scaffold, the proper mechanical properties, scaffold pore morphology, porosity, and pore size

are important to maintain the shape of scaffold against stress and for determining cell ingrowth [7, 8]. The mechanical strength of the scaffolds was usually low due to high porosity, large pore size, and interconnected structures. To overcome these limitations, a number of studies have been focused on improving mechanical strength of β-TCP bioceramics [9–11]. Several fabrication techniques, such as replication of polymer foams [12], freeze casting [13], gel-casting foaming [14, 15], and foaming with employment of several pore creating additives [16, 17], have been developed to control pore sizes, porosity, pores interconnectivity, and mechanical strength of β-TCP scaffold. Numbers of studies have been focused on the sintering properties of β-TCP bioceramics to improve their mechanical strength [18, 19]. Additionally, the scaffold structures must have sufficient porosity to allow for cellular infiltration and proper cell function. A porosity of 90% was recommended for optimum diffusive transport within a cell-scaffolds construct under *in vitro* conditions [20].

Supplemented osseointegration and shortened healing time are desired to guarantee a direct bone to implant adherence. Meanwhile, scaffolds employed for bone tissue regeneration should have highly porous structures with a well-interconnected 3D pore network to encourage cell growth, vascularization, and transport of nutrient and metabolic waste [21, 22]. Many pores enlarge to the surface and can be vascularized with an adequate diameter (>approx. 100 μm) [23, 24]. Smaller pore diameters are more gainful in the adhesion and integration of mineralized tissues, contraction of the cell to implant, and the absorption of extracellular liquids [25, 26].

The wound infection is one of extremely severe issues in tissue regeneration and biomaterial implantation. An acute wound arising from a cut, surgical procedure, provides an opportunity for such bacteria to attack and colonize the underlying tissue, inclusive of connective tissue, muscle, and bone tissue. Bacterial contamination of implant results in tissue breakdown and degradation round the implant material. Recently, oxidative stress caused by excessive generation of intracellular reactive oxygen species (ROS) has been recommended to be associated with the adverse biological effects of such materials [27, 28]. N-acetyl-L-cysteine (NAC) is an antioxidant amino acid derivative and a sulfhydryl group, the functional half of NAC, directly resists ROS [29–33]. NAC can be harmonized into a cell and deacetylated into L-cysteine, a leader of glutathione [32, 34], which plays a key part in intracellular redox balance [35]. In fact, NAC prevented mastering of cell viability and function in fibroblasts and dental pulp cells created by resin [36, 37].

In this study, the effects of β-TCP contents and coating time on the mechanical properties and porosity of scaffolds developed by polyurethane replica method were investigated and then the effect of NAC loaded to β-TCP scaffold on biocompatibility was identified by various in vitro and in vivo tests.

2. Materials and Methods

2.1. Preparation of β-TCP Scaffolds.
To fabricate interconnected porous β-TCP scaffold by the replicating, polyurethane (PU) foam (Joon Ang sponge, 45 PPI, Republic of Korea) cut off with dimension of 4 mm in height and width and 25 mm in length. This PU foam was cleaned in distilled water using ultrasonic cleaner (Branson 2210, USA) for 10 min and dried at 80°C for 24 h. The coating slurry was prepared using different concentrations (35, 40, 45, and 50 w/v%) of commercial β-TCP [Ca$_3$(PO$_4$)$_2$, Sigma-Aldrich, Germany] powder, respectively. Briefly 5 g of polyvinyl butyral (PVB) (Sigma-Aldrich, USA) as binder was added in 100 ml of ethanol (C$_2$H$_6$O, Mallinckrodt Baker Inc, Malaysia) and vigorously stirred for 3 h, and then the given weight of β-TCP and 5 g of triphenyl phosphate (TEP) [(C$_6$H$_5$O)$_3$PO, Sigma-Aldrich, USA] additive for increase in the fluidity of material were added. The mixture was again stirred for an additional 24 h.

The cleaned PU foam was immersed in the coating slurry for 1 min and then it was dried at 80°C for 5 min after blowing up by the compressed air through coated foam with airgun

to disperse the slurry uniformly throughout the porous scaffolds without blocking the pores; dipping-and-drying steps were repeated twice and thrice in each concentration. Then the prepared samples were denoted as 35TCP2, 35TCP3, 40TCP2, 40TCP3, 45TCP2, 45TCP3, 50TCP2, and 50TCP3, respectively. The ceramic slurry coated foam was dried in drying oven at 60°C for 24 h. Then the prepared scaffolds were heat-treated to burn out the sponge and binder at 600°C for 3 h with the heating rate of 2°C/min and again solidified at 1300°C for 3 h with heating rate of 5°C /min.

2.2. The Scaffold Preparation for NAC Loading.
The scaffolds were cleaned in 70% of ethanol three times and fully dried before loading of the NAC solution. Briefly, NAC (C$_5$H$_9$NO$_3$S, Sigma-Aldrich, USA) solutions with different concentration (1, 5, 10, and 15 mM) were prepared with distilled water. NAC was loaded on each scaffold by dipping method and dried in vacuum oven at 37°C for 48 h, and the loading step was repeated five times.

2.3. Characterization of Scaffold

2.3.1. Surface Morphology and Crystallinity.
The surface morphology of the scaffolds was observed by scanning electron microscope (SEM, SUPRA 40 VP, Carl Zeiss, Germany). The average pore diameter of the scaffolds was calculated from SEM images by selecting five arbitrary areas. X-ray diffraction (XRD) was used to analyze the crystallographic structure and chemical composition of scaffolds after sintering using X-ray diffraction spectroscopic machine (Rigaku, Model D/mas 2550 v, USA) at 40 kV and 40 mA Cu Kα monochromatic radiation from $2\theta = 20° \sim 60°$ scanning angle with a step size of 0.02° and step scan rate of 1.8°/min.

2.3.2. Total Porosity.
The pores in the β-TCP scaffolds and the micro-tubes in the scaffold were responsible for the total porosity. The total porosity of the sintered β-TCP scaffold was determined by using the following equations [38]. Three identical specimens were used to determine the total porosity.

Total Porosity (%) = [1- (M / V × ρ)] × 100,
in which

M is the mass of the sintered sample,

V is the volume of the sintered sample,

ρ is the density of the sintered ceramic which is 3.07 g/cm^3 [39].

2.3.3. Mechanical Test.
The flexural strength of the scaffolds was investigated using an Instron machine (Model: 4201, USA). Four-point bending test was performed with fabricated scaffolds of dimensions 3.5 × 3.5 × 28 mm^3. The tests were carried out according to the procedure described in ASTM C1674-11, with the span in the bend test either 21 mm. The speed of the moving support was set at 0.126 mm/min in ambient conditions. Eight identical samples for each group were tested for each condition. The peak force recorded was used to calculate the flexural strength (S) from [40].

S = 3 PL / 4 bd^2,
in which

P is the break force,

L is the outer (support) span,

b is the specimen width,

d is the specimen thickness.

2.4. Assessment of the Scaffold Bioactivity. β-TCP scaffold was immersed in simulated body fluid (SBF) solution at 37°C for 14 days; after that, it was distinguished for the formation of a calcium phosphate layer on the surfaces as one indication of bioactivity. The complement of the SBF was identical to that described by Kokubo et al. [41]. The scaffolds sizes of $8 \times 8 \times 4$ mm^3 were sterilized by soaking in 70% ethanol for 1 h and then left 30 min in UV light followed by being rinsed with phosphate buffered saline (PBS) three times and dried in clean bench for 24 h and then placed in falcon tubes containing 5 ml of SBF at 37 ± 2°C. The SBF was replaced every 24 h. After immersion test, the scaffolds were gently washed in distilled water and dried at room temperature for 24 h. Surface morphologies and chemical composition of byproducts formed on the surface were characterized by SEM, energy dispersive X-ray (EDX).

2.5. Cell Viability

2.5.1. Cell Seeding and Culture. Mouse osteoblast cells (MC3T3-E1) were purchased from ATCC (American Type Culture Collection). 10% fetal bovine (FBS, Gibco Co., USA), 500 units/ml of penicillin (Gibco Co., USA), and 500 units/ml of streptomycin (Gibco Co., USA) were added to α-MEM (Gibco Co., USA) to prepare the culture medium. Incubation was carried out at 37°C in an atmosphere containing 5 vol. % CO_2 according to ISO 10993-5:1999 [42]. The medium was replaced after 24 h then every 2 days. After the cells reached 80% confluency, the cells were harvested using a commonly procedure. The harvested cells were resuspended in the culture medium at a density of 2.4×10^4 cells/ml.

Subsequently, the cell suspension was mixed with the medium and then dropped into the β-TCP scaffolds (1 ml/scaffold) and culture in 24-well culture plates, where α-MEM medium was used as the negative control; the cells were cultured in atmosphere at 37°C under a 5% CO_2. The cell medium was changed every 2 days during the cell culture of 5 days. At the specified time endpoints, cell-scaffold media were removed and characterized for cell viability.

2.5.2. Water-Soluble Tetrazolium Salt Assay Test. Cell viability was evaluated using the water-soluble tetrazolium salt (WST) assay (Fluka Co., Milwaukee, USA). The scaffold constructs were placed in 24-well plates with 1,000 μl of cell medium for 5 days of culture. After culture time, the 100 μl of media in each well was removed out, and then 100 μl of WST solution was added newly for color expression. The scaffold samples were removed for absorbance evaluation. 200 μl of supernatant from each well was transferred to a 96-well plate and the optical density (OD) measurements were conducted at a wavelength of 450 nm using a microplate spectrophotometer (EMax, Molecular Devices, USA). Six

specimens for each group were tested, and each test was repeated three times.

2.6. Animal Experiment. We used skeletally two mature male beagles of which body weight was about 35 kg (HUVET, Icsan, South Korea). Housing and feeding of the animals were performed according to the NIH guidelines for the care and use of laboratory animals. The ethical approval of animal study was accepted by the Institutional Animal Care and Use Committee of the Chonbuk National University Laboratory Animal Center (CBNU 2017-0005).

Cylindrical scaffolds with 4 mm in diameter and 8 mm in length were made for *in vivo* test. For the implanted surgery, the beagle was premedicated with ketamine hydrochloride. General anesthesia of the animal was guided with thiopental and, after endotracheal intubation, maintained with halothane. Supplementary analgesia was obtained by local administration of 2% xylocaine containing epinephrine.

The left and right humeral and femoral diaphysis of animals were shaved and cleaned with ethanol-iodine and draped for sterile surgical procedure. The bones were exposed after transcutaneous incision and reflection of the periosteum. Eight corticocancellous defects ($4 \times 4 \times 8$ mm in height) at all the implant sites were prepared using slow drill speeds and plenty saline irrigation to minimize mechanical and thermal trauma to cortical bone. The final drill hole gave a slight press-fit of the transcortical implants which were inserted into the holes with 2 to 3 mm left proud of the periosteal surface. Two implants were carefully inserted into each humeral and femoral created defects of each animal side, and the 40TCP3 group in which NAC drug was not loaded was implanted into each femoral side for control group. 15mM40TCP3 group with the lowest cytotoxicity at *in vitro* was excluded for *in vivo* test.

After a healing period of 24 weeks, beagle was sacrificed. Immediately thereafter the implants were exposed and removed. After formalin fixation for 48 h, all bone samples were kept in 70% ethanol and scanned with micro-CT. The samples were thereafter stained in Villaneuva osteochrome bone stain (Polyscience, Inc.) and embedded in methyl methacrylate, monomer (Yakuri Pure Chemical So., LTD./Kyoto Japan). Histological analysis was performed to examine for new bone formation by optical microscopy (DM 2500M, Leica, Germany).

2.7. Statistical Analysis. All results of the control and experimental groups were analyzed independently. Statistical analysis was performed using Student's *t*-test at a significance level of P value < 0.05 to examine the difference in the variables among different experimental conditions.

3. Results

3.1. Characterization of β-TCP Scaffolds

3.1.1. Microstructure and Crystallinity. Figure 1 shows the typical SEM morphologies of the porous β-TCP scaffolds by varying the concentrations of β-TCP and coating times. When compared to the scaffolds at the same concentration,

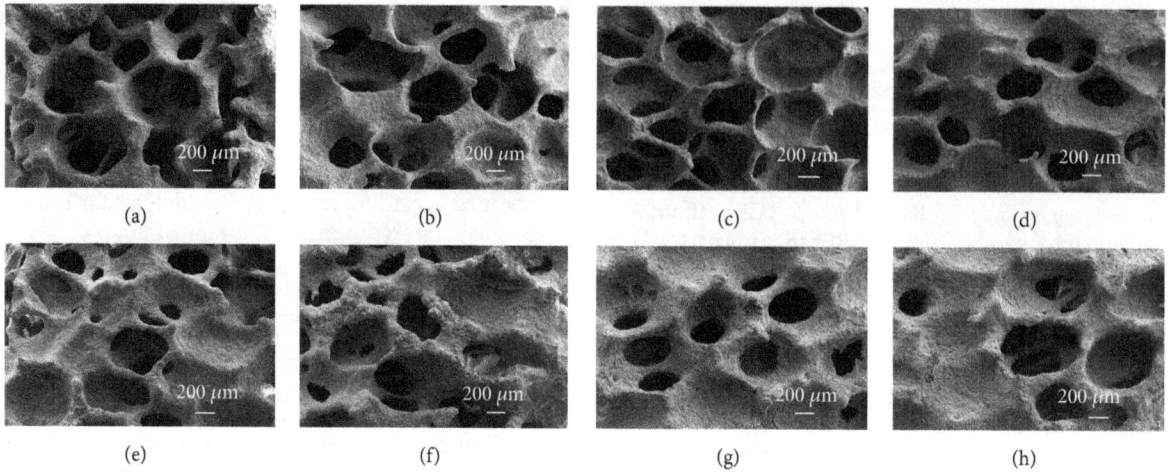

FIGURE 1: SEM morphologies of the β-TCP porous scaffolds with different concentration and coating times: (a) 35TCP2, (b) 35TCP3, (c) 40TCP2, (d) 40TCP3, (e) 45TCP2, (f) 45TCP3, (g) 50TCP2, and (h) 50TCP3. In the abbreviation, the number in front signifies the concentration (w/v%) of β-TCP powder included in slurry, and the number in back signifies the repetition number of immersion in slurry for fabrication of scaffold.

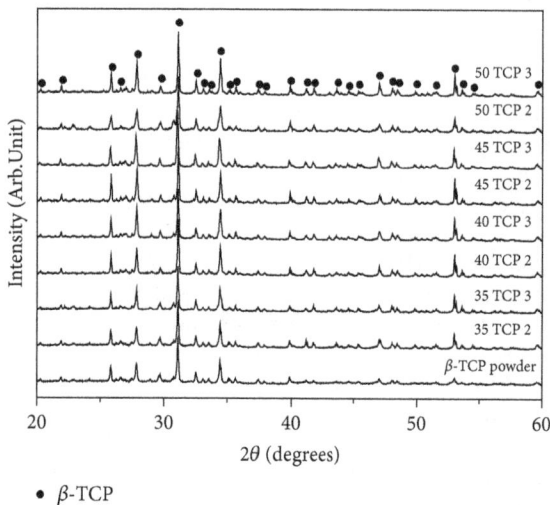

FIGURE 2: XRD pattern of the as-received β-TCP powder and scaffolds with different w/v% and coating times after sintering.

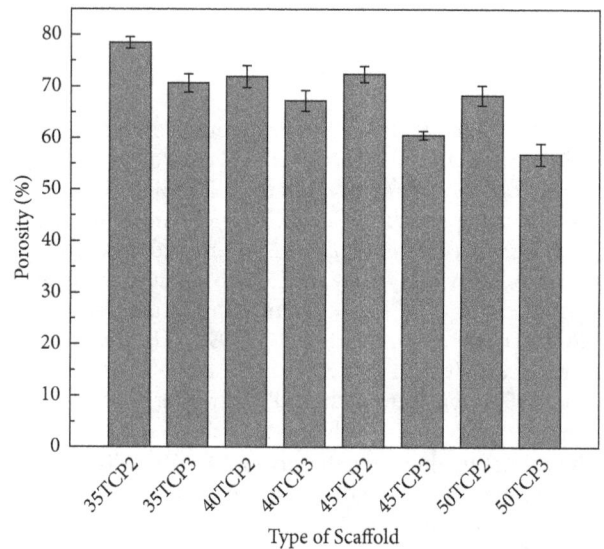

FIGURE 3: Porosity of β-TCP scaffold with TCP content and coating times. Error bar represents mean ± Standard Deviation (SD); $n = 5$.

the stems of the scaffold became thicker and some pores were partly closed as the coating time increased. As the concentration of TCP increased, the stem became much thicker and some pores were completely clogged as shown in Figures 1(e)–1(h). Good porous structure with strong interconnected pores was observed in 40 TCP 2 (Figure 1(c)). It was revealed that β-TCP concentration and number of coating times visibly affected the interconnectivity and size of the pores.

The XRD patterns of as-received β-TCP powders and sintered scaffolds prepared by different concentration of β-TCP and coating times were presented in Figure 2. The XRD pattern is composed of major phases of β-TCP (JCPDS # 09-0169), and it can be seen that the scaffolds sintered at 1300°C were mostly composed of highly crystalline and single phase

β-TCP, and the intensity of β-TCP peaks was increased with increase in concentration and coating time.

3.1.2. Porosity. Appropriate porosity and porous structure is important for cells penetration and ingrowth. Figure 3 shows the total porosity of the scaffolds with different concentrations of TCP and coating times. The total porosity was decreased with increase in the concentrations of β-TCP and coating time.

3.1.3. Mechanical Properties. Figure 4 shows the influence of β-TCP concentrations and coatings time on the flexural strength of the sintered scaffolds. Increasing the concentrations and coating times, the flexural strength was increased.

TABLE 1: Element composition of the scaffold after immersion in SBF for 14 days.

Sample abbreviation	Ca/P ratio	Chemical composition (at. %)		
		Ca	P	O
40TCP3 (Before immersion)	1.12	19.56	17.40	57.92
40TCP3	1.30	18.74	16.59	59.33
1 mM 40TCP3	1.35	19.59	14.55	58.34
5 mM 40TCP3	1.38	22.15	16.09	55.97
10 mM 40TCP3	1.65	25.65	16.10	53.52
15 mM 40TCP3	1.18	21.14	17.86	52.83

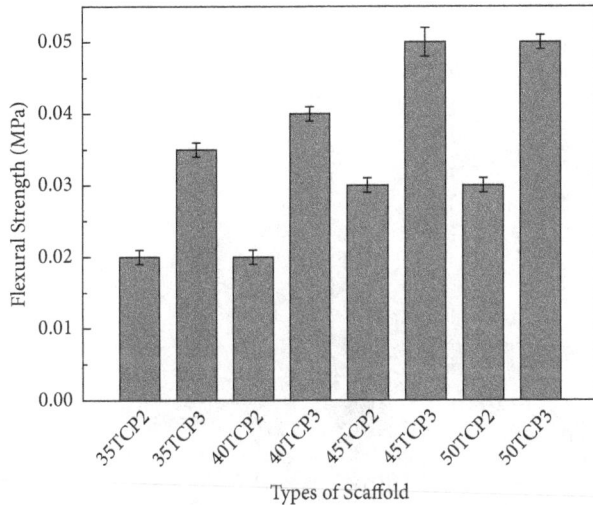

FIGURE 4: Flexural strength of β-TCP scaffold with concentration and coating times. Error bar represents mean ± Standard Deviation (SD); $n = 5$.

Results demonstrated that the flexural strength of the scaffold containing 35, 40, and 45 w/v% β-TCP with three times coating was steadily increased while 50 w/v% β-TCP showed the same strength as 45 w/v% β-TCP.

3.2. In Vitro Bioactivity of NAC Loaded Scaffolds

3.2.1. Immersion Test. Figure 5 represents surface morphologies of NAC loaded β-TCP scaffolds before and after immersion in SBF for 3 and 14 days. In comparison with the surface of 40 TCP 3, more and denser particulate layers were formed on the surface of NAC loaded 40 TCP 3 scaffolds within 3 days, and the densest particulate layer was observed on the surface of 5 mM 40 TCP 3 after immersion for 14 days. Porous network of needle-like crystals with nanometer size was identified in high-resolution FE-SEM images of all scaffolds after immersion.

Table 1 shows EDX results before and after immersion in SBF solution for 14 days. Compared with Ca/P ratio before immersion of 40 TCP 3, Ca/P ratio on 40 TCP 3 after immersion was gently increased, and the highest Ca/P ratio was confirmed on 10 mM 40TCP3, which was of 1.65.

3.2.2. Cell Viability. Figure 6 presents the viability of MC3T3-E1 cells cultured for 5 days in the experimental groups to compare the effect of NAC loaded on β-TCP scaffolds on the cell proliferation. After 5 days of culture, higher cell growth was shown in the β-TCP scaffolds loading the lower concentration of NAC (below 5 mM) than 40TCP3 group (the control group). But in the groups loading NAC concentration over 10 mM, the cell viability showed a significant decrease ($p < 0.05$) compared to that in the control group.

3.3. Evaluation of Bone-Forming Effects.

The surgical procedure and external radiation delivery were completed without major complications. Histological specimens were retrieved after 24 weeks following β-TCP scaffold implants, and tissue ingrowth to the periphery of the scaffold was evaluated.

Figure 7 shows micro-CT images at 24 weeks after implantation of scaffolds, where the light blue in images is TCP blocks embedded in hole defect of bone. It was identified that the new bone penetrated into all porous TCP blocks and the degradation rate of 40 TCP 3 scaffold was decreased as the amount of NAC loading increased.

Figure 8 presents the histological results, where the positions of the scaffolds embedded in the cancellous bone are indicated by white squares. As the empty space of the bone was filled with scaffold, a new bone grew inside the scaffold from the boundary of the scaffold and bone. Defects in the compact bone at the upper part of the scaffold were seen in several groups and no inflammation was observed. The residual scaffold of the unloaded NAC (a) was observed to have many pores (blue ∗ marks) due to the porous form, and new bone cell was not formed inside the pores. In the group loaded with 1~5 mM NAC, dense binding of scaffold with new bone and bone marrow cell was observed. In the group loaded with 10 mM NAC(d), the pores of the scaffold were almost filled, so the new cartilage and bone marrow cells did not grow to the inside, and empty space was formed inner core.

4. Discussion

The objective of this study was to alter the pore structures of scaffolds for enhancing the mechanical properties with controlling the porosity and pore size by changing the concentrations of TCP slurry and number of coating times during the preparation.

Adequate porosity and mechanical strength of scaffolds are a prerequisite to allow its use in bone tissue engineering without rigid support. Pore size and porosity of scaffolds play a critical role in bone formation *in vitro* and *in vivo*.

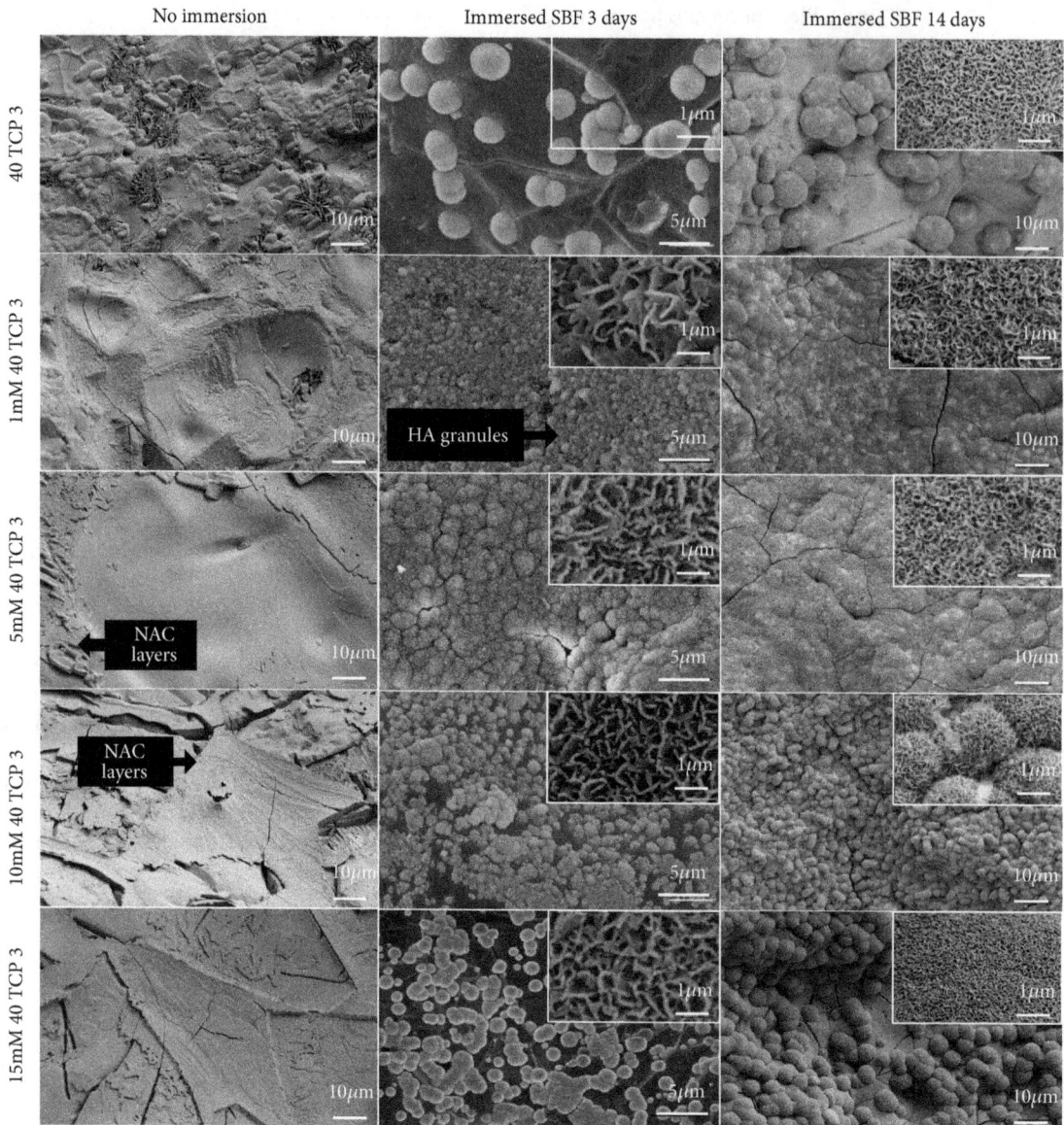

FIGURE 5: FE-SEM images of the surface of the β-TCP and NAC loaded scaffold before and after immersion for 3 days and 14 days in SBF.

The porous structure of scaffold can provide good condition for cell growth and be easy to transport the nutrients and metabolic waste while the dense scaffold can impart excellent mechanical strength when implanted in body. The optimal pore size for bone tissue engineering required for bone ingrowth has been suggested in the range of 100-800 μm [43]. Pores above 100 μm allow to scaffold locating by osteoblasts [44]. And scaffolds must have sufficient porosity for nutrient and gas exchange. Previous research reported that a porosity of more than 80% was characteristic of an ideal scaffold [45, 46]. 40 TCP 3 scaffold had a pore size of 150-950 μm and porosity of 67.22 ± 2% in this study. When compared to the prior study, it was similar to that observed on 13-93 bioactive glass with a pore size of 100-500 μm and porosity was slightly different of 85 ± 2%, which was fabricated with

a microstructure using a polymer foam replication technique [47].

The mechanical properties of the different concentration of β-TCP were examined by flexural strength measurements. As expected, the results indicated that the flexural strength of the 45 and 50 wt.% TCP scaffold was higher compared to that of 35 and 40 wt% TCP scaffold. Previous study also reported that the porosity would influence the mechanical properties of porous scaffolds [48]. Figure 4 shows that the flexural strength decreases progressively with the increase in porosity of the scaffolds. The mechanical properties of a scaffold used for tissue engineering are very important due to the need for the structural stability to resist the stresses that occurred during implantation *in vivo*. Even though 40TCP3 scaffold had a low flexural strength compared to a compressive

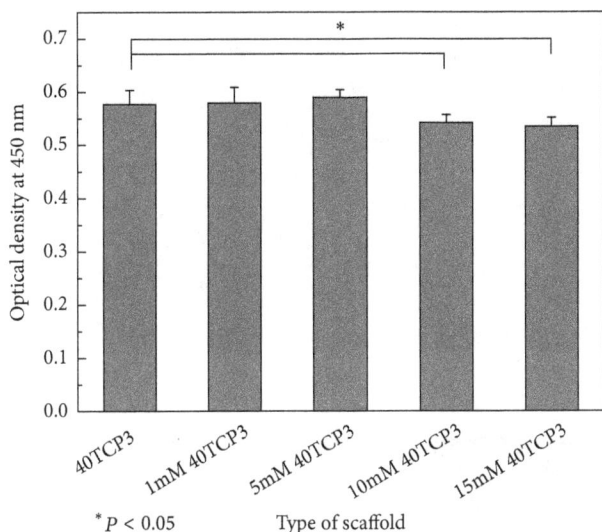

FIGURE 6: WST assay for viability of MC3T3-E1 at 5-day culture. Mean ± SD; $n = 4$. $^*P < 0.05$ means significantly different.

strength in the range of 1 to 100 MPa of the cancellous bone [49], it had a good porous and interconnected pore in this study.

Mainly, TCP, whose calcium/phosphate (Ca/P) ratio is 1.5, has three crystal forms: α, β, and γ, which are thermodynamically determined by the sintering temperature of amorphous TCP. Despite a lower solubility than that of the α-form, β-TCP is markedly more soluble than synthetic HA that is verified by its solubility product constant (α-TCP: 3.16×10^{-26}; β-TCP: 2.19×10^{-30}: synthetic HA: 2.13×10^{-59}). This suggests that β-TCP releases Ca cation (Ca^{2+}) and P anion (P^i), which play important role on bone homeostasis. Clearly, Ca^{2+} modulate osteoblastic viability, motility, proliferation, and differentiation through activation of calcium-sensing receptors, enhancement of Ca^{2+} influx into cells, and subsequent intracellular calcium signaling pathway [50, 51].

It was confirmed that application of NAC significantly alleviated cytotoxicity of β-TCP in the previous studies. A number of studies have indicated that NAC detoxifies oxidative stress-inducing cytotoxic material, particularly on the scaffold with NAC 1 mM and 5 mM loaded. Several presumable pathways of NAC-mediated detoxification had been presented. In the previous study, NAC alleviated intracellular level of ROS on/around β-TCP [31, 52]. Also, it was reported that NAC could directly displace extracellular ROS [30] and toxic compounds [53]. Thus, NAC encourages an increase in osteoblasts and reducing of inflammatory molecules [52, 54, 55]. We would need the further investigation for the most effective method to apply NAC on β-TCP scaffold material on the basis of the NAC-mediated detoxification mechanism and the physiochemical property of β-TCP scaffold. The results of this study suggested that application of antioxidant amino acid derivative NAC may offer a guide for the biofunctionalization of β-TCP scaffold-based material for bone regeneration. Likewise, this may

provide important information for future development of synthetic bone substitutes.

Generally it has been known that the carbonated apatite is precipitated on the surface of biomaterial by immersion in a SBF. The precipitates are grown by extermination of the calcium, phosphate, carbonate, and hydroxide ions in the SBF [56]. The formation of a biologically active HA layer, which is chemically and structurally equivalent to the main mineral composition of bone, is a key requirement for developing a strong interfacial bond between bioactive ceramics and tissue in vivo [57]. The formation of HA surface layer in vitro is indicative of a material's bioactive potential in vivo [58]. In this study, the whole groups of the β-TCP scaffold illustrated that the formation of HA surface layer was observed within 3 days (Figure 6), which means that the β-TCP scaffold has the good bioactivity when considering that the formation of an amorphous calcium phosphate layer on the surface of 13-93 bioactive glass and the porous 45S5 glass-ceramic constructs was observed within 7 days and after 28 days in an SBF [47, 59]. More rapid formation of HA layer is adaptable for obtaining early biological fixation of a bioactive implant in bone repair and regeneration. The results provided useful information for designing the β-TCP scaffold with NAC of various concentrations. In this result, HA layer with Ca/P ratio of 1.65 was formed in 10 mM 40TCP3 group after immersion in SBF for 14 days, which was highest compared with other gourps. Nevertheless, the estimation of the biocompatibility and bioactivity using in vitro test in SBF must be cogitated very carefully as it may differ from that of the practical in vivo reaction [60].

The MC3T3-E1 cells are a well-characterized mouse preosteoblastic cell line and have been used largely for in vitro cytotoxicity testing of biomaterials [61, 62]. As shown in Figure 6, good cell proliferation was identified in the β-TCP scaffolds loading the lower concentration of NAC below 5 mM, but less cell proliferation was confirmed in the β-TCP scaffolds loading the lower concentration of NAC over 10 mM than 40TCP3. It indicates that the proper amount of NAC loading on the β-TCP scaffold can increase biocompatibility of scaffold, but excess loading of NAC can cause cytotoxicity and reduce the site where the cells can attach and grow due to blocking the pores.

To investigate of the bone formation, histological samples were harvested 24 weeks after implantation of the β-TCP and NAC loaded scaffold. In all groups, empty defects of the bone were filled with the osteoblast and marrow cell based on the scaffold. The osteoblasts come from the differentiation of osteogenic cells in the tissue that covers the outside of the bone, or the periosteum and the bone marrow. The osteoblast is trapped inside of the bone after growth once it hardens, and it changed to be known as the osteocyte (the red * marks) [63]. Through the loading of 1~5 mM of NAC, osteocyte areas including osteoblasts grew widely in healed defect areas. Furthermore, micro-CT 3D images and histological analysis revealed new bone formation and osteoblasts were plenty which occurred on the whole group of scaffold; at the same time the degradation of scaffold seemed slowly decreasing by increasing the ratio of NAC. The porous β-TCP scaffold served tissue ingrowth, indicating that connective tissue and

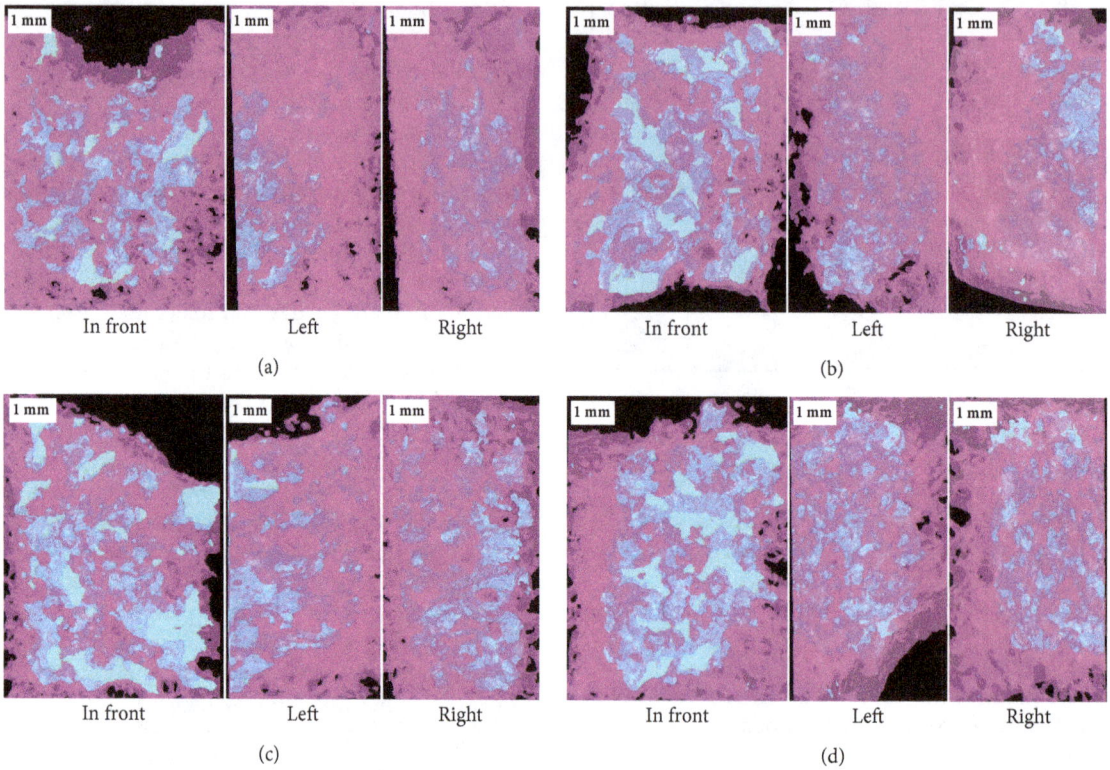

FIGURE 7: Micro-CT images at 24 weeks after implantation: (a) 40 TCP 3, (b) 1 mM 40 TCP 3, (c) 5 mM 40 TCP 3, and (d) 10 mM 40 TCP 3.

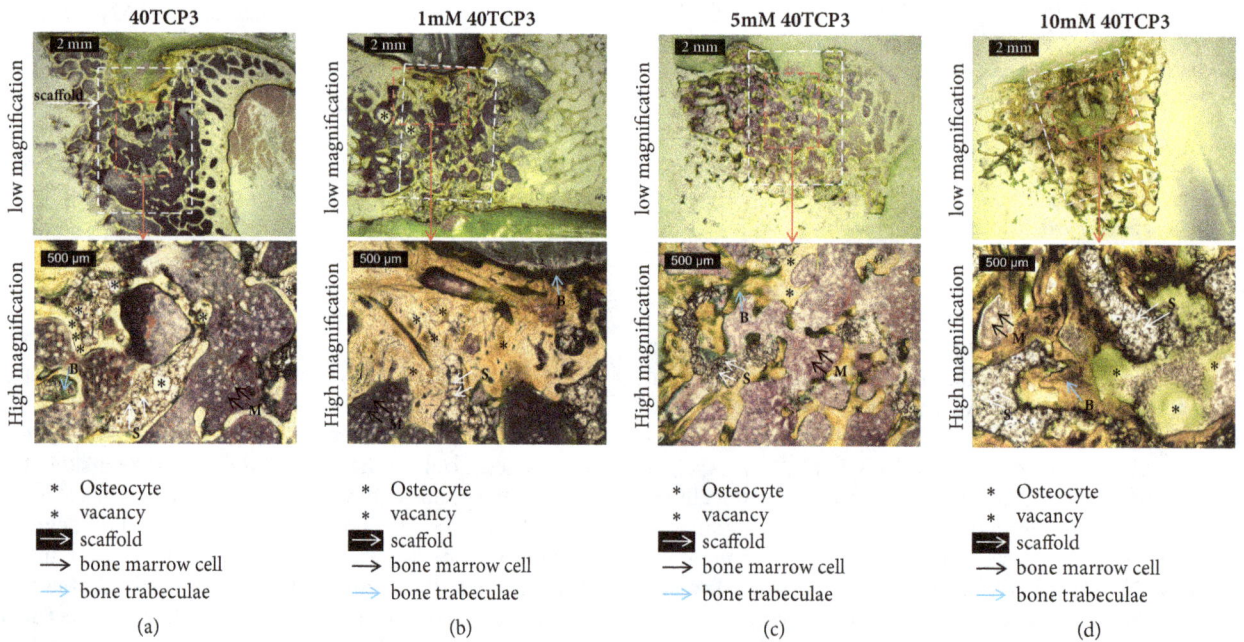

FIGURE 8: Histological analyses of humeral, femoral, and tibia bone by Villanueva Osteochrome bone staining at 24 weeks after implantation of each group: (a) 40 TCP 3, (b) 1 mM 40 TCP 3, (c) 5 mM 40 TCP 3, and (d) 10 mM 40 TCP 3. The number in front of 40 TCP 3 signifies the concentration of NAC in coating solution for loading NAC on the scaffold.

bone-like tissue entered the block interior. The scaffold of the spongy structure plays a role of supporting the connection of the osteocytes, but the cells and the blood are hardly penetrated into the pores in the scaffold structure as 40TCP3 group. In previous study, it was reported that the percentage of viable osteoblasts was increased to 94% from 88% under β-TCP granules at 24 h by preaddition of NAC [64]. On the other hand, a high amount of NAC loading as 10 mM 40TCP3 in this study occurred to fill the outer pores of the scaffold, which make it difficult to grow inside of osteoblasts and circulate blood. The effect of scaffold pore size is the impact on bone regeneration and vascularization. The larger pore size around 350 μm diameters can be partially responsible for the enhanced bone formation [65]. In this histological study, we found that the 1~5 mM 40TCP3 scaffolds elicit a higher level of osteoblastic activity, leading to higher mature bone formation.

5. Conclusions

This study focused on evaluating the biocompatibility of the NAC loaded β-TCP bone scaffold which was fabricated by replica method using slurries containing various concentrations of β-TCP and different coating times. The flexural strength of the interconnected β-TCP bone scaffolds was increased and its porosity was decreased as TCP content and coating times increased. Among the group, the porous 40TCP3 scaffold had the most appropriate interconnected porous structure and porosity. NAC loading within 10 mM on β-TCP scaffold significantly improved the osteoblastic response to the material. 15 mM NAC loaded β-TCP scaffold showed low cell viability and cell spreading on the surface of scaffolds. After implantation, the NAC loaded β-TCP scaffolds more effectively enhance the long-term bone remodeling and bone augmentation for bone formation than the scaffold without NAC loading *in vivo*.

Disclosure

This research was carried out without funding. The current manuscript is an extended and updated version of the conference paper in the following link: http://www.dbpia.co.kr/Journal/ArticleDetail/NODE02511894.

Conflicts of Interest

The authors declare that they have no conflicts of interest.

Authors' Contributions

Yong-Seok Jang and Phonelavanh Manivong contributed equally to this work and Phonelavanh Manivong is considered as joint first author.

References

[1] D. Hutmacher, "Scaffolds in tissue engineering bone and cartilage," *Biomaterials*, vol. 21, no. 24, pp. 2529–2543, 2000.

[2] C. N. Cornell, "Osteoconductive materials and their role as substitutes for autogenous bone grafts," *Orthopedic Clinics of North America*, vol. 30, no. 4, pp. 591–598, 1999.

[3] S. Samavedi, A. R. Whittington, and A. S. Goldstein, "Calcium phosphate ceramics in bone tissue engineering: a review of properties and their influence on cell behavior," *Acta Biomaterialia*, vol. 9, no. 9, pp. 8037–8045, 2013.

[4] J. Elliot, *Structure and Chemistry of the Apatites and Other Calcium Phosphates*, Elsevier, Amsterdam, Netherlands, 1994.

[5] S. V. Dorozhkin, "Calcium orthophosphates," *Journal of Materials Science*, vol. 42, no. 4, pp. 1061–1095, 2007.

[6] Y.-H. Kim, M. A. Jyoti, M.-H. Youn et al., "In vitro and in vivo evaluation of a macro porous β-TCP granule-shaped bone substitute fabricated by the fibrous monolithic process," *Biomedical Materials*, vol. 5, no. 3, Article ID 035007, 2010.

[7] L. Indolfi, A. B. Baker, and E. R. Edelman, "The role of scaffold microarchitecture in engineering endothelial cell immunomodulation," *Biomaterials*, vol. 33, no. 29, pp. 7019–7027, 2012.

[8] J. A. A. Hendriks, L. Moroni, J. Riesle, J. R. de Wijn, and C. A. van Blitterswijk, "The effect of scaffold-cell entrapment capacity and physico-chemical properties on cartilage regeneration," *Biomaterials*, vol. 34, no. 17, pp. 4259–4265, 2013.

[9] F. H. Perera, F. J. Martínez-Vázquez, P. Miranda, A. L. Ortiz, and A. Pajares, "Clarifying the effect of sintering conditions on the microstructure and mechanical properties of β-tricalcium phosphate," *Ceramics International*, vol. 36, no. 6, pp. 1929–1935, 2010.

[10] H. R. R. Ramay and M. Zhang, "Biphasic calcium phosphate nanocomposite porous scaffolds for load-bearing bone tissue engineering," *Biomaterials*, vol. 25, no. 21, pp. 5171–5180, 2004.

[11] K. Lin, J. Chang, J. Lu, W. Wu, and Y. Zeng, "Properties of β-Ca3(PO4)2 bioceramics prepared using nano-size powders," *Ceramics International*, vol. 33, no. 6, pp. 979–985, 2007.

[12] I. Sopyan, M. Mel, S. Ramesh, and K. A. Khalid, "Porous hydroxyapatite for artificial bone applications," *Science and Technology of Advanced Materials*, vol. 8, no. 1-2, pp. 116–123, 2007.

[13] E.-J. Lee, Y.-H. Koh, B.-H. Yoon, H.-E. Kim, and H.-W. Kim, "Highly porous hydroxyapatite bioceramics with interconnected pore channels using camphene-based freeze casting," *Materials Letters*, vol. 61, no. 11-12, pp. 2270–2273, 2007.

[14] H. R. Ramay and M. Zhang, "Preparation of porous hydroxyapatite scaffolds by combination of the gel-casting and polymer sponge methods," *Biomaterials*, vol. 24, no. 19, pp. 3293–3302, 2003.

[15] M. Potoczek, A. Zima, Z. Paszkiewicz, and A. Ślósarczyk, "Manufacturing of highly porous calcium phosphate bioceramics via gel-casting using agarose," *Ceramics International*, vol. 35, no. 6, pp. 2249–2254, 2009.

[16] B. Kundu, D. Sanyal, and D. Basu, "Physiological and elastic properties of highly porous hydroxyapatite potential for integrated eye implants: Effects of SIRC and L-929 cell lines," *Ceramics International*, vol. 39, no. 3, pp. 2651–2664, 2013.

[17] M. Descamps, T. Duhoo, F. Monchau et al., "Manufacture of macroporous β-tricalcium phosphate bioceramics," *Journal of the European Ceramic Society*, vol. 28, no. 1, pp. 149–157, 2008.

[18] Y. Li, W. Weng, and K. C. Tam, "Novel highly biodegradable biphasic tricalcium phosphates composed of α-tricalcium phosphate and β-tricalcium phosphate," *Acta Biomaterialia*, vol. 3, no. 2, pp. 251–254, 2007.

[19] F. Laurent, A. Bignon, J. Goldnadel et al., "A new concept of gentamicin loaded HAP/TCP bone substitute for prophylactic action: In vitro release validation," *Journal of Materials Science: Materials in Medicine*, vol. 19, no. 2, pp. 947–951, 2008.

[20] L. E. Freed, G. Vunjak-Novakovic, R. J. Biron et al., "Biodegradable polymer scaffolds for tissue engineering," *Bio/Technology*, vol. 12, no. 7, pp. 689–693, 1994.

[21] A. Tampieri, G. Celotti, F. Szontagh, and E. Landi, "Sintering and characterization of HA and TCP bioceramics with control of their strength and phase purity," *Journal of Materials Science: Materials in Medicine*, vol. 8, no. 1, pp. 29–37, 1997.

[22] M. Akao, H. Aoki, K. Kato, and A. Sato, "Dense polycrystalline β-tricalcium phosphate for prosthetic applications," *Journal of Materials Science*, vol. 17, no. 2, pp. 343–346, 1982.

[23] J. Handschel, K. Berr, R. Depprich et al., "Compatibility of embryonic stem cells with biomaterials," *Journal of Biomaterials Applications*, vol. 23, no. 6, pp. 549–560, 2009.

[24] C. Naujoks, F. Langenbach, K. Berr et al., "Biocompatibility of osteogenic predifferentiated human cord blood stem cells with biomaterials and the influence of the biomaterial on the process of differentiation," *Journal of Biomaterials Applications*, vol. 25, no. 5, pp. 497–512, 2011.

[25] P. S. Eggli, W. Muller, and R. K. Schenk, "Porous hydroxyapatite and tricalcium phosphate cylinders with two different pore size ranges implanted in the cancellous bone of rabbits. A comparative histomorphometric and histologic study of bone ingrowth and implant substitution," *Clinical Orthopaedics and Related Research*, no. 232, pp. 127–138, 1988.

[26] K. A. Hing, B. Annaz, S. Saeed, P. A. Revell, and T. Buckland, "Microporosity enhances bioactivity of synthetic bone graft substitutes," *Journal of Materials Science: Materials in Medicine*, vol. 16, no. 5, pp. 467–475, 2005.

[27] H. K. Varma and S. Sureshbabu, "Oriented growth of surface grains in sintered β tricalcium phosphate bioceramics," *Materials Letters*, vol. 49, no. 2, pp. 83–85, 2001.

[28] R. Tsaryk, M. Kalbacova, U. Hempel et al., "Response of human endothelial cells to oxidative stress on Ti6Al4V alloy," *Biomaterials*, vol. 28, no. 5, pp. 806–813, 2007.

[29] M. Yamada, T. Ueno, H. Minamikawa et al., "N-acetyl cysteine alleviates cytotoxicity of bone substitute," *Journal of Dental Research*, vol. 89, no. 4, pp. 411–416, 2010.

[30] M. Yamada and T. Ogawa, "Chemodynamics underlying N-acetyl cysteine-mediated bone cement monomer detoxification," *Acta Biomaterialia*, vol. 5, no. 8, pp. 2963–2973, 2009.

[31] G. Spagnuolo, V. D'Antò, C. Cosentino, G. Schmalz, H. Schweikl, and S. Rengo, "Effect of N-acetyl-L-cysteine on ROS production and cell death caused by HEMA in human primary gingival fibroblasts," *Biomaterials*, vol. 27, no. 9, pp. 1803–1809, 2006.

[32] M. Zafarullah, W. Q. Li, J. Sylvester, and M. Ahmad, "Molecular mechanisms of N-acetylcysteine actions," *Cellular and Molecular Life Sciences*, vol. 60, no. 1, pp. 6–20, 2003.

[33] H. Schweikl, G. Spagnuolo, and G. Schmalz, "Genetic and cellular toxicology of dental resin monomers," *Journal of Dental Research*, vol. 85, no. 10, pp. 870–877, 2006.

[34] N. Tsukimura, M. Yamada, H. Aita et al., "N-acetyl cysteine (NAC)-mediated detoxification and functionalization of poly(methyl methacrylate) bone cement," *Biomaterials*, vol. 30, no. 20, pp. 3378–3389, 2009.

[35] N. Kojima, M. Yamada, A. Paranjpe et al., "Restored viability and function of dental pulp cells on poly-methylmethacrylate

[36] M. Yamada, N. Kojima, A. Paranjpe et al., "N-acetyl cysteine (NAC)-assisted detoxification of PMMA resin," *Journal of Dental Research*, vol. 87, no. 4, pp. 372–377, 2008.

[37] W. Att, M. Yamada, N. Kojima, and T. Ogawa, "N-Acetyl cysteine prevents suppression of oral fibroblast function on poly(methylmethacrylate) resin," *Acta Biomaterialia*, vol. 5, no. 1, pp. 391–398, 2009.

[38] X. Miao, D. M. Tan, J. Li, Y. Xiao, and R. Crawford, "Mechanical and biological properties of hydroxyapatite/tricalcium phosphate scaffolds coated with poly(lactic-co-glycolic acid)," *Acta Biomaterialia*, vol. 4, no. 3, pp. 638–645, 2008.

[39] R. Narayan, P. Colombo, S. Widjaja, and D. Singh, *Advances in Bioceramics and Porous Ceramics IV: Ceramic Engineering and Science Proceedings, Volume 32*, John Wiley & Sons, 2011.

[40] ASTM C1674-16, *Standard Test Method for Flexural Strength of Advanced Ceramics with Engineered Porosity (Honeycomb Cellular Channels) at Ambient Temperatures*, ASTM International, West Conshohocken, Pa, USA, 2016.

[41] T. Kokubo, S. Ito, Z. T. Huang et al., "Ca, P-rich layer formed on high-strength bioactive glass-ceramic A-W," *Journal of Biomedical Materials Research Part B: Applied Biomaterials*, vol. 24, no. 3, pp. 331–343, 1990.

[42] "Biological evaluation of medical devices. Part 5. Tests for cytotoxicity: in vitro methods Arlington, VA: ANSI/AAMI ISO 10993-5. International organization for standardization," 1999.

[43] M. Sous, R. Bareille, F. Rouais et al., "Cellular biocompatibility and resistance to compression of macroporous β-tricalcium phosphate ceramics," *Biomaterials*, vol. 19, no. 23, pp. 2147–2153, 1998.

[44] K. Kieswetter, Z. Schwartz, T. W. Hummert et al., "Surface roughness modulates the local production of growth factors and cytokines by osteoblast-like MG-63 cells," *Journal of Biomedical Materials Research Part B: Applied Biomaterials*, vol. 32, no. 1, pp. 55–63, 1996.

[45] M. A. Attawia, K. M. Herbert, K. E. Uhrich, R. Langer, and C. T. Laurencin, "Proliferation, morphology, and protein expression by osteoblasts cultured on poly(anhydride-co-imides)," *Journal of Biomedical Materials Research Part B: Applied Biomaterials*, vol. 48, no. 3, pp. 322–327, 1999.

[46] Z. She, B. Zhang, C. Jin, Q. Feng, and Y. Xu, "Preparation and in vitro degradation of porous three-dimensional silk fibroin/chitosan scaffold," *Polymer Degradation and Stability*, vol. 93, no. 7, pp. 1316–1322, 2008.

[47] Q. Fu, M. N. Rahaman, B. Sonny Bal, R. F. Brown, and D. E. Day, "Mechanical and in vitro performance of 13–93 bioactive glass scaffolds prepared by a polymer foam replication technique," *Acta Biomaterialia*, vol. 4, no. 6, pp. 1854–1864, 2008.

[48] K. R. Butler Jr. and H. A. Benghuzzi, "Morphometric analysis of the hormomal effect on tissue-implant response associated with TCP bioceramic implants," *Biomedical Sciences Instrumentation*, vol. 39, pp. 535–540, 2003.

[49] D. T. Reilly and A. H. Burstein, "The elastic and ultimate properties of compact bone tissue," *Journal of Biomechanics*, vol. 8, no. 6, pp. 393–405, 1975.

[50] H. C. Blair, P. H. Schlesinger, C. L. Huang, and M. Zaidi, "Calcium signalling and calcium transport in bone disease," *Subcellular Biochemistry*, vol. 45, pp. 539–562, 2007.

[51] M. Zayzafoon, "Calcium/calmodulin signaling controls osteoblast growth and differentiation," *Journal of Cellular Biochemistry*, vol. 97, no. 1, pp. 56–70, 2006.

[52] M. Yamada, H. Minamikawa, T. Ueno, K. Sakurai, and T. Ogawa, "N-acetyl cysteine improves affinity of beta-tricalcium phosphate granules for cultured osteoblast-like cells," *Journal of Biomaterials Applications*, vol. 27, no. 1, pp. 27–36, 2012.

[53] Y.-H. Lee, N.-H. Lee, G. Bhattarai et al., "Enhancement of osteoblast biocompatibility on titanium surface with Terrein treatment," *Cell Biochemistry & Function*, vol. 28, no. 8, pp. 678–685, 2010.

[54] Y.-F. Feng, L. Wang, Y. Zhang et al., "Effect of reactive oxygen species overproduction on osteogenesis of porous titanium implant in the present of diabetes mellitus," *Biomaterials*, vol. 34, no. 9, pp. 2234–2243, 2013.

[55] T. Hanawa, M. Kon, H. Ukai, K. Murakami, Y. Miyamoto, and K. Asaoka, "Surface modification of titanium in calcium-ion-containing solutions," *Journal of Biomedical Materials Research Part B: Applied Biomaterials*, vol. 34, no. 3, pp. 273–278, 1997.

[56] L. L. Hench, "Bioactive materials: The potential for tissue regeneration," *Journal of Biomedical Materials Research Part B: Applied Biomaterials*, vol. 41, no. 4, pp. 511–518, 1998.

[57] P. Banerjee, D. J. Irvine, A. M. Mayes, and L. G. Griffith, "Polymer latexes for cell-resistant and cell-interactive surfaces," *Journal of Biomedical Materials Research Part B: Applied Biomaterials*, vol. 50, no. 3, pp. 331–339, 2000.

[58] Q. Z. Chen, I. D. Thompson, and A. R. Boccaccini, "45S5 Bioglass^5-derived glass-ceramic scaffolds for bone tissue engineering," *Biomaterials*, vol. 27, no. 11, pp. 2414–2425, 2006.

[59] L. M. Hirakata, M. Kon, and K. Asaoka, "Evaluation of apatite ceramics containing α-tricalcium phosphate by immersion in simulated body fluid," *Bio-Medical Materials and Engineering*, vol. 13, no. 3, pp. 247–259, 2003.

[60] S. Foppiano, S. J. Marshall, G. W. Marshall, E. Saiz, and A. P. Tomsia, "The influence of novel bioactive glasses on in vitro osteoblast behavior," *Journal of Biomedical Materials Research Part A*, vol. 71, no. 2, pp. 242–249, 2004.

[61] M. Inoue, R. Z. LeGeros, M. Inoue et al., "In vitro response of osteoblast-like and odontoblast-like cells to unsubstituted and substituted apatites," *Journal of Biomedical Materials Research Part A*, vol. 70, no. 4, pp. 585–593, 2004.

[62] Takeshi Ikeda, Kahori Ikeda, Kouhei Yamamoto et al., "Fabrication and Characteristics of Chitosan Sponge as a Tissue Engineering Scaffold," *BioMed Research International*, vol. 2014, Article ID 786892, 8 pages, 2014.

[63] H. Robert Dudley and S. David, "The fine structure of bone cells," *The Journal of Cell Biology*, vol. 11, no. 3, pp. 627–649, 1961.

[64] L. Galois and D. Mainard, "Bone ingrowth into two porous ceramics with different pore sizes: An experimental study," *Acta Orthopædica Belgica*, vol. 70, no. 6, pp. 598–603, 2004.

[65] Y. Kuboki, Q. Jin, M. Kikuchi, J. Mamood, and H. Takita, "Geometry of artificial ECM: Sizes of pores controlling phenotype expression in BMP-induced osteogenesis and chondrogenesis," *Connective Tissue Research*, vol. 43, no. 2-3, pp. 529–534, 2002.

Permissions

List of Contributors

Viswanathan Gayathri, Varma Harikrishnan and Parayanthala Valappil Mohanan
Biomedical Technology Wing, Sree Chitra Tirunal Institute for Medical Sciences and Technology, Poojappura, Thiruvananthapuram, Kerala 695 012, India

Girdhari Rijal, Chandra Bathula and Weimin Li
Department of Biomedical Sciences, Elson S. Floyd College of Medicine, Washington State University, Spokane, WA 99210, USA

Morshed Khandaker, Albert Orock, Stefano Tarantini and Jeremiah White
Department of Engineering and Physics, University of Central Oklahoma, Edmond, OK 73034, USA

Ozlem Yasar
Department of Mechanical Engineering, New York City College of Technology, Brooklyn, NY 11201, USA

Georgios S. Theodorou, Konstantinos Chrissafis, Konstantinos and M. Paraskevopoulos
Department of Physics, Aristotle University of Thessaloniki, 54124 Thessaloniki, Greece

Eleana Kontonasaki, Anna Theocharidou, Athina Bakopoulou, Maria Bousnaki, Christina Hadjichristou, Eleni Papachristou and Petros T. Koidis
Dentistry Department, Laboratory of Fixed Prosthesis and Implant Prosthodontics, Aristotle University ofThessaloniki, 54124Thessaloniki, Greece

Lambrini Papadopoulou and Nikolaos A. Kantiranis
Department of Geology, Aristotle University ofThessaloniki, 54124Thessaloniki, Greece

Pawan Kumar
Department of Materials Science and Nanotechnology, Deenbandhu Chhotu Ram University of Science and Technology, Murthal-131039, India

Ana M. Cortizo, Flavia N. Mazzini and M. Silvina Molinuevo
LIOMM, Dto. Ciencias Biológicas, Facultad de Ciencias Exactas, Universidad Nacional de La Plata, 1900 La Plata, Argentina

Graciela Ruderman and Ines G. Mogilner
IFLYSIB, CONICET, Facultad de Ciencias Exactas, Universidad Nacional de La Plata, 1900 La Plata, Argentina

Davood Almasi and Izman Sudin
Department of Materials, Manufacturing and Industrial Engineering, Faculty of Mechanical Engineering, Universiti Teknologi Malaysia, 81310 Skudai, Johor, Malaysia

Nida Iqbal and Mohammed Rafiq Abdul Kadir
Medical Implant Technology Group (MEDITEG), Faculty of Bioscience and Medical Engineering, Universiti Teknologi Malaysia, 81310 Skudai, Johor, Malaysia

Maliheh Sadeghi
Faculty of Chemical and Energy Engineering, Universiti Teknologi Malaysia, 81310 Skudai, Johor, Malaysia

Tunku Kamarul
Department of Orthopaedic Surgery, NOCERAL, Faculty ofMedicine, University of Malaya, 50603 Kuala Lumpur, Malaysia

Maryam Tamaddon, Sorousheh Samizadeh, Gordon Blunn and Chaozong Liu
Institute of Orthopaedic & Musculoskeletal Science, University College London, Royal National Orthopaedic Hospital, Stanmore HA7 4LP, UK

Ling Wang
State Key Laboratory for Manufacturing System Engineering, Xi'an Jiaotong University, Xi'an, Shanxi Province 710049, China

Brooke McClarren and Ronke Olabisi
Department of Biomedical Engineering, Rutgers University, 599 Taylor Rd, Piscataway, NJ 08854, USA

Magdalini Tsintou and Kyriakos Dalamagkas
Centre for Nanotechnology & Regenerative Medicine, Division of Surgery and Interventional Science, University College of London, London, UK

Alexander Seifalian
Nanotechnology & Regenerative Medicine
Commercialisation Centre Ltd., The London
BioScience Innovation Centre, London, UK

**R. T. De Silva, M. M. M. G. P. G. Mantilaka and
S. P. Ratnayake**
Nanotechnology and Science Park, Sri Lanka
Institute of Nanotechnology (SLINTEC), Pitipana,
Homagama, Sri Lanka

G. A. J. Amaratunga
Nanotechnology and Science Park, Sri Lanka
Institute of Nanotechnology (SLINTEC), Pitipana,
Homagama, Sri Lanka
Electrical Engineering Division, Department
of Engineering, University of Cambridge, 9 J.
J.Thomson Avenue, Cambridge CB3 0FA, UK

K. M. Nalin de Silva
Nanotechnology and Science Park, Sri Lanka
Institute of Nanotechnology (SLINTEC), Pitipana,
Homagama, Sri Lanka
Department of Chemistry, University of Colombo,
Colombo 3, Sri Lanka

K. L. Goh
School of Mechanical and Systems Engineering,
Newcastle University, Newcastle Upon Tyne, UK

**P. V. Popryadukhin, I. P. Dobrovolskaya, E. M.
Ivan'kova and V. E. Yudin**
Institute of Macromolecular Compounds, Russian
Academy of Sciences, Bolshoy Pr. 31, Saint-
Petersburg 199004, Russia
Peter the Great Saint-Petersburg State Polytechnical
University, Polytechnicheskaya Str. 29, Saint-
Petersburg 194064, Russia

G. I. Popov, G. Yu. Yukina and V. N. Vavilov
Pavlov First Saint-Petersburg State Medical
University, Leo Tolstoy Str. 6-8, Saint-Petersburg
197022, Russia

Daniyal Jamal and Roche C. de Guzman
Bioengineering Program, Department of
Engineering, Hofstra University, Hempstead, NY
11549, USA

Ziyu Ge, Yani Wu and Yanzhen Zhang
Department of General Dentistry, The Second
Affiliated Hospital, Zhejiang University School of
Medicine, 310052, China

**Luming Yang, Fang Xiao, Tingting Yu, Jing Chen
and Jiexin Lin**
Zhejiang University, 310058, China

Julio Bissoli and Homero Bruschini
Hospital das Clínicas da Faculdade de Medicina da
Universidade de São Paulo (HCFMUSP), 05410-020
São Paulo, SP, Brazil

Yingge Zhou
Department of Industrial, Manufacturing and
SystemEngineering, Texas Tech University,
Lubbock, TX, USA

Joanna Chyu and Mimi Zumwalt
Department of Orthopedic Surgery and Rehabilitation,
Texas Tech University Health Sciences Center,
Lubbock, TX, USA

**Maidy Rehder Wimmers Ferreira, Roger Rodrigo
Fernandes and Karina Fittipaldi Bombonato-Prado**
Cell Culture Laboratory, Department of Morphology,
Physiology and Basic Pathology, School of Dentistry
of Ribeirão Preto, University of São Paulo, 14040-904
Ribeirão Preto, SP, Brazil

Geraldo A. Passos
Cell Culture Laboratory, Department of
Morphology, Physiology and Basic Pathology,
School of Dentistry of Ribeirão Preto, University
of São Paulo, 14040-904 Ribeirão Preto, SP, Brazil
Molecular Immunogenetics Group, Department of
Genetics, Ribeirão Preto Medical School, University
of São Paulo, 14049-900 Ribeirão Preto, SP, Brazil

Amanda Freire Assis and Janaína A. Dernowsek
Molecular Immunogenetics Group, Department of
Genetics, Ribeirão Preto Medical School, University
of São Paulo, 14049-900 Ribeirão Preto, SP, Brazil

Fabio Variola
Faculty of Engineering, Department of Mechanical
Engineering, University of Ottawa, Ottawa, ON,
Canada K1N 6N5

Kanchan Maji and Sudip Dasgupta
Department of Ceramic Engineering, National
Institute of Technology, Rourkela 769008, India

Krishna Pramanik and Akalabya Bissoyi
Department of Biotechnology and Medical
Engineering, National Institute of Technology,
Rourkela 769008, India

Yong-Seok Jang, Phonelavanh Manivong, Yu-Kyoung Kim, Tae-Sung Bae and Min-Ho Lee
Department of Dental Biomaterials and Institute of Biodegradable Materials, Institute of Oral Bioscience and School of Dentistry (Plus BK21 Program), Chonbuk National University, Jeonju, Republic of Korea

Kyung-Seon Kim
Department of Dental Hygiene, Jeonju Kijeon College, Jeonju, Republic of Korea

Sook-Jeong Lee
Department of Bioactive Material Science, Chonbuk National University, Jeonju, Republic of Korea

Index

A

Adipose Derived Mesenchymal Stem Cells, 1, 3

Alkaline Phosphatase, 47, 60, 81, 83-84, 158-161, 163, 165

Apoptosis, 1, 4-6, 8-9, 132, 161, 163-167

Articular Cartilage Tissue Engineering, 148, 150, 152, 155

B

Bioactive Ceramic Scaffolds, 35

Biomimetic Scaffolds, 29-30, 33, 51, 149, 154

Biomineralization, 27-28, 69, 185

Biopolymer Scaffolds, 103

Bone Marrow Progenitor Cells, 45-46

Bone Tissue Engineering, 1-2, 7, 9, 36-38, 43-44, 46, 51, 54, 68-69, 74, 77, 109, 135, 139, 154, 171, 173, 185-187, 191-192, 195, 197

Breast Cancer, 10-14, 17-18, 138

Brunauer-emmett-teller Analyzer, 38

C

Cancer Cells, 10-18, 138-139

Carbon Nanotubes, 121-122, 126-127, 129-131, 152, 156

Ceramic Scaffolds, 27-28, 30-37, 197

Chitosan Scaffold, 38-39, 41, 174, 185, 196

Collagen Fibers, 46, 49, 88, 115, 117, 119

Collagen Scaffold, 45-47, 54, 87, 91-92, 134, 137

Compressed Collagen Hydrogels, 93, 96, 98-99

Confocal Microscopy, 1, 3, 6, 14

Cyclic Tensile Strain, 79-80, 85, 87-88, 90-91

Cytotoxicity, 1, 3-5, 22, 34, 36, 38, 40, 42, 45-47, 51, 53-55, 65, 93, 98, 125, 132-134, 139-140, 186, 189, 193, 196

D

Dental Pulp Stem Cells, 29-30, 36, 134, 139

Dental Tissue Regeneration, 27-28, 30

Dexamethasone, 47, 71, 79, 87, 91, 121-122, 126, 128, 136, 140, 159

Diamond-like Carbon, 60, 67

Direct Metal Laser Sintering, 69, 77

E

Electrospinning, 102-104, 109-113, 115, 117, 119, 142-143, 147-152, 154-156

Electrospun Alginate, 102-109

Endothelium, 112-113, 115-117, 119

Estrogen Receptor, 10, 14

Extracellular Matrix, 6, 10, 17, 27, 35, 38, 46, 51, 83, 91-92, 100, 102, 109-110, 125, 145, 156, 167, 173, 184

F

Fluorescein Diacetate, 3

Fluorescence Microscopy, 12, 22, 38, 41

Fluorinated Graphene, 131, 140

Fourier Transform Infrared Spectroscopy, 28, 175

G

Glass-ceramic Scaffolds, 27-28, 30-34, 36-37, 197

Graphene Family Nanomaterials, 130, 138

Graphene Nanosheets, 130, 136, 140

H

Hydrogel Systems, 92, 98-99

Hydroxyapatite, 1-2, 4, 7, 9, 27, 36, 38, 56, 59-60, 62-63, 65-69, 78, 103, 109-110, 129, 136-137, 140, 165, 171, 173-174, 185-186, 195-196

Hydroxyapatite Burr Hole Button Device, 1-2

I

Injectable Hydrogel, 92-93, 100

M

Mesenchymal Stem Cells, 1, 3, 6-7, 9, 45, 54, 58-59, 61, 66, 71, 77, 79, 82, 89-91, 121-122, 124, 128, 134, 139-140, 149-150, 152, 154-155, 167, 172, 176, 179

Methoxyphenyl Tetrazolium Salt, 60

Mgo Nanoparticles, 102-105, 107-109

Microfibers, 111-115, 117, 119, 151

Msc Osteodifferentiation, 80, 87, 121

N

Nanofibrous Scaffolds, 43, 102-104, 107, 109, 150, 154-156

Nanoribbons, 130

Nutrient Conduit Networks, 19-20, 24

O

Osteoblastic Cells, 67, 75, 137, 158-159, 164-165, 167-170

Osteochondral Tissue Engineering, 45, 54, 77, 152, 156

Osteoconduction, 5, 69, 77, 173

Oxidative Nanopatterning, 158-159, 161, 163, 165, 168
Oxidative Stress, 1, 3, 6, 9, 17, 55, 132-133, 137, 139-140, 188, 193, 196

P
Pelvic Organ Prolapse, 142, 145-147
Photopolymerization System, 20
Plastically Compressed Collagen Hydrogel, 92, 96
Polyether Ether Ketone, 56, 66
Polyethylene Glycol Diacrylate, 19-20, 25
Polymeric Nanofibres, 102
Porous Scaffold, 2, 16, 60, 173, 175
Porous Titanium Scaffold, 68-69
Progenitor Cells, 1, 45-46, 91, 100, 128, 187

R
Reduced Graphene Oxide, 130-131, 133, 138
Resorption Rate, 119

S
Selective Laser Melting, 69, 77-78
Selective Laser Sintering, 62-63, 67-69, 74
Synthetic Polymeric Scaffolds, 10

T
Tissue Engineering Scaffolds, 24, 43, 68, 109, 130, 148, 151-152, 155
Tissue-engineered Vascular Grafts, 111, 119-120
Titanium Dioxide, 56, 60

V
Vanadium-loaded Collagen Scaffold, 45
Vascular Graft, 111, 119-120
Vascular Surgery, 111, 119-120

W
Wet-spinning, 102

www.ingramcontent.com/pod-product-compliance
Lightning Source LLC
Chambersburg PA
CBHW082018190326
41458CB00010B/3222